CLASSICAL MECHANICS
FOR PHYSICS GRADUATE STUDENTS

CLASSICAL MECHANICS
FOR PHYSICS GRADUATE STUDENTS

Ernesto Corinaldesi

Department of Physics
Boston University

World Scientific
Singapore • New Jersey • London • Hong Kong

Published by

World Scientific Publishing Co. Pte. Ltd.
P O Box 128, Farrer Road, Singapore 912805
USA office: Suite 1B, 1060 Main Street, River Edge, NJ 07661
UK office: 57 Shelton Street, Covent Garden, London WC2H 9HE

Library of Congress Cataloging-in-Publication Data
Corinaldesi, E.
Classical mechanics for physics graduate students / Ernesto Corinaldesi.
p. cm.
Includes index.
ISBN 9810236255
1. Mechanics. I. Title.
Q125.C655 1998
531--dc21 98-48670
CIP

British Library Cataloguing-in-Publication Data
A catalogue record for this book is available from the British Library.

This book is printed on acid-free paper.

Printed in Singapore by Uto-Print

to Abner Shimony, gratefully

Preface

This book evolved from my occasional teaching of graduate Classical Mechanics at Boston University.

In the late 1960's I recommended the well-known texts by H. Goldstein, L.D. Landau and E.M. Lifshitz (?), and A. Sommerfeld. In the late 1980's I learned some modern differential geometry while catching up with gauge field theory. Thus I became able to appreciate the elegance of V.I. Arnold's *Mathematical Methods of Classical Mechanics.*

I took the educational risk of presenting Hamiltonian mechanics expressed in Cartan's notation as a kind of appendix following the traditional treatment. Instructors in other universities also seem to have recognized that the graduate teaching of classical mechanics in physics departments should be updated. In this book I have tried to satisfy both traditionalists and modernists by approaching each subject at successive levels of abstraction.

My work is designed for a first-year physics graduate student at Boston University, who has taken "Intermediate Mechanics". Knowledge of curvilinear coordinates, vector analysis, advanced calculus etc. is assumed. The wealth of detail offered should not lull the reader into thinking that the material can be learned by leafing through the book. As much time as possible should be devoted to going through the calculations and solving problems.

Chapter 1 is an introduction meant to establish a rapport between the reader and my way of presenting the material. Chapter 2 builds up a stock of formulae to be drawn from in later chapters. Chapter 3 is mostly devoted to oscillations. It ends with a sketchy account of chaos for discrete maps to introduce the reader to a field founded a century ago by Poincaré and cultivated intensively in recent years.

Chapters 4 and 5 cover coordinate systems, inertial forces, and rigid bodies.

Analytical mechanics begins with chapter 6, Lagrangians. Problems already worked out in previous chapters are again solved by using Lagrangian methods. The reader is invited to compare the amount of labor involved in the two versions.

Chapter 7 presents Hamiltonian mechanics. Sections 7.1 to 7.4 give a simple account of the traditional treatment. Section 7.5 introduces Cartan's notation, which is used throughout the following sections as well as in chapter 8, action-angle variables and adiabatic invariants.

Chapter 9 is on classical perturbation theory with emphasis on the similarity with the perturbation theory of quantum mechanics and field theory.

Relativistic dynamics is discussed in chapter 10, with attention to the "spinor connection", the spin, and the Thomas precession.

Chapter 11 illustrates Lagrangian and Hamiltonian methods for continuous systems by discussing two case studies, the vibrating string and the ideal incompressible non-viscous fluid. The Lagrangian description of the latter is not widely known. I worked unsuccessfully trying to formulate it, until I was lucky enough to find D.E. Soper's *Classical Field Theory* and reference to the original 1911 paper by G. Herglotz.

Acknowledgements

I wish to express my gratitude to my students, especially to Nathaniel R. Greene and Dinesh Loomba, for asking challenging questions and bringing to my attention interesting material. After taking my course Gregg Jaeger and John B. Ross became my graders and helped me in the preparation of the monthly exams from which some of the problems in this text originated.

Adriana Ruth Corinaldesi painstakingly checked parts of my work pointing out errors and lack of clarity. Abner Shimony kindly read the final version of the manuscript and suggested improvements. Neither Abner nor Adriana are responsible for any residual errors and imperfections, which are indeed possible because of the unusual number of formulae presented.

Anthony P. French, John Stachel, and Edwin F. Taylor kindly helped me to dispel doubts about one of the relativity problems.

In addition to the books mentioned in the Preface I learned from many others, but I wish to mention J.D. Jackson *Classical Electrodynamics*, C.W. Kilmister and J.E. Reeves *Rational Mechanics*, R.A. Mann *The Classical Dynamics of Particles - Galilean and Lorentz relativity*, B.F. Schutz *Geometrical methods of mathematical physics*, W.Thirring *Classical Dynamical Systems.*

I was able to prepare the printed manuscript only because of the generous coaching in Latex given me by João Leao. I am also indebted to Jinara Reyes and Guoan Hu for frequent help.

Wei Chen and Lakshmi Narayanan of World Scientific have been an example of amicable and helpful editorship.

Contents

List of Figures

Chapter 1

INTRODUCTION

This chapter covers informally material that will be treated extensively in the book. Some of it may have already been learned in undergraduate mechanics. Introduction or review as they may be, the following pages are meant to lead into the substance of analytical mechanics.

1.1 Motion in phase space

It is known from General Physics that the dynamics of particles and rigid bodies is governed by second order differential equations. For computational purposes, it is convenient to replace these by systems of first order equations. Thus Newton's second law in the form

$$m\ddot{x} = f(x, \dot{x}, t) \tag{1.1}$$

is replaced by the system

$$\dot{p} = f(q, p, t) \quad , \quad \dot{q} = p \quad , \tag{1.2}$$

where we have renamed the coordinate ($q = x$), denoted by p the momentum, and taken the mass equal to unity.

This technical artifice is epitomized by Hamilton's equations for the generalized coordinates $q_1, \ldots q_n$ and their conjugate momenta $p_1, \ldots p_n$,

$$\dot{p}_k = -\partial H(p, q, t)/\partial q_k \quad , \tag{1.3}$$

$$\dot{q}_k = \partial H(p, q, t)/\partial p_k \quad , \tag{1.4}$$

where H is the "Hamiltonian" (see chapter 7).

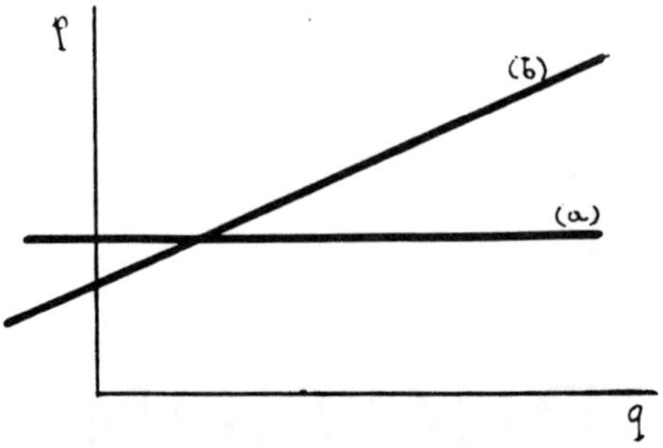

Figure 1.1: Arbitrariness of p

The above process is not unique. For example, for a free particle ($\ddot{q} = 0$) rather than

$$\dot{p} = 0 \quad , \quad \dot{q} = p \quad , \tag{1.5}$$

according to which p is constant and $q = pt + q_0$ (figure 1.1(a)), we might equivalently have written (figure 1.1(b))

$$\dot{p} = 1 \quad , \quad \dot{q} = p - t \quad , \quad p = t + \alpha \quad , \quad q = \alpha t + \beta \quad . \tag{1.6}$$

As we shall see in Chapter 6, this arbitrariness is connected with one for the Lagrangian.

Once the second order equations have been replaced by systems of first order ones, it is natural to describe the evolution of a system as the motion of a point in a $2n$-dimensional "phase space" for the coordinates $q_1, \ldots q_n$ and the momenta $p_1, \ldots p_n$.

1.2 Motion of a particle in one dimension

It is instructive to study the motion of a particle in one dimension subject to the elastic force $-kq$ and the repulsive force $+kq$ ($k > 0$), respectively.

In the former case, the equations $\dot{p} = -kq$, $\dot{q} = p/m$ have the general solution ($\omega = \sqrt{k/m}$)

$$q = a\cos(\omega t + \alpha) \quad , \quad p = -m\omega a \sin(\omega t + \alpha) \quad , \tag{1.7}$$

so that the representative point in the phase plane (p, q) moves on an ellipse of semi-axes $a = \sqrt{2E/k}$ and $m\omega a = \sqrt{2mE}$ (figure 1.2), where E is the energy. The point $q = 0$, $p = 0$ describes a state of stable equilibrium ($a = 0$, $E = 0$).

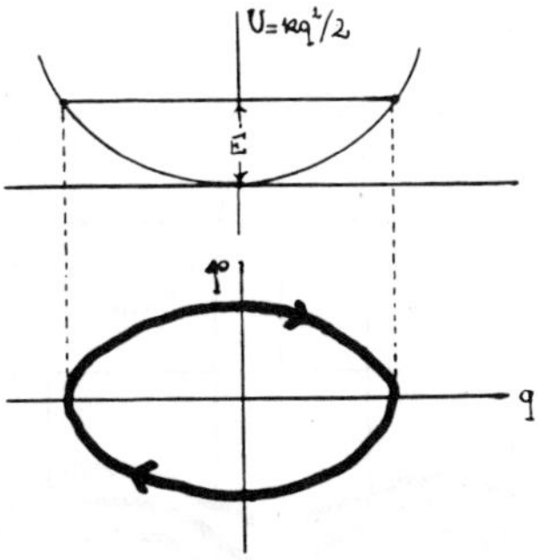

Figure 1.2: Motion under the force $-kq$

In the case of the repulsive force $+kq$, the trajectories in the phase plane are hyperbolae. The solutions

$$q = \pm a \cosh(\omega t) \quad , \quad p = \pm m\omega a \sinh(\omega t) \quad , \tag{1.8}$$

correspond to a particle with initial velocity towards the center of repulsion coming from $q = \pm\infty$ respectively, and stopping at $\pm a$ before rebounding (figure 1.3). In these two cases the energy $E = (p^2/2m) - (kq^2/2)$ is negative ($E = -ka^2/2$).

The solutions

$$q = \pm a \sinh(\omega t) \quad , \quad p = \pm m\omega a \cosh(\omega t) \quad , \tag{1.9}$$

correspond to the positive energy $E = +ka^2/2$. For $t = 0$ one has $q = 0$, $p = \pm m\omega a$, i.e. a particle right on the center of repulsion with a finite velocity (figure 1.3).

The asymptotes $p = \pm m\omega q$ separate the regions of positive and negative energies. They intersect at the point of unstable equilibrium $p = 0$, $q = 0$.

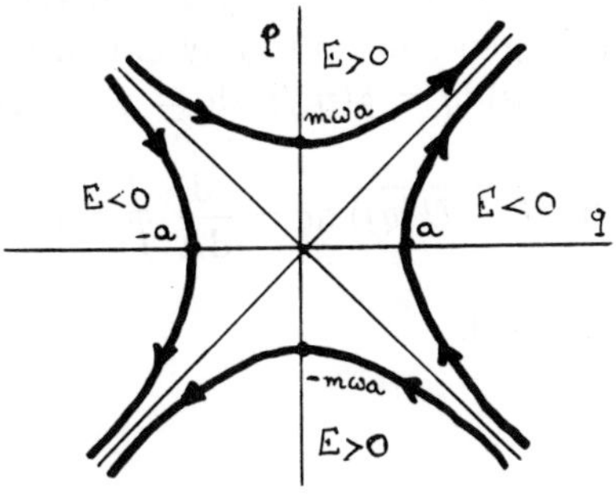

Figure 1.3: Motion under the force $+kq$

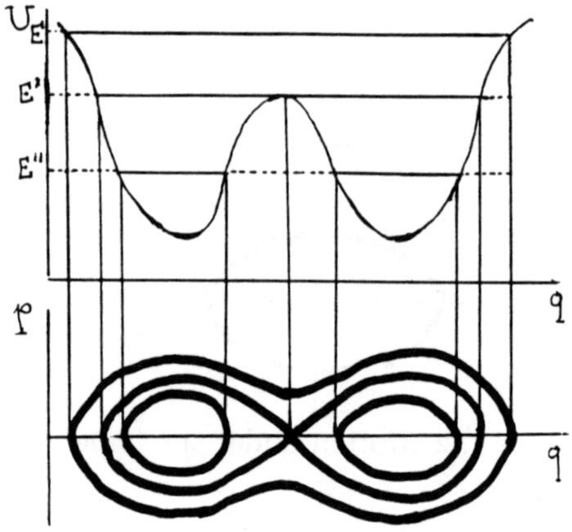

Figure 1.4: Motion near potential energy extrema

Consider now the one-dimensional motion of a particle subject to a conservative force $f(q) = -\mathrm{d}U(q)/\mathrm{d}q$. In the vicinity of a potential energy minimum $q = q_0$, one has $f(q) = -k(q - q_0)$ with $k = (\mathrm{d}^2U/\mathrm{d}q^2)_{q=q_0} > 0$.

If the energy E differs little from $U(q_0)$, the particle moves on an ellipse of center $q = q_0$, $p = 0$ in the (p, q) plane with the period $T = 2\pi\sqrt{m/k}$.

In the vicinity of a potential energy maximum the motion resembles that of a particle subject to the repulsive force $f(q) = +k(q - q_0)$ with $k = -(\mathrm{d}^2U/\mathrm{d}q^2)_{q=q_0} > 0$. Passing through the point $(q = q_0$, $p = 0)$ there will be a critical curve with tangents $p = \pm\sqrt{mk}\,(q - q_0)$, corresponding to the asymptotes of figure 1.3. Such a curve separates regions with $E > U(q_0)$ and $E < U(q_0)$.

Returning to the elastic force $f(q) = -kq$, we notice that the area enclosed by an elliptical trajectory in the (p, q) plane is

$$A = \pi a(m\omega a) = 2\pi E\sqrt{m/k} \ ,$$

while the period is $T = 2\pi\sqrt{m/k}$. Clearly $T = \mathrm{d}A/\mathrm{d}E$.

This relation is easily generalized to any closed trajectory in the phase plane: $T = \oint \mathrm{d}t = \oint (\mathrm{d}t/\mathrm{d}q)\mathrm{d}q = \oint (m/p)\mathrm{d}q = \oint m\ \mathrm{d}q/\sqrt{2m(E - U(q))}$,

$$T = \frac{\mathrm{d}}{\mathrm{d}E}\oint \sqrt{2m(E - U(q))}\,\mathrm{d}q = \frac{\mathrm{d}}{\mathrm{d}E}\oint p\ \mathrm{d}q = \frac{\mathrm{d}A}{\mathrm{d}E} \quad , \tag{1.10}$$

where A is the area enclosed.

1.3 Flow in phase space

Consider a region in the (p, q) plane and assume that each point in the region represents a possible position and momentum of the particle at the time $t = 0$. If each point moves according to the dynamics of the particle, those points will in general cover a different region at a later time $t > 0$. We can visualize this as the flow of a "fluid" in two dimensions.

If the forces are conservative, the fluid is incompressible. (See problem 1.5 for a dissipative case.)
For example, for the vertical motion of particles under gravity one has $q = (gt^2/2) + (p_0 t/m) + q_0$, $p = mgt + p_0$. The points of the rectangular region $0 \leq q \leq Q$, $0 \leq p \leq P$ at time $t = 0$, at a later time $t > 0$ will cover the parallelogram with vertices $a = (gt^2/2, mgt)$, $b = (gt^2/2 + Q, mgt)$, $c = (gt^2/2 + Pt/m + Q, P + mgt)$, $d = (gt^2/2 + Pt/m, P + mgt)$ (see figure 1.5). It can be seen by inspection that the area of this parallelogram is equal to the area PQ of the rectangle at time $t = 0$.

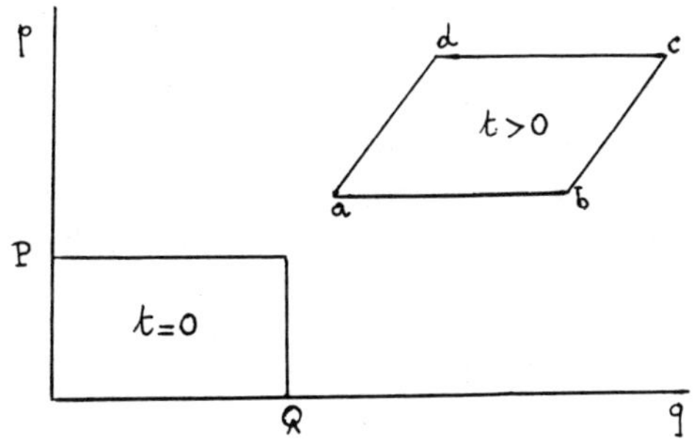

Figure 1.5: Vertical motion under gravity

As a prelude to the general case, we can also express the area at time $t > 0$ in the form

$$A_t = \int_{D_t} dp_t \, dq_t = \int_{D_0} \left| \frac{\partial(p_t, q_t)}{\partial(p_0, q_0)} \right| dp_0 \, dq_0 \; . \qquad (1.11)$$

But the Jacobian determinant

$$\frac{\partial(p_t, q_t)}{\partial(p_0, q_0)} = \begin{vmatrix} \partial p_t/\partial p_0 & \partial p_t/\partial q_0 \\ \partial q_t/\partial p_0 & \partial q_t/\partial q_0 \end{vmatrix} = \begin{vmatrix} 1 & 0 \\ t/m & 1 \end{vmatrix} = 1$$

and so

$$A_t = \int_{D_0} dp_0 \, dq_0 = A_0 \quad . \qquad (1.12)$$

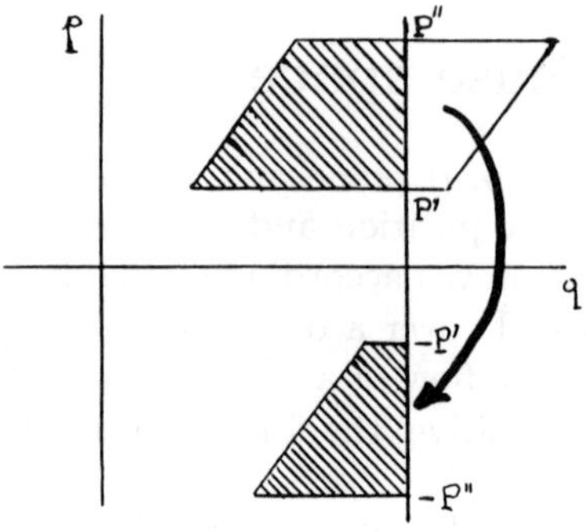

Figure 1.6: Elastic collision against a wall

The same property applies to particles undergoing elastic collision against a "wall" (see figure 1.6).

For the harmonic oscillator $q_t = q_0 \cos(\omega t) + p_0 \sin(\omega t)/m\omega$, $p_t = -m\omega q_0 \sin(\omega t) + p_0 \cos(\omega t)$ one has similarly

$$\frac{\partial(p_t, q_t)}{\partial(p_0, q_0)} = \begin{vmatrix} \cos(\omega t) & -m\omega \sin(\omega t) \\ \sin(\omega t)/m\omega & \cos(\omega t) \end{vmatrix} = 1 \quad ,$$

and so the area conservation applies also in this case.

Can this property be generalized to any conservative force? Since $p_{t+\mathrm{d}t} = p_t - (\mathrm{d}U/\mathrm{d}q)_t \, \mathrm{d}t$, $q_{t+\mathrm{d}t} = q_t + (p_t/m) \, \mathrm{d}t$, one has

$$\frac{\partial(p_{t+\mathrm{d}t}, q_{t+\mathrm{d}t})}{\partial(p_t, q_t)} = \begin{vmatrix} 1 & -(\mathrm{d}^2U/\mathrm{d}q^2)_t \mathrm{d}t \\ \mathrm{d}t/m & 1 \end{vmatrix} = 1 + O((\mathrm{d}t)^2) \quad .$$

Hence

$$\frac{\mathrm{d}A_t}{\mathrm{d}t} = 0 \quad . \tag{1.13}$$

This is a special case of Liouville's theorem, which will be proved in Chapter 7.

For a one-dimensional system with conservative forces, by Stokes' theorem area conservation in the (p, q) plane is equivalent to conservation of a line integral along the curve enclosing the area,

$$\oint_{C_t} p \, \mathrm{d}q = \oint_{C_0} p \, \mathrm{d}q \quad , \tag{1.14}$$

where C_0 and C_t are the boundaries of D_0 and D_t, respectively.

Direct proof: If $p = p(\lambda, t)$, $q = q(\lambda, t)$ are the parametric equations of C_t, where the domain of λ does not depend on t, one has

$$\oint_{C_t} p \, \mathrm{d}q = \oint \mathrm{d}\lambda \; p(\lambda, t) \; \partial q(\lambda, t)/\partial\lambda \quad ,$$

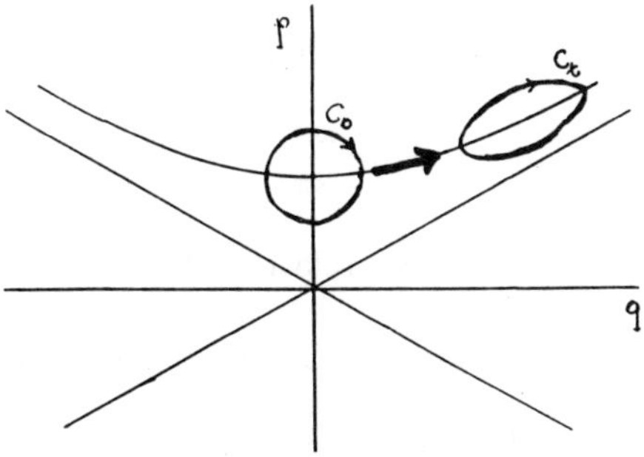

Figure 1.7: Example of equation (1.14)

$$\frac{\mathrm{d}}{\mathrm{d}t}\oint_{C_t} p\ \mathrm{d}q = \oint \left(\frac{\partial p}{\partial t}\frac{\partial q}{\partial \lambda} + p\frac{\partial^2 q}{\partial \lambda \partial t}\right) \mathrm{d}\lambda = \oint \left(\frac{\partial p}{\partial t}\frac{\partial q}{\partial \lambda} - \frac{\partial p}{\partial \lambda}\frac{\partial q}{\partial t}\right) \mathrm{d}\lambda$$

$$= \oint \left[-\frac{\partial U}{\partial q}\frac{\partial q}{\partial \lambda} - \frac{\partial}{\partial \lambda}\left(\frac{p^2}{2m}\right)\right] \mathrm{d}\lambda = -\oint \frac{\mathrm{d}}{\mathrm{d}\lambda}\left(\frac{p^2}{2m} + U\right) \mathrm{d}\lambda = 0 \quad .$$

As an example, let C_0 be the circle of equations

$$q(\lambda) = r\ \cos\lambda,\ p(\lambda) = p_0 - r\ \sin\lambda \ ,$$

and C_t the curve

$$p(\lambda, t) = m\omega q(\lambda)\ \sinh(\omega t) + p(\lambda)\ \cosh(\omega t) \ ,$$
$$q(\lambda, t) = q(\lambda)\ \cosh(\omega t) + (p(\lambda)/m\omega)\ \sinh(\omega t) \ ,$$

which is the mapping of C_0 under the flow induced by the repulsive force $f(q) = kq$ ($k = m\omega^2$). One has (see figure 1.7)

$$\oint_{C_t} p\ \mathrm{d}q = \oint_0^{2\pi} [m\omega r\ \cos\lambda\ \sinh(\omega t) + (p_0 - r\ \sin\lambda) \cosh\ (\omega t)]$$
$$[-r\ \sin\lambda\ \cosh(\omega t) - r\ \cos\lambda\ \sinh(\omega t)/m\omega]\ \mathrm{d}\lambda$$
$$= -r^2\pi[\sinh^2(\omega t) - \cosh^2(\omega t)] = +\pi r^2 = \oint_{C_0} p\ \mathrm{d}q \quad .$$

If C_0 is a (p, q)-orbit of a periodic motion, then $C_t = C_0$ at all times. In the phase flow, the points of C_0 pursue one another along C_0 itself. In this case,

$$J = \oint_{C_0} p\ \mathrm{d}q \tag{1.15}$$

is called a "cyclic action variable".

For the harmonic oscillator

$$q(\lambda, t) = a\ \cos(\omega t + \lambda),\ p(\lambda, t) = -m\omega a\ \sin(\omega t + \lambda),$$

$$J = \oint m\omega^2 a^2\ \sin^2(\omega t + \lambda)\ \mathrm{d}t = m\omega a^2 \int_0^{2\pi} \sin^2\theta\ \mathrm{d}\theta = 2\pi\frac{E}{\omega} \quad . \tag{1.16}$$

The energy can be expressed in terms of J,

$$E = \frac{\omega J}{2\pi} \quad . \tag{1.17}$$

Since $J = A$, this agrees with the relation $A = 2\pi E\sqrt{m/k} = 2\pi E/\omega$ on page 4.

In the "old quantum mechanics", Sommerfeld's quantization condition

$$J = nh \qquad (n = 0, 1 \ldots) \tag{1.18}$$

yielded the energy levels for the harmonic oscillator

$$E = n\hbar\omega \tag{1.19}$$

instead of the correct $E = (n + 1/2)\hbar\omega$.

For a multiply periodic system with n degrees of freedom, the energy can be expressed as a function of n cyclic action variables, $E = E(J_1, J_2 \ldots J_n)$. An example will be presented in the next chapter (Keplerian motion) and the general case will be discussed in chapter 8.

1.4 The action integral

The cyclic action variable J originates from Hamilton's "characteristic function"

$$S(q_2, q_1, E) = \int_{q_1}^{q_2} p \, \mathrm{d}q \quad . \tag{1.20}$$

This is a very useful function. We notice first that the momenta at q_1 and q_2 are given by $p_2 = \partial S/\partial q_2$ and $p_1 = -\partial S/\partial q_1$.

The time between two positions q_1 and q_2 is given by

$$t_2 - t_1 = \partial S(q_2, q_1, E)/\partial E \tag{1.21}$$

In fact

$$t_2 - t_1 = \int_{q_1}^{q_2} \frac{\mathrm{d}t}{\mathrm{d}q}\mathrm{d}q = m\int_{q_1}^{q_2} \frac{\mathrm{d}q}{p} = m\int_{q_1}^{q_2} \frac{\mathrm{d}q}{\sqrt{2m(E - U(q))}}$$

$$= \frac{\partial}{\partial E}\int_{q_1}^{q_2} \sqrt{2m(E - U(q))} \, \mathrm{d}q = \frac{\partial}{\partial E} S(q_2, q_1, E) \quad .$$

The formula $T = \mathrm{d}J/\mathrm{d}E$ for the period of a closed orbit is a special case for $q_2 = q_1$.

For the harmonic oscillator one has

$$S(q_2, q_1, E = ka^2/2) = \int_{q_1}^{q_2} \sqrt{2m(E - kq^2/2)}\ \mathrm{d}q = m\omega \int_{q_1}^{q_2} \sqrt{a^2 - q^2}\ \mathrm{d}q$$

$$= (m\omega/2)\left[q_2\sqrt{a^2 - q_2^2} + a^2 \sin^{-1}(q_2/a) - \text{same for } q_1\right] \quad , \qquad (1.22)$$

while (1.21) gives

$$t_2 - t_1 = [\sin^{-1}(q_2/a) - \sin^{-1}(q_1/a)]/\omega \quad . \qquad (1.23)$$

Taking $q_1 = a$, $t_1 = t_0$, $q_2 = q$, $t_2 = t$, we have

$$t - t_0 = [\sin^{-1}(q/a) - \pi/2]/\omega \ , \ q = a\cos(\omega(t - t_0)) \quad . \qquad (1.24)$$

The action $S(q, q_0, E)$ obeys the Hamilton-Jacobi equation

$$\frac{1}{2m}\left(\frac{\partial S}{\partial q}\right)^2 + U(q) = E \quad , \qquad (1.25)$$

as can be seen by substituting $p = \partial S/\partial q$ in the expression for the energy, $E = p^2/2m + U(q)$.

As an example, verify that the action for a particle of positive energy E subject to the force $f(q) = +kq$, $U = -kq^2/2$, is

$$S(q, q_0, E) = \int_{q_0}^{q} \sqrt{2m(E - U(q'))}\ \mathrm{d}q'$$

$$= \frac{m\omega}{2}\left[q\sqrt{q^2 + 2E/k} - q_0\sqrt{q_0^2 + 2E/k} + \frac{2E}{k}\ln\left(\frac{q + \sqrt{q^2 + 2E/k}}{q_0 + \sqrt{q_0^2 + 2E/k}}\right)\right] \quad , \qquad (1.26)$$

and that it satisfies the Hamilton-Jacobi equation for $U(q) = -kq^2/2$.

All this may seem somewhat futile. However, the usefulness of action integrals becomes evident as soon as we extend their definition to more than one dimension, say three dimensions.

Let

$$S(\mathbf{r}, \mathbf{r}_0, E, \alpha) = \int_{\mathbf{r}_0}^{\mathbf{r}} \mathbf{p} \cdot \mathrm{d}\mathbf{r} = \int_{\mathbf{r}_0}^{\mathbf{r}} \sqrt{2m(E - U(\mathbf{r}))}\ \mathrm{d}s \quad , \qquad (1.27)$$

where the line integral is along a trajectory characterized by the energy E and some integrals of motion summarily referred to by the symbol α. (In one dimension, the only integral of motion was E, hence no α.) Note that along a trajectory $\mathbf{p} \cdot \mathrm{d}\mathbf{r} = |\mathbf{p}|\ \mathrm{d}s$, $\mathrm{d}s = |\mathrm{d}\mathbf{r}|$.

For the two-dimensional case of projectile motion, with the x-axis horizontal and the y-axis vertically upwards, and $\mathbf{r}_0 = 0$, in the ascending branch of the trajectory one finds

$$S(\mathbf{r}, 0, E, \alpha = p_x) = \int_0^x p_x \, \mathrm{d}x' + \int_0^y \sqrt{2m(E - E_x - mgy')} \, \mathrm{d}y'$$

$$= p_x x + [(2m(E - E_x))^{\frac{3}{2}} - (2m(E - E_x - mgy))^{\frac{3}{2}}]/3m^2 g \quad , \qquad (1.28)$$

where $E_x = p_x^2/2m$.

Consider the normal to the curve $S = 0$ at $\mathbf{r} = 0$. Since $\partial S/\partial x = p_x = mv_{0x}$ and $(\partial S/\partial y)_{y=0} = \sqrt{2m(E - E_x)} = mv_{0y}$, it is clear that the trajectory is normal to the curve $S = 0$ at the $\mathbf{r} = 0$ intersection.

This is not surprising. In the three-dimensional case, the normal to the surface $S(\mathbf{r}, \mathbf{r}_0, E, \alpha) = 0$ at $\mathbf{r}_0$ is parallel to the gradient $(\nabla S)_{\mathbf{r}_0}$, and this in turn is equal to the momentum $\mathbf{p}$ at $\mathbf{r}_0$.

A method for finding trajectories suggests itself. The action integral satisfies the Hamilton-Jacobi equation

$$(\nabla S)^2/(2m) + U = E \quad . \qquad (1.29)$$

If $S(\mathbf{r})$ is a solution of this equation, consider the family of surfaces $S(\mathbf{r}) = \text{const}$. The trajectories form a family of curves orthogonal to these surfaces. The problem of finding trajectories is similar to that of finding lines of force in electrostatics once the equipotential surfaces are known.

Let us re-examine the problem of projectile motion from this viewpoint. The Hamilton-Jacobi equation

$$\frac{1}{2m}\left[\left(\frac{\partial S}{\partial x}\right)^2 + \left(\frac{\partial S}{\partial y}\right)^2\right] + mgy = E \qquad (1.30)$$

can be solved by separation of variables, $S(x, y) = S_x(x) + S_y(y)$, $(\mathrm{d}S_x/\mathrm{d}x)^2/(2m) = E_x$, $(\mathrm{d}S_y/\mathrm{d}y)^2/(2m) + mgy = E_y$, with $E_x + E_y = E$.

The function

$$S(x, y) = \sqrt{2mE_x}\, x - [(2m(E - E_x - mgy))^{\frac{3}{2}} - (2m(E - E_x))^{\frac{3}{2}}]/3m^2 g \qquad (1.31)$$

is a solution of equation (1.30), corresponding to a curve passing through the origin.

To find the trajectory through the origin corresponding to the energies E_x and E_y, we write

$$\mathrm{d}y/\mathrm{d}x = (\partial S/\partial y)/(\partial S/\partial x) = \sqrt{E - E_x - mgy}/\sqrt{E_x} \quad ,$$

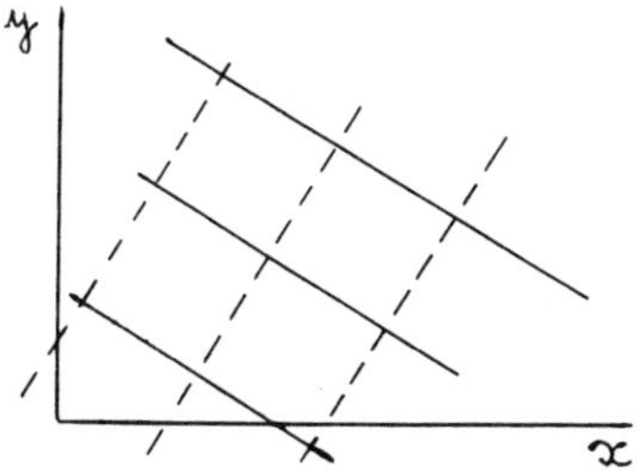

Figure 1.8: All trajectories in this figure correspond to the same values of p_x and p_y.

where we have chosen the positive sign for $\mathrm{d}y/\mathrm{d}x$ (ascending branch of the trajectory). This expresses the fact that a displacement $(\mathrm{d}x, \mathrm{d}y)$ along the trajectory is parallel to the normal to the curve represented by $S(x, y)$.

Separating variables, we have

$$\mathrm{d}y/\sqrt{E - E_x - mgy} = \mathrm{d}x/\sqrt{E_x} \quad ,$$

$$-(2/mg)\sqrt{E - E_x - mgy} = x/\sqrt{E_x} - (2/mg)\sqrt{E - E_x} \quad .$$

Putting $E_x = mv_{0x}^2/2$ and $E - E_x = mv_{0y}^2/2$, and assuming that v_{0x} and v_{0y} are both positive, this reduces to the familiar formula

$$y = \frac{v_{0y}}{v_{0x}}x - \frac{gx^2}{2v_{0x}^2} \quad . \tag{1.32}$$

This, of course, can be derived by equating the two expressions for t, x/v_{0x} and $(v_{0y} - \sqrt{v_{0y}^2 - 2gy}\,)/g$.

The Hamilton-Jacobi equation for a free particle of energy E in two dimensions with momentum component p_x has the solution

$$S = p_x x \pm \sqrt{2m(E - E_x)}\; y \quad . \tag{1.33}$$

The curves $S = \text{const}$ (solid) and the trajectories (broken) are shown in figure 1.8 for $p_x > 0$ and the plus sign. Of course, the roles of x and y can be interchanged.

However, the Hamilton-Jacobi equation for a free particle of energy E in two dimensions has also the solution

$$S = \pm\sqrt{2mE}\,|\mathbf{r} - \mathbf{r}_0| \quad . \tag{1.34}$$

The curves S=const are circles, the trajectories are straight lines through $\mathbf{r}_0$ (see figure 1.9). The angular momentum about $\mathbf{r}_0$, $l = 0$, plays the role of the constant of motion α other than the energy (see equation (1.27)).

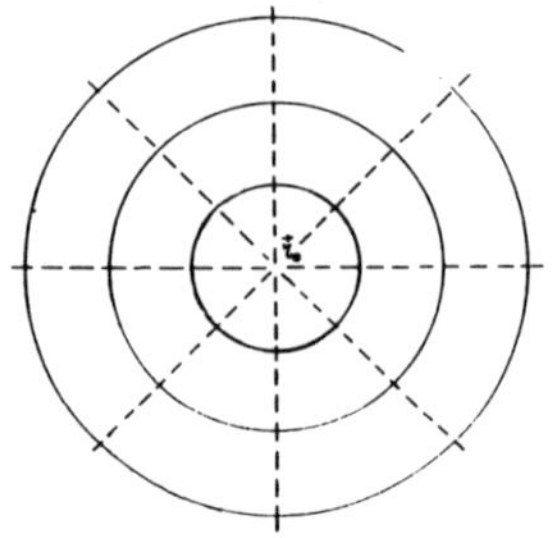

Figure 1.9: Equation (1.34)

1.5 The Maupertuis principle

We have seen that

$$S(B) - S(A) = \left(\int_{\mathrm{A}}^{\mathrm{B}} \mathbf{p} \cdot \mathrm{d}\mathbf{r}\right)_{\mathrm{t}} = \left(\int_{\mathrm{A}}^{\mathrm{B}} |\mathbf{p}| \mathrm{d}s\right)_{\mathrm{t}} \quad , \tag{1.35}$$

where $\mathbf{p} = \nabla S$, $|\mathbf{p}| = |\nabla S| = \sqrt{2m(E - U(\mathbf{r})}$ and "t" is the actual trajectory from A to B.

Now if "c" is a curve from A to B adjacent to the trajectory, we see that

$$S(B) - S(A) = \left(\int_{\mathrm{A}}^{\mathrm{B}} \mathbf{p} \cdot \mathrm{d}\mathbf{r}\right)_{\mathrm{c}} \leq \left(\int_{A}^{B} \sqrt{2m(E - U(\mathbf{r}))}\, \mathrm{d}s\right)_{\mathrm{c}} \quad , \tag{1.36}$$

since $\mathbf{p}$ and $\mathrm{d}\mathbf{r}$ are not parallel on "c".

Hence the Maupertuis principle or principle of "least action": The action integral from a point A to a point B, $\int_A^B \sqrt{2m(E - U(\mathbf{r})}\, \mathrm{d}s$, is minimal along a trajectory. Less restrictively, Euler's formulation of the principle states that

$$\delta \int_A^B \sqrt{E - U(\mathbf{r})}\, \mathrm{d}s = 0 \quad , \tag{1.37}$$

where δ indicates variation of the integral when the coordinates $\mathbf{r} = (x, y, z)$ of each point of a trajectory from A to B are changed to $\mathbf{r} + \delta\mathbf{r} = (x + \delta x, y + \delta y, z + \delta z)$, with $\delta\mathbf{r}$ vanishing at A and B.

It must be possible to derive the equations of trajectories from this variational principle. Let us do so in two dimensions and assuming that $U = U(y)$. With $y' \equiv \mathrm{d}y/\mathrm{d}x$, we have

$$\delta \int_A^B \sqrt{E - U} \mathrm{d}s = \delta \int_A^B \sqrt{E - U(y)}\, \sqrt{1 + y'^2}\, \mathrm{d}x$$

$$= \int_A^B \left[-\frac{1}{2}\sqrt{\frac{1+y'^2}{E-U}}\frac{\mathrm{d}U}{\mathrm{d}y}\delta y + \sqrt{\frac{E-U}{1+y'^2}}\, y'\frac{\mathrm{d}\delta y}{\mathrm{d}x} \right] \mathrm{d}x = 0 \quad .$$

Partial integration using $\delta y_A = \delta y_B = 0$ yields the trajectory equation

$$\frac{\mathrm{d}}{\mathrm{d}x}\left(y'\sqrt{\frac{E-U}{1+y'^2}} \right) + \frac{1}{2}\sqrt{\frac{1+y'^2}{E-U}}\frac{\mathrm{d}U}{\mathrm{d}y} = 0 \quad .$$

Multiplying by $y'\sqrt{(E-U)/(1+y'^2)}$ we obtain

$$\frac{\mathrm{d}}{\mathrm{d}x}\left(\frac{y'^2}{1+y'^2}(E-U) + U \right) = 0 \quad ,$$

so that the quantity in parentheses is a constant, which we denote by C. Hence

$$y'^2 = \frac{C-U}{E-C} \quad .$$

The value of the constant can be determined from that of y' at a point of the trajectory. For a projectile ($U = mgy$), we have $E = mv_0^2/2$, $(y')_{x=y=0} = v_{0y}/v_{0x}$, and so $C = mv_{0y}^2/2$, $y' = (\sqrt{v_{0y}^2 - 2gy})/v_{0x}$. Solution by separation of variables yields equation (1.32) if the value of the integration constant is chosen so that the trajectory may pass through the origin.

1.6 The time

What about the time? Equation (1.21) expressed the time interval during which a particle moves from a position to another as the derivative of the action integral with respect to the energy. That formula is easily generalized to three dimensions. In fact

$$\frac{\partial}{\partial E}S(\mathbf{r}_2, \mathbf{r}_1, E, \alpha) = \frac{\partial}{\partial E}\int_{\mathbf{r}_1}^{\mathbf{r}_2} \mathbf{p}\cdot\mathrm{d}\mathbf{r} = \frac{\partial}{\partial E}\int_{\mathbf{r}_1}^{\mathbf{r}_2} \sqrt{2m(E-U(\mathbf{r}))}\,\mathrm{d}s$$

$$= \int_{\mathbf{r}_1}^{\mathbf{r}_2} m\,\mathrm{d}s/\sqrt{2m(E-U(\mathbf{r}))} = \int_{\mathbf{r}_1}^{\mathbf{r}_2} \mathrm{d}s/v = t_2 - t_1 \quad , \qquad (1.38)$$

where v denotes the velocity.

It is interesting to compare this time interval with the time it would take a particle of the same mass, total energy E, and potential energy $U(\mathbf{r})$, to

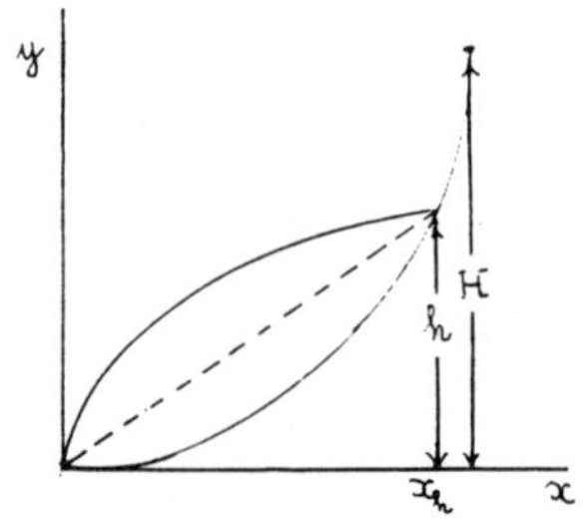

Figure 1.10: Real and varied motion under gravity

go from $\mathbf{r}_1$ to $\mathbf{r}_2$ along a path other than the trajectory. This latter motion requires a frictionless constraint.

Consider as an example the motion of a projectile. Since

$$S(x, y, 0, E, p_x = mv_{0x}) = p_x x$$

$$+[(2E - mv_{0x}^2)^{\frac{3}{2}} - (2E - mv_{0x}^2 - 2mgy)^{\frac{3}{2}}]/3\sqrt{m}g \quad ,$$

the time t from $(0,0)$ to (x,y) is

$$t = \partial S/\partial E = \left[\sqrt{2E - mv_{0x}^2} - \sqrt{2E - mv_{0x}^2 - 2mgy}\right] / g\sqrt{m}$$

$$= \left(v_{0y} - \sqrt{v_{0y}^2 - 2gy}\right) / g \quad ,$$

a formula known to high-school students. The time taken to reach the highest point of the trajectory, $y = h = v_{0y}^2/2g$, $x = x_h = v_{0x}v_{0y}/g$, is, of course, $t = v_{0y}/g$.

Note that instead of choosing p_x as a constant of motion α other than the total energy E, we might have chosen the energy of the y-motion, $E_y = mv_y^2/2 + mgy$. Then we would have had
$t = \partial S_x/\partial E = \partial(\sqrt{2m(E - E_y)}\, x)/\partial E = x\sqrt{m/2(E - E_y)} = x/v_{0x}$.

Suppose now that the body, with the same total energy $E = m(v_{0x}^2 + v_{0y}^2)/2$, travels from $(0,0)$ to (x_h, h) along the straight line of equation $y = hx/x_h$ (see figure 1.10) under the influence of gravity and a frictionless constraint.

The time t' taken will be shorter than $t = v_{0y}/g$,

$$t' = \int m \, \mathrm{d}s/\sqrt{2m(E - mgy)} = \int_0^h \mathrm{d}y\sqrt{1 + (x_h/h)^2}/\sqrt{v_0^2 - 2gy}$$

$$= (1/g)\sqrt{1 + 4v_{0x}^2/v_{0y}^2}\ \left(\sqrt{v_{0x}^2 + v_{0y}^2} - v_{0x}\right) < v_{0y}/g = t \quad .$$

Does a shortest-time curve from $(0, 0)$ to (x_h, h) exist for that energy E? Yes, it is Bernoulli's "brachistochrone".

Its equation is obtained by requiring that

$$\delta \int_A^B \frac{\mathrm{d}s}{v} = 0 \tag{1.39}$$

for variations $\delta \mathbf{r}$ restricted to vanish at A and B.

A calculation similar to that presented in section 1.5 yields the differential equation

$$\frac{\mathrm{d}}{\mathrm{d}x}\left(\frac{y'^2}{(E-U)(1+y'^2)}\right) = \frac{1}{(E-U)^2}\frac{\partial U}{\partial y}y' \quad .$$

For $U = mgy$ one obtains by a first integration

$$y' = \sqrt{\frac{C - (H - y)}{H - y}} \quad ,$$

where C is an integration constant and $H = E/mg$. Note that y' is infinite (tangent is vertical) for $y = H$, indicating that $v_x = 0$ for $y = H$. Thus H is indeed the maximum height reached by a body of energy E.

A second integration putting $H - y = C(1 + \cos\theta)/2$, and a suitable choice of integration constants, yield

$$x = H(\theta + \sin\theta)/2 \quad , \quad y = H(1 - \cos\theta)/2 \quad . \tag{1.40}$$

These are the parametric equations of a cycloid. Thus the brachistochrone is also "tautochrone" (see problem 1.8).

Note that $x = y = 0$ for $\theta = 0$, and $(x = \pi H/2, y = H)$ for $\theta = \pi$. The time taken by the projectile of energy $E = mgH$ to go from $(0, 0)$ to the point characterized by the value θ of the parameter is

$$t'' = \int \frac{\mathrm{d}s}{v} = \sqrt{\frac{H}{2g}}\int_0^y \frac{\mathrm{d}y}{\sqrt{y(H-y)}}$$

$$= \sqrt{\frac{H}{2g}}\int_0^\theta \frac{\sin\theta\ \mathrm{d}\theta}{\sqrt{(1+\cos\theta)(1-\cos\theta)}} = \sqrt{\frac{H}{2g}}\,\theta \quad ,$$

where we have used $\mathrm{d}s = \sqrt{1 + (\mathrm{d}x/\mathrm{d}y)^2}\mathrm{d}y = \sqrt{H/y}\ \mathrm{d}y$, $v = \sqrt{2g(H-y)}$, and the parametric equations.

We now want to find a brachistochrone through the points $(0,0)$ and (x_h, h) of the projectile trajectory. It is easy to show that a brachistochrone for $E = mv_0^2/2 = m(v_{0x}^2 + v_{0y}^2)/2$ and $H = v_0^2/2g > v_{0y}^2/2g = h$ will intersect the projectile trajectory at (x_h, h) for the value $\theta = 2v_{0x}v_{0y}/v_0^2$ of the parameter. The time from $(0,0)$ to (x_h, h) will be

$$t'' = \frac{1}{\sqrt{1 + (v_{0y}/v_{0x})^2}} \frac{v_{0y}}{g} < \frac{v_{0y}}{g} = t \quad .$$

It is also easy to show that $t'' < t'$, see problem 1.9.

1.7 Fermat's principle

We must briefly mention the remarkable similarity between geometrical optics and mechanics of a point-mass. In the former, Fermat's principle

$$\delta \int_A^B \frac{n(\mathbf{r})}{c}\, \mathrm{d}s = 0 \tag{1.41}$$

expresses the fact that the time taken by light to travel from A to B along a light ray ("optical path length") is less than the time it would take along any adjacent path from A to B.

In mechanics, Maupertuis' principle is the analogue of Fermat's principle. The action (not the time) is the analogue of the optical path length.

Fermat's principle is the consequence of the existence of families of "iconal surfaces" to which the families of light rays are orthogonal, in the same way as trajectories are orthogonal to $S =$const surfaces.

An "iconal equation" is obtained as first approximation of the wave equation for a given frequency ν,

$$\nabla^2 \psi + \frac{4\pi^2}{\lambda^2}\psi = 0 \quad , \tag{1.42}$$

where $\lambda = \lambda_0/n = c/n\nu$, λ_0 is the vacuum wavelength, and $n(\mathbf{r})$ is the refractive index.

One begins by expressing ψ in the form

$$\psi = \exp(2\pi \mathrm{i} S(\mathbf{r})/\lambda_0) \quad , \tag{1.43}$$

where $S = S_0 + (\lambda_0/2\pi)S_1 + (\lambda_0/2\pi)^2 S_2 + \ldots$

Substituting in the wave equation, one finds in first approximation the iconal equation

$$(\nabla S_0)^2 = n^2 \quad . \tag{1.44}$$

This is the analogue of the Hamilton-Jacobi equation.

Wave mechanics (de Broglie) is to classical mechanics what wave optics is to geometrical optics. In fact, comparing the Schrödinger equation

$$\nabla^2\psi + \frac{2m}{\hbar^2}(E-U)\psi = 0 \tag{1.45}$$

with equation (1.42) one sees that $\sqrt{2m(E-U)}$ plays the role of a refractive index, while the wave-mechanical wavelength is

$$\lambda = h/\sqrt{2m(E-U)} = h/p(\mathbf{r}).$$

1.8 Chapter 1 problems

1.1 The damped oscillator equation $\ddot{q} = -q - \dot{q}$ is equivalent to the system $\dot{q} = p$, $\dot{p} = -q - \gamma p$. Show that these equations cannot be expressed in the Hamiltonian form (1.3-4).

1.2 The damped oscillator equation of the preceding problem is also equivalent to $\dot{q} = p\exp(-\gamma t)$, $\dot{p} = -q\ \exp(\gamma t)$. Verify that these equations are Hamilton equations, with $H = [p^2\exp(-\gamma t) + q^2\exp(\gamma t)]/2$.

1.3 The equation $\ddot{q} = -(q - q_0)$ for a harmonic oscillator with equilibrium point $q = q_0$, can be converted into the system of equations $\dot{q} = p - p_0$, $\dot{p} = -q + q_0$, where p_0 is an arbitrary constant. These are Hamilton equations, since $\partial(p - p_0)/\partial q = 0$ and $\partial(-q + q_0)/\partial p = 0$. (i) Find the Hamiltonian. (ii) What is the (p, q) trajectory for energy E?

1.4 Show that the 4-dimensional phase space volume is conserved in the elastic collision of two particles in one dimension.

1.5 Non-conservative forces invalidate Liouville's theorem. Show that, if $\dot{p} = -\gamma\dot{q}$ $(\gamma > 0)$, $\dot{q} = p/m$, areas in the (p, q)-plane decrease exponentially with time.

1.6 According to Liouville's theorem, conservative forces cannot change the small-scale phase space density of a system of particles. Yet they can change the large-scale density producing accumulation.

Give a simple one-dimensional example (two-dimesnional phase space). Look up also S. van der Meer, "Stochastic cooling and the accumulation of antiprotons", *Revs. Mod. Phys.*,**57**(1985)689.

1.7 For the two-dimensional oscillator with $T = m(\dot{x}^2 + \dot{y}^2)/2$, $U = m\omega^2(x^2 + y^2)/2$, and energy $E = E_x + E_y$ with $E_x = m\omega^2 a^2/2$, $E_y = m\omega^2 b^2/2$, the action integral is given by $S = S_x + S_y$, with S_x obtained from equation (1.22) with the replacement $q_2 \to x$, $q_1 \to x_0$ and S_y with the replacements $a \to b$, $q_2 \to y$, $q_1 \to y_0$.

Show that

$$t - t_0 = [\sin^{-1}(x/a) - \sin^{-1}(x_0/a)]/\omega = [\sin^{-1}(y/b) - \sin^{-1}(y_0/b)]/\omega \quad .$$

Equating these two expressions for $t - t_0$, show that the trajectory equation is

$$b^2x^2 + a^2y^2 - 2ab\cos(\alpha - \beta)xy = a^2b^2\sin^2(\alpha - \beta) \quad \text{(an ellipse)} \quad ,$$

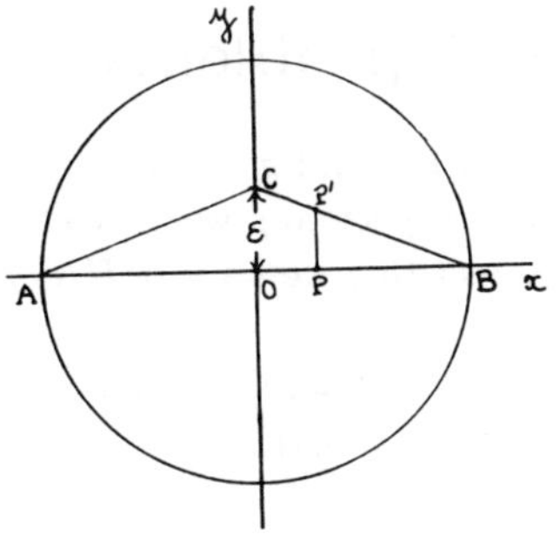

Figure 1.11: Illustration of Maupertuis principle

where $\alpha = \cos^{-1}(x_0/a)$ and $\beta = \cos^{-1}(y_0/b)$.

1.8 A particle, initially at rest, slides down without friction along the brachistochrone (1.40). Show that the time required to reach the bottom is $\pi\sqrt{H/2g}$, independent of the initial position. Thus the brachistochrone is also tautochrone.

1.9 With the notation of section 1.6, show that $t'' < t'$.

1.10 In figure 1.11, the circle represents an earth meridian, A and B are the initial and final positions of an object released from rest at A and travelling to B along a smooth straight tunnel.

Assuming ϵ infinitesimal, show that

$$D = \left(\int \sqrt{2m(E - U(P'))}\,\mathrm{d}s'\right)_{ACB} - \left(\int \sqrt{2m(E - U(P))}\,\mathrm{d}s\right)_{AOB} > 0$$

as expected from the Maupertuis principle.

Solutions to ch. 1 problems

S1.1 $\dot{p} = -\partial H/\partial q$, $\dot{q} = \partial H/\partial p$ would give $-q - \gamma p = -\partial H/\partial q$, $p = \partial H/\partial p$. Differentiating the first with respect to p we have $\partial^2 H/\partial p \partial q = \gamma$, while differentiating the second with respect to q we have $\partial^2 H/\partial q \partial p = 0$.

S1.3 (i) $H = [(p-p_0)^2 + (q-q_0)^2]/2$ (ii) $q = q_0 + \sqrt{2E}\ \cos t$, $p = p_0 - \sqrt{2E}\ \sin t$, circle of radius $\sqrt{2E}$ and center (p_0, q_0).

S1.4 $Mp_1 = (m_1 - m_2)p_{10} + 2m_1 p_{20}$, $Mp_2 = 2m_2 p_{10} + (m_2 - m_1)p_{20}$,
$M = m_1 + m_2$
$\int \mathrm{d}p_1 \mathrm{d}q_1 \mathrm{d}p_2 \mathrm{d}q_2 = \int \mathrm{d}q_{10} \mathrm{d}q_{20} \int |\partial(p_1, p_2)/\partial(p_{10}, p_{20})| \mathrm{d}p_{10} \mathrm{d}p_{20} = \int \mathrm{d}p_{10} \mathrm{d}q_{10} \mathrm{d}p_{20} \mathrm{d}q_{20}$,
since

$$\frac{\partial(p_1, p_2)}{\partial(p_{10}, p_{20})} = \begin{vmatrix} M^{-1}(m_1 - m_2) & 2M^{-1}m_1 \\ 2M^{-1}m_2 & M^{-1}(m_2 - m_1) \end{vmatrix} = -1$$

S1.5

$$\frac{\partial(p_{t+\mathrm{d}t}, q_{t+\mathrm{d}t})}{\partial(p_t, q_t)} = \begin{vmatrix} 1 - m^{-1}\gamma \mathrm{d}t & 0 \\ m^{-1}\mathrm{d}t & 1 \end{vmatrix} = 1 - \frac{\gamma}{m}\,\mathrm{d}t \quad ,$$

and so $\mathrm{d}A_t/\mathrm{d}t = -m^{-1}\gamma A_t$, $A_t = A_0 \exp(-m^{-1}\gamma t)$.

S1.6 In figure 1.2, think of two small phase space regions around $(q = a, p = 0)$ and $(q = -a, p = 0)$ filled with particles. If the p-axis of the ellipse is much smaller than the q-axis (k very small), after a quarter of a period the two regions will have moved to $(q = 0, p = \pm m\omega a)$ with resulting accumulation around $(q = 0, p = 0)$.

In an antiproton accumulator the empty (phase) spaces between the particles are squeezed outwards, while each antiproton is pushed towards the center of the distribution.

S1.7 Regarding E_y (and, therefore, b) as constant of motion other than E, since $a = \omega^{-1}\sqrt{2(E - E_y)/m}$, the first of the two expressions for $t - t_0$ is given by $t - t_0 = \partial S_x/\partial E = (\partial S_x/\partial a)(\partial a/\partial E)$. Also $t - t_0 = \partial S_y/\partial E = (\partial S_y/\partial b)(\partial b/\partial E)$.

The trajectory equation is obtained by equating the two expressions for $t - t_0$. This gives $\sin^{-1}(x/a) - \sin^{-1}(y/b) = \sin^{-1}(x_0/a) - \sin^{-1}(y_0/b)$, which yields the desired result by using the identity

$$\sin^{-1}\epsilon - \sin^{-1}\eta = \sin^{-1}\left(\epsilon\eta + \sqrt{(1-\epsilon^2)(1-\eta^2)}\right).$$

S1.8 It is convenient to use the parametric equations of the cycloid in terms of the arc length s from the lowest point, $x = \ldots$, $y = s^2/4H$. Then

$$m\ddot{s} = -mg\ \mathrm{d}y/\mathrm{d}s = -mgs/2H,\ \ddot{s} + \omega^2 s = 0 \text{ with } \omega = \sqrt{g/2H},$$

harmonic motion of the cycloidal pendulum.
The time sought is $T/4 = \pi/2\omega = \pi\sqrt{H/2g}$.

S1.9 With $x = v_{0y}/v_{0x} > 0$, $t''?t'$ is equivalent to

$$\frac{1}{\sqrt{1+x^2}} ? \sqrt{1+\frac{4}{x^2}}\left(\sqrt{1+\frac{1}{x^2}}-\frac{1}{x}\right) \quad .$$

By simple algebra this yields $0?x^8 + 8x^6 + 20x^4 + 12x^2$. Thus "?" is "<".

S1.10

$$P(x,0),\ P'(x,\delta y) \text{ with } \delta y = \epsilon\eta(x+\eta R)/R,\ \eta = -x/|x|,$$
$$U(P) = mg(x^2-3R^2)/2R,\ U(P') = mg(x^2+(\delta y)^2-3R^2)/2R,$$
$$E = -mgR,\ \sqrt{E-U(P')} - \sqrt{E-U(P)} = -mg(\delta y)^2/(4R\sqrt{E-U(P)}\,),$$
$$\mathrm{d}s' - \mathrm{d}s \simeq \epsilon^2 \mathrm{d}x/(2R^2),$$

$$D \simeq \int_{-R}^{R} [\sqrt{2m(E-U(P'))} - \sqrt{2m(E-U(P))} + \frac{\epsilon^2}{2R^2}\sqrt{2m(E-U(P))}\,]\,\mathrm{d}x$$

$$\simeq -\frac{m^{\frac{3}{2}} g\epsilon^2}{\sqrt{2}R^3}\int_{-R}^{R}\frac{x(x+\eta R)\mathrm{d}x}{\sqrt{E-U(P)}} > 0 \quad ,$$

since $x(x+\eta R) < 0$ in the integration interval.

Chapter 2

EXAMPLES OF PARTICLE MOTION

We collect a body of notions to be used in the more formal parts of the book.

2.1 Central forces

If a particle moves under the action of a central force $\mathbf{f} = f(r)\mathbf{r}/r$, its angular momentum $\mathbf{l}$ with respect to the center of force $\mathbf{r} = 0$ is conserved. This follows from the vanishing of the torque τ due to $\mathbf{f}$ and $\mathbf{r}$ being along the same line. The motion is confined to a plane normal to $\mathbf{l}$. Expressing Newton's second law in polar coordinates, we have

$$ma_r = f(r) \quad \text{and} \quad ma_\theta = 0 \quad , \tag{2.1}$$

where

$$a_r = \ddot{r} - r\dot{\theta}^2 \quad \text{and} \quad a_\theta = r\ddot{\theta} + 2\dot{r}\dot{\theta} = \frac{1}{r}\frac{\mathrm{d}}{\mathrm{d}t}(r^2\dot{\theta}) \quad . \tag{2.2}$$

These second-order equations can be replaced by the following first-order system:

$$\dot{p}_r = f(r) + \frac{p_\theta^2}{mr^3} \quad , \quad \dot{r} = \frac{p_r}{m} \quad , \quad \dot{p}_\theta = 0 \quad , \quad \dot{\theta} = \frac{p_\theta}{mr^2} \quad , \tag{2.3}$$

where $p_\theta = mr^2\dot{\theta}$ has the physical dimensions of an angular momentum. In fact, $p_\theta = l = |\mathbf{l}|$ <u>is</u> the magnitude of the conserved angular momentum.

If the force is conservative, the energy E is a constant of motion. Denoting by $U(r)$ the potential energy ($f(r) = -U'(r)$), E can be expressed in the form:

$$E = m(\dot{r}^2 + r^2\dot{\theta}^2)/2 + U(r) = m\dot{r}^2/2 + U_e(r) \quad , \tag{2.4}$$

where

$$U_e(r) = U(r) + \frac{l^2}{2mr^2} \tag{2.5}$$

is an "effective potential energy" for the radial motion. A second-order equation for r can be written in the form

$$m\ddot{r} = -U'(r) + \frac{l^2}{mr^3} = -U_e'(r) \quad . \tag{2.6}$$

Note the important formulae

$$t = t_0 + \int_{r_0}^{r} \frac{\mathrm{d}r}{\sqrt{2(E - U_e(r))/m}} \tag{2.7}$$

and

$$\theta = \theta_0 + \frac{l}{m}\int_{t_0}^{t} \frac{\mathrm{d}t}{(r(t))^2} = \theta_0 + \frac{l}{m}\int_{r_0}^{r} \frac{\mathrm{d}r}{\dot{r}r^2} \quad ,$$

$$\theta = \theta_0 + l\int_{r_0}^{r} \frac{\mathrm{d}r}{r^2\sqrt{2m(E - U_e(r))}} \quad , \tag{2.8}$$

where the square roots must be taken with the signs of $\mathrm{d}r(t)/\mathrm{d}t$ and $\mathrm{d}r(\theta)/\mathrm{d}\theta$, respectively.

The study of the r-motion is similar to that of the one-dimensional motion of chapter 1 but for the presence of the "centrifugal term" in the effective potential energy.

Thus if $f(r)$ is attractive but $\lim_{r\to 0} r^2|U(r)| = 0$, the centrifugal term prevails at short distances, and there is a mimimum distance of approach to the force center.

On the other hand, if $\lim_{r\to 0} r^2|U(r)| = \infty$ (e.g. $U = -a/r^3$, $a > 0$) the particle falls into the center of force.

2.2 Circular and quasi-circular orbits

Equilibrium points of one-dimensional motion correspond to circular orbits. There will be a circular orbit of radius R if $(\mathrm{d}U_e(r)/\mathrm{d}r)_{r=R} = 0$, namely $f(R)+l^2/(mR^3) = 0$, which is nothing but the elementary $mv^2/R = -f(R)$ ($f(R) < 0$). The period will be $T = 2\pi\sqrt{mR/|f(R)|}$.

We now study small oscillations about a stable circular orbit. Denoting by R the radius of a circular orbit, for r differing little from R we have

$$m\,\mathrm{d}^2(r-R)/\mathrm{d}t^2 = m\,\ddot{r} = -U_e'(r)$$

$$\simeq -[U_e'(R) + U_e''(R)(r-R)] = -U_e''(R)(r-R) \quad . \tag{2.9}$$

The circular orbit is stable if $U_e''(R) > 0$. The angular frequency of small oscillations about a stable circular orbit is

$$\omega_{\mathrm{osc}} = \sqrt{U_e''(R)/m} = \sqrt{(U''(R) + 3l^2/mR^4)/m} \tag{2.10}$$

or, using $l^2/mR^3 = -f(R) = U'(R)$,

$$\omega_{\mathrm{osc}} = \sqrt{(U''(R) + 3U'(R)/R)/m} \quad , \tag{2.11}$$

and the period is

$$T_{\mathrm{osc}} = 2\pi\sqrt{m/(U''(R) + 3U'(R)/R)} \quad , \tag{2.12}$$

while that of the circular orbit is

$$T = 2\pi\sqrt{mR/U'(R)} \quad . \tag{2.13}$$

For an almost circular orbit, the angle $\Delta\theta$ between two successive pericenters is approximately

$$\Delta\theta = \dot{\theta}T_{\mathrm{osc}} = \frac{2\pi l}{R\sqrt{m(R^2U''(R) + 3RU'(R))}} = 2\pi\sqrt{\frac{U'(R)}{RU''(R) + 3U'(R)}} \quad . \tag{2.14}$$

For the Newton force ($f(r) = -k/r$) this gives $\Delta\theta = 2\pi$, and for the isotropic oscillator ($f(r) = -kr$) $\Delta\theta = \pi$. In both cases the orbits are closed.

Note that for $U(r) = Cr^\alpha$ and $U(r) \coloneqq C\ln(r/r_0)$, $\Delta\theta$ does not depend on R.

For $U(r) = -(k/r) - h/r^2$ the orbits are not closed (rosettes). If h is small

$$\Delta\theta = 2\pi\sqrt{1 + \frac{2h}{kR}} \simeq 2\pi\left(1 + \frac{h}{kR}\right) \quad . \tag{2.15}$$

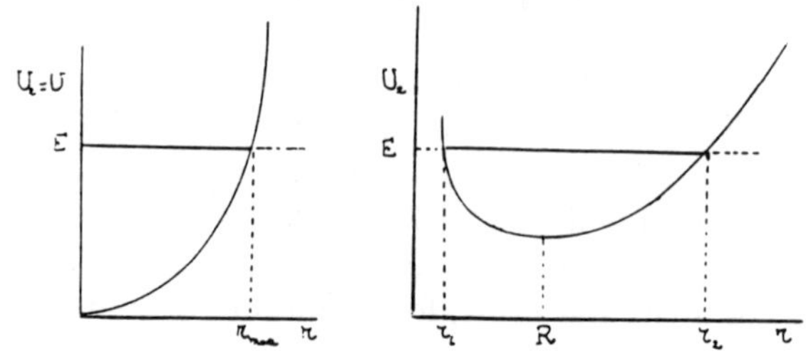

Figure 2.1: Isotropic harmonic oscillator

The average angular velocity of precession is

$$\omega_{\mathbf{pr}} = \frac{\Delta\theta - 2\pi}{T} = \frac{2\pi h}{kRT} = \frac{2\pi hm}{l^2 T} \quad . \tag{2.16}$$

For $h > 0$ (attractive cubic force), the precession is in the direction of motion.

2.3 Isotropic harmonic oscillator

The force is central and elastic ($f(r) = -kr$), $U_e(r) = (kr^2 + l^2/mr^2)/2$ is shown in figures 2.1 (left $l = 0$, right $l > 0$).

It is convenient to introduce the variable $s = 1/r^2$. The turning points r_1 and $r_2 > r_1$ are then given by $r_1 = 1/\sqrt{s_1}$ and $r_2 = 1/\sqrt{s_2}$, ($s_1 > s_2$) with

$$\begin{pmatrix} s_1 \\ s_2 \end{pmatrix} = \frac{mE}{l^2}\left(1 \pm \sqrt{1 - \frac{l^2\omega^2}{E^2}}\right) \quad . \tag{2.17}$$

The minimum of U_e is for $r = R$, where R is the radius of the stable circular orbit $R = \sqrt{l/m\omega}$. Since $U_e(R) = \omega l$, for given l must be $E > \omega l$.

Equation (2.8) gives

$$\theta = \theta_0 - \frac{1}{2}\int_{s_0}^{s} \frac{\mathrm{d}s}{\sqrt{(s_1 - s)(s - s_2)}} \quad , \tag{2.18}$$

$$\theta = \theta_0 + \frac{1}{2}\left[\cos^{-1}\left(\frac{2s - s_1 - s_2}{s_1 - s_2}\right) - \cos^{-1}\left(\frac{2s_0 - s_1 - s_2}{s_1 - s_2}\right)\right] \quad . \tag{2.19}$$

Taking $s_0 = (s_1 + s_2)/2$ and $\theta_0 = \pi/4$, so that

$$\theta = \frac{1}{2}\cos^{-1}\left(\frac{2s - s_1 - s_2}{s_1 - s_2}\right) \quad , \tag{2.20}$$

we find the ellipse.

$$s = [s_1 + s_2 + (s_1 - s_2)\cos(2\theta)]/2 \quad , \tag{2.21}$$

$$\frac{1}{r^2} = \frac{mE}{l^2}\left(1 + \sqrt{1 - \frac{\omega^2 l^2}{E^2}}\cos(2\theta)\right) \quad . \tag{2.22}$$

Note that $r = r_1$ for $\theta = 0$ and $r = r_2$ for $\theta = \pi/2$, in accordance with what was noted after (2.14). See also problems 2.3 and 2.5.

We end this section with the calculation of the cyclic action integral $J_r = \oint p_r \mathrm{d}r$, where $p_r = m\dot{r} = (l/r^2)\mathrm{d}r/\mathrm{d}\theta$:

$$J_r = l\left(1 - \frac{\omega^2 l^2}{E^2}\right)\int_0^{2\pi} \frac{\sin^2(2\theta)\mathrm{d}\theta}{\left[1 + \sqrt{1 - \omega^2 l^2/E^2}\cos(2\theta)\right]^2} \quad ,$$

$$J_r = -2\pi l + \frac{2\pi E}{\omega} \quad . \tag{2.23}$$

But $2\pi l = \int_0^{2\pi} l\mathrm{d}\theta = \int_0^{2\pi} p_\theta \mathrm{d}\theta$, with p_θ as defined in (2.3), and so

$$2\pi E/\omega = \oint p_r \mathrm{d}r + \oint p_\theta \mathrm{d}\theta = J_r + J_\theta \quad . \tag{2.24}$$

Since $\oint \mathbf{p}\cdot \mathrm{d}\mathbf{r} = m\oint(\dot{x}\ \mathrm{d}x + \dot{y}\ \mathrm{d}y) = m\oint(\dot{r}\ \mathrm{d}r + r^2\dot{\theta}\ \mathrm{d}\theta)$, we have $J_r + J_\theta = J_x + J_y$. Therefore by Sommerfeld's quantization the quantized energy would be $E = \hbar\omega(n_x + n_y)$ or $E = \hbar\omega(n_r + n_\theta)$.

2.4 The Kepler problem

We discuss negative-energy orbits for

$$U(r) = -\frac{k}{r} - \frac{h}{r^2} \tag{2.25}$$

with $h \ll l^2/2m$.

Note that the $-h/r^2$ potential <u>cannot</u> be compensated by changing the value of l. In fact, l appears not only under the square root, but also as a factor multiplying the integral in equation (2.8). This gives

$$\theta = \theta_0 + l\int_{r_0}^{r} \frac{\mathrm{d}r}{r^2\sqrt{2m[E + (k/r) + (2mh - l^2)/2mr^2]}} \quad . \tag{2.26}$$

Introducing the variable $s = 1/r$, θ can be expressed in the form

$$\theta = \theta_0 - \frac{1}{\alpha}\int_{s_0}^{s} \frac{\mathrm{d}s}{\sqrt{(s_1 - s)(s - s_2)}} \quad , \tag{2.27}$$

where

$$\alpha = \sqrt{1 - 2mh/l^2} \quad , \tag{2.28}$$

$$\begin{pmatrix} s_1 \\ s_2 \end{pmatrix} = \frac{mk}{\alpha^2 l^2}\left(1 \pm \sqrt{1 - \frac{2|E|l^2\alpha^2}{mk^2}}\right) \quad . \tag{2.29}$$

The integral is elementary. We find

$$\theta = \theta_0 - \frac{1}{\alpha}\left[\sin^{-1}\left(\frac{2s - s_1 - s_2}{s_1 - s_2}\right) - \sin^{-1}\left(\frac{2s_0 - s_1 - s_2}{s_1 - s_2}\right)\right] \quad . \tag{2.30}$$

Taking $s_0 = s_1$ and $\theta_0 = 0$, we have

$$\theta = \frac{1}{\alpha}\cos^{-1}\left(\frac{2s - s_1 - s_2}{s_1 - s_2}\right) \quad , \tag{2.31}$$

$$s = [s_1 + s_2 + (s_1 - s_2)\cos(\alpha\theta)]/2 \quad . \tag{2.32}$$

We see at once that the orbit is not closed if $\alpha \neq 1$ ($h \neq 0$). The angle of precession per revolution is then $2\pi(\alpha^{-1} - 1) \simeq 2\pi mh/l^2$ as anticipated in equation (2.16). For $\alpha = 1$ ($h = 0$) the orbit is an ellipse of equation

$$\frac{1}{r} = \frac{1 + \epsilon\cos\theta}{a(1 - \epsilon^2)} \quad . \tag{2.33}$$

Here

$$a = \frac{1}{2}\left(\frac{1}{s_1} + \frac{1}{s_2}\right) = \frac{r_1 + r_2}{2} = \frac{k}{2|E|} \quad , \tag{2.34}$$

$$E = -k/2a \; , \; \epsilon = \sqrt{1 - 2|E|l^2/mk^2} \; , \; a(1 - \epsilon^2) = l^2/mk \quad . \tag{2.35}$$

We now calculate the action integral $J_r = \oint p_r \mathrm{d}r$, where $p_r = (l/r^2)\mathrm{d}r/\mathrm{d}\theta = l\epsilon\sin\theta/a(1-\epsilon^2)$, $\mathrm{d}r = a\epsilon(1-\epsilon^2)\sin\theta/(1 + \epsilon\cos\theta)^{-2}\mathrm{d}\theta$. A partial integration yields

$$J_r = \oint p_r \mathrm{d}r = -2\pi l + l\int_0^{2\pi} \frac{\mathrm{d}\theta}{1 + \epsilon\cos\theta} \quad . \tag{2.36}$$

The evaluation of the integral is a well-known exercise in complex integration: $\int_0^{2\pi} \mathrm{d}\theta/(1+\epsilon\cos\theta) = (2/\mathrm{i}\epsilon)\oint \mathrm{d}z/(z^2+2\epsilon^{-1}z+1)$, where the path integral is along a circle of unit radius and center at the origin. The pole $z = (-1 + \sqrt{1-\epsilon^2})/\epsilon$ falls inside this circle.

We find

$$J_r = -J_\theta + \frac{2\pi l}{\sqrt{1 - \epsilon^2}} = -J_\theta + 2\pi k\sqrt{\frac{m}{2|E|}} \quad . \tag{2.37}$$

Hence

$$|E| = \frac{2\pi^2 mk^2}{(J_r + J_\theta)^2} \quad , \tag{2.38}$$

and, by Sommerfeld's quantization and $k = e^2$,

$$E = -\frac{me^4}{2\hbar^2 (n_r + n_\theta)^2} \tag{2.39}$$

for the hydrogen energy levels.

Note that both here and in equation (2.24) for the isotropic harmonic oscillator, we would expect to see $J_r + J_\theta + J_\phi$ and $n_r + n_\theta + n_\phi$ instead of $J_r + J_\theta$ and $n_r + n_\theta$. The absence of J_ϕ is due to our having confined the orbit to a meridian plane so that $\dot{\phi} = 0$.

Schwinger (see G. Baym,*Lectures on Quantum Mechanics* (Addison-Wesley,1969) problem 3, p. 179) noticed that the radial Schrödinger equation for the hydrogen atom can be reduced to the equation for the two-dimensional isotropic harmonic oscillator. In the present context, let us compare the equation for the electron orbit

$$\frac{1}{r} = \frac{me^2}{l^2}\left(1 + \sqrt{1 - \frac{2|E_{\text{hyd}}|l^2}{me^4}}\cos\theta\right) \tag{2.40}$$

with that for the isotropic harmonic oscillator

$$\frac{1}{\rho^2} = \frac{mE_{\text{osc}}}{l^2}\left(1 + \sqrt{1 - \frac{\omega^2 l^2}{E_{\text{osc}}^2}}\cos(2\varphi)\right) \quad , \tag{2.41}$$

where we have renamed the polar angle and the distance from the origin. With the correspondence $r \to \rho^2$ and $\theta \to 2\varphi$, the hydrogen orbit becomes that of an oscillator of energy $E_{\text{osc}} = e^2$ and frequency $\omega = \sqrt{2|E_{\text{hyd}}|/m}$.

Therefore, if in $E_{\text{osc}} = \hbar\omega(n_r + n_\theta)$ we replace $E_{\text{osc}} \to e^2$ and $\omega \to \sqrt{2|E_{\text{hyd}}|/m}$, we find $e^2 = \hbar\sqrt{2|E_{\text{hyd}}|/m}\,(n_r + n_\theta)$,

$$|E_{\text{hyd}}| = \frac{me^4}{2\hbar^2 (n_r + n_\theta)^2} \quad . \tag{2.42}$$

The period is given by

$$T = \frac{\mathrm{d}J_r}{\mathrm{d}E} = \pi k\sqrt{\frac{m}{2|E|^3}} = 2\pi\sqrt{\frac{ma^3}{k}} \quad , \tag{2.43}$$

yielding Kepler's third law ($k = GmM$)

$$T^2 = \frac{4\pi^2 a^3}{GM} \quad . \tag{2.44}$$

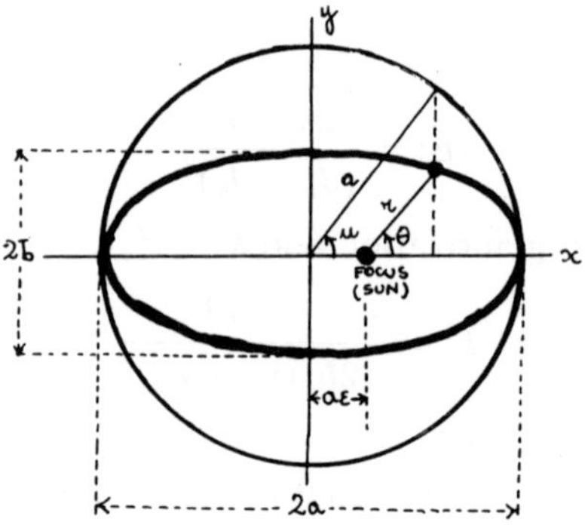

Figure 2.2: Eccentric anomaly

In contrast to θ, the angle u ("eccentric anomaly") shown in figure (2.2) is simply related to the time t.

We see from the figure that $a\cos u = a\epsilon + r\cos\theta$. Substituting $r\cos\theta = [a(1-\epsilon^2) - r]/\epsilon$ (from equation (2.33)), we find $r = a(1-\epsilon\cos u)$, $s = 1/a(1-\epsilon\cos u)$. Now

$$s_1 - s = \frac{1}{a(1-\epsilon)} - \frac{1}{a(1-\epsilon\cos u)} = \frac{\epsilon(1-\cos u)}{a(1-\epsilon)(1-\epsilon\cos u)} \quad ,$$

$$s - s_2 = \frac{1}{a(1-\epsilon\cos u)} - \frac{1}{a(1+\epsilon)} = \frac{\epsilon(1+\cos u)}{a(1+\epsilon)(1-\epsilon\cos u)} \quad ,$$

and so

$$t = t_0 + \int_{r_0}^{r} \frac{\mathrm{d}r}{\sqrt{2(E - U_e(r))/m}}$$

$$= \frac{m}{l}\int_{r_0}^{r} \frac{\mathrm{d}r}{\sqrt{(s_1 - s)(s - s_2)}} = \frac{ma^2\sqrt{1-\epsilon^2}}{l}(u - \epsilon\sin u) \quad , \tag{2.45}$$

where we have taken $t_0 = 0$ for $u_0 = 0$. Hence

$$u - \epsilon\sin u = nt \tag{2.46}$$

with

$$n = \frac{l}{ma^2\sqrt{1-\epsilon^2}} = \sqrt{\frac{k}{ma^3}} \quad . \tag{2.47}$$

Comparing (2.47) with (2.43) we see that $n = 2\pi/T$ is the average angular velocity. Of course, we could have seen this by taking $t = T$ and $u = 2\pi$ in (2.46).

This equation is useful for the calculation of time averages, since

$$\mathrm{d}t/T \;=\; (1 - \epsilon\cos u)\mathrm{d}u/2\pi \quad .$$

For example, with $x = r\ \cos\theta$ equation (2.33) gives $r + \epsilon x = a(1-\epsilon^2)$. Denoting by $\langle r\rangle$ and $\langle x\rangle$ the time averages of r and x, we have

$$\epsilon\langle x\rangle - a(1-\epsilon^2) = -\langle r\rangle = -\frac{1}{T}\int_0^T a(1-\epsilon\cos u)dt$$

$$= -\frac{1}{2\pi}\int_0^{2\pi} a(1-\epsilon\cos u)^2 du = -a(1+\epsilon^2/2) \quad .$$

Therefore $\epsilon\langle x\rangle = -3a\epsilon^2/2$. Since $\langle y\rangle = 0$, we find

$$\langle \mathbf{r}\rangle = -(3a\epsilon/2)\hat{\mathbf{e}}_x = -(3a/2)\mathbf{A} \quad ,$$

where $\mathbf{A}$ is a vector of magnitude ϵ pointing from the center of force to the pericenter, the L-R-L vector discussed in the next section.

2.5 The L-R-L vector

For $U = -k/r$, the Laplace-Runge-Lenz (L-R-L) vector

$$\mathbf{A} = -\frac{\mathbf{r}}{r} + \frac{\mathbf{v}\times\mathbf{l}}{k} \tag{2.48}$$

is a constant of motion ($d\mathbf{A}/dt = 0$) as is easy to prove. [1] At the point of nearest approach, $\mathbf{A} = (-1 + l^2/kmr)\mathbf{r}/r$ for both closed and open orbits. Using (2.33), the last of (2.35), and anticipating (2.54) and the last of (2.55), it is easy to show that at the point of nearest approach $\mathbf{A} = \epsilon k\mathbf{r}/|k|r$ for both closed and open orbits. Thus $|\mathbf{A}| = \epsilon$, and $\mathbf{A}$ points from the center of force to the point of nearest approach for $k > 0$ and E negative or positive, and in the opposite direction for $k < 0$ and E, of course, positive.

Multiplying vectorially both sides of the equation $k\mathbf{r}/r = \mathbf{v}\times\mathbf{l} - k\mathbf{A}$ by $\mathbf{l}$ and squaring, one finds

$$(\mathbf{v} - k\mathbf{l}\times\mathbf{A}/l^2)^2 = k^2/l^2 \quad . \tag{2.49}$$

Thus the tip of $\mathbf{v}$ is on a circle of center $k\mathbf{l}\times\mathbf{A}/l^2$ and radius k/l. Note that $|\mathbf{l}\times\mathbf{A}| = l\epsilon$. For closed orbits $\mathbf{v}$ describes the whole circle (2.49) and one has

$$\frac{v_{\max}}{v_{\min}} = \frac{1+\epsilon}{1-\epsilon} = \frac{r_{\max}}{r_{\min}} \quad . \tag{2.50}$$

[1] The "accidental degeneracy" of the energy levels of the hydrogen atom, i.e. their independence of the angular momentum quantum number, is due to the inverse-square form of the Coulomb force. This endows the Lagrangian with a special symmetry property (see problem 6.24), which is responsible for the constancy of the L-R-L vector.

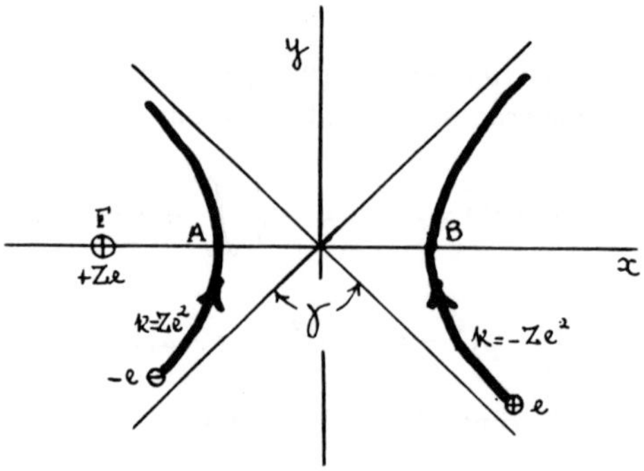

Figure 2.3: Rutherford orbits

Therefore $v_{\max} r_{\min} = v_{\min} r_{\max}$ as one knows from General Physics. For open orbits (Kepler-Rutherford trajectories) see section 2.6.

If the force acting on the particle is $\mathbf{f} = -k\mathbf{r}r^{-3} + \delta\mathbf{f}$, one finds easily

$$\frac{\mathrm{d}\mathbf{A}}{\mathrm{d}t} = \frac{1}{km}\,\delta\mathbf{f} \times \mathbf{l} + \frac{1}{k}\,\mathbf{v} \times (\mathbf{r} \times \delta\mathbf{f}) \quad . \tag{2.51}$$

If $\delta\mathbf{f}$ is central ($\delta\mathbf{f} = \delta f(r)\,\mathbf{r}/r$), this reduces to

$$\frac{\mathrm{d}\mathbf{A}}{\mathrm{d}t} = \frac{1}{kmr}\,\delta f(r)\,\mathbf{r} \times \mathbf{l} \quad . \tag{2.52}$$

A derivation of equation (2.15) based on this formula is presented in problem 2.6.

The L-R-L vector is "Laplace's second vector", the first being the angular velocity. It was used by W. Lenz who took it from a textbook by C. Runge. Neither of these two authors claimed it as an original finding. This historical question was discussed by H. Goldstein, *Am. J. Phys.*,**43**(1975)737 and **44**(1976)1123.

2.6 Open Kepler-Rutherford orbits

For $U = -k/r$ we consider the case $E > 0$, but allow k to be positive (attractive force) or negative (repulsive force). We find

$$\frac{1}{r} = \frac{m|k|}{l^2}\left(\frac{k}{|k|} + \sqrt{1 + \frac{2El^2}{mk^2}}\,\cos\theta\right) \quad , \tag{2.53}$$

$$\frac{1}{r} = \frac{(k/|k|) + \epsilon\cos\theta}{a(\epsilon^2 - 1)} \tag{2.54}$$

with

$$\epsilon = \sqrt{1 + 2El^2/mk^2} > 1 \;,\; a(\epsilon^2 - 1) = l^2/m|k| \quad . \tag{2.55}$$

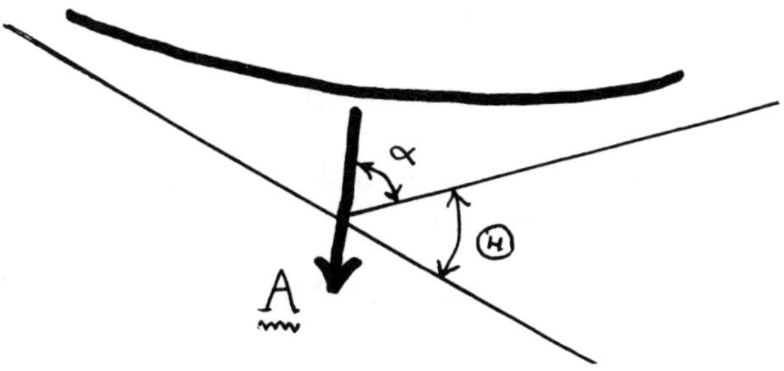

Figure 2.4: L-R-L vector for Rutherford scattering

The orbits are hyperbolae (see figure 2.3) $(x^2/a^2) - (y^2/b^2) = 1$, $a = l^2/m|k|(\epsilon^2 - 1)$, $b = a\sqrt{\epsilon^2 - 1} = |k|/2E$.

	A	B	F
x	$-a$	a	$-a\epsilon$
r	$a(\epsilon - 1)$	$a(\epsilon + 1)$	0
θ	0	0	

The positron travels from $r = \infty$, $\theta^+_{\text{in}} = -\cos^{-1}(1/\epsilon) < 0$ to $r = \infty$, $\theta^+_{\text{out}} = \cos^{-1}(1/\epsilon) > 0$, $|\theta^+_{\text{in}}| = |\theta^+_{\text{out}}| < \pi/2$.

The electron travels from $r = \infty$, $\theta^-_{\text{in}} = \theta^+_{\text{in}} - \gamma < 0$ to $r = \infty$, $\theta^-_{\text{out}} = \theta^+_{\text{out}} + \gamma > 0$, $|\theta^-_{\text{in}}| = |\theta^-_{\text{out}}| > \pi/2$.

As one knows from analytical geometry, b is the "impact parameter", shortest distance of F from the asymptotes $y/x = \pm b/a$. For $k > 0$ (attractive force), $E = mv_0^2/2$, where v_0 is the velocity of the incident particle at infinite distance from F. A simple calculation gives $l = bmv_0$. In this case the tip of $\mathbf{v}$ does not cover the whole of the circle (2.49). One finds $v_{\max} = k(1+\epsilon)/l$ as for closed orbits. In order to find $v_{\min} = v_0$, one must substitute $v_x = -v_0 \cos\theta^+_{\text{out}} = -v_0/\epsilon$, $v_y = v_0 \sin\theta^+_{\text{out}}$ in (2.49). This gives $v_0 = (k/l)\sqrt{\epsilon^2 - 1} = b/r_{\min}$, $v_{\max} r_{\min} = v_{\min} b$.

Zero-energy orbits: These are possible only for $k > 0$. They are parabolas ($\epsilon = 1$)

$$\frac{1}{r} = \frac{mk}{l^2}(1 + \cos\theta) \quad . \tag{2.56}$$

The L-R-L vector is a useful tool also in treating open orbits. In order to clarify its meaning in this case, we define

$$\mathbf{I} = \frac{km\mathbf{A}}{|\mathbf{p}|^2} = \left(1 - \frac{km}{|\mathbf{p}|^2 r}\right)\mathbf{r} - \frac{\mathbf{r}\cdot\mathbf{p}}{|\mathbf{p}|^2}\mathbf{p} \quad .$$

If the center of force is inactive, then

$$\lim_{k\to 0} \mathbf{I} = \mathbf{r} - (\mathbf{r}\cdot\mathbf{p})\mathbf{p}/|\mathbf{p}|^2 = \mathbf{r}_\perp \quad .$$

Using the L-R-L vector, it is easy to show that the scattering angle Θ in Rutherford's scattering is given by

$$\tan(\Theta/2) = qQ/2Eb \quad , \tag{2.57}$$

where q is the charge of the scattered particle, Q that of the fixed scattering center ($qQ > 0$), E is the energy and b the impact parameter.

In fact, $\mathbf{A}\cdot\mathbf{p} = -\mathbf{r}\cdot\mathbf{p}/r$, $\lim_{r\to\infty}\mathbf{A}\cdot\mathbf{p} = -|\mathbf{p}_{\text{in}}|$, where $\mathbf{p}_{\text{in}}$ is the momentum of the incident particles, $\cos\alpha = -\lim_{r\to\infty}\mathbf{A}\cdot\mathbf{p}/|\mathbf{A}||\mathbf{p}| = 1/|\mathbf{A}| = 1/\epsilon$, $\Theta = \pi - 2\alpha$, $\sin(\Theta/2) = 1/\epsilon$, $\tan(\Theta/2) = 1/\sqrt{\epsilon^2 - 1} = k/2Eb$, $k = qQ$.

2.7 Integrability

The isotropic harmonic oscillator and the Kepler motion share two distinct properties. They are integrable and exactly solvable.

The former property, integrability, is apparent from equations (2.7) and (2.8) giving $t = t(r)$ and $\theta = \theta(r)$ in the form of definite integrals. Once these formulae have been written, the remaining task is "to do" the integrals. This could be done exactly and equations (2.22) and (2.33) were established.

The equations of motion of a particle in a conservative central field with potential energy $U(r)$ are integrable because equations (2.7) and (2.8) are available. If, however, a search in a table of integrals proves unsuccessful, and one must resort to numerical integration, the equations are obviously not solvable exactly.

This definition of exact solvability is a little loose. For instance, an integral may be expressed analytically as a slowly converging series whose use is less convenient than calculating it numerically.

Integrability results from the existence of a sufficient number of integrals of motion, and the existence of an integral of motion is connected with the invariance of the equations under a transformation. This will be the leitmotif of Chapters 6 and 7.

Let us consider a particle in three dimensions acted upon by a force $\mathbf{f} = -\nabla U(\mathbf{r})$ which does not depend explicitly on the time. The energy E is an integral of motion connected with the invariance of $\mathbf{f} = m\mathbf{a}$ under time translations $t' = t + \delta t$ (δt constant).

If $U = U(\rho)$ ($\rho = \sqrt{x^2 + y^2}$), the z-components of the momentum and of the angular momentum, p_z and l_z, are integrals of motion. They result from the invariance of the equations under translations $z' = z + \delta z$ (δz constant) and rotations $\phi' = \phi + \delta\phi$ ($\delta\phi$ constant). The three integrals of motion, E, p_z, and l_z make the equations integrable.

If $U = U(r)$ ($r = \sqrt{x^2 + y^2 + z^2}$), l_x, l_y, and l_z are integrals of motion resulting from the invariance of the equations under rotations about each of the coordinate axes. They amount to the magnitude of the angular momentum, and a direction which we took as the z axis. Together with E, $l_z = |\mathbf{l}| = l$ made the equations integrable.

A non-trivial example of integrable two-dimensional problem is the motion with the Toda potential energy

$$U_{\mathrm{T}} = \frac{1}{24}[\mathrm{e}^{2(x-y\sqrt{3})} + \mathrm{e}^{2(x+y\sqrt{3})} + \mathrm{e}^{-4x}] - \frac{1}{8} \quad . \tag{2.58}$$

Together with the energy E, the integral of motion

$$I = 8\dot{y}(\dot{y}^2 - 3\dot{x}^2) + (\dot{y} + \sqrt{3}\dot{x})\mathrm{e}^{2(x-y\sqrt{3})} + (\dot{y} - \sqrt{3}\dot{x})\mathrm{e}^{2(x+y\sqrt{3})} - 2\dot{y}\mathrm{e}^{-4x} \tag{2.59}$$

found by Hénon makes the problem integrable.

Expanding U_{T} to the third order in x and y one has

$$U_{\mathrm{T}} \simeq \frac{1}{2}(x^2 + y^2) - \frac{1}{3}x^3 + xy^2 = U_{\mathrm{HH}} \quad , \tag{2.60}$$

the Hénon-Heiles potential energy. It seems that the equations of motion with the potential energy U_{HH} have only one integral of motion, the energy. Therefore, they are not integrable.

2.8 Chapter 2 problems

2.1 For a constant attractive central force ($U = Cr,\ C > 0$), equation (2.12) gives $T_{\rm osc} = 2\pi\sqrt{mR/3C} = T/\sqrt{3}$. Compare this result with the following.

A puck of mass m on a frictionless horizontal table is pulled by an inextensible string passing through a small hole in the table and attached to a hanging mass M.

2.2 Study the stability of circular orbits for
(i) $U = Cr^\alpha$, and (ii) $U = C\ln(r/r_0)$, where C, $\alpha \neq 0$, and r_0 are constants.

2.3 Substituting $U = Cr^\alpha$ in equation (2.12), we find

$$T_{\rm osc} = 2\pi\sqrt{mR^{2-\alpha}/\alpha(\alpha+2)C} \quad .$$

For the isotropic oscillator ($C = k/2, \alpha = 2$) and for the Newton-Coulomb force ($C = -k < 0, \alpha = -1$), compare $T_{\rm osc}$ with the period T of the circular orbit, and draw illustrative graphs.

2.4 Study the negative- and zero-energy orbits of a particle acted upon by the attractive force $f_r = -2h/r^3$ ($h > 0$). Consider only the case $2mh > l^2$. For other cases, as well as for a combination of an inverse-square and an inverse-cube force, see G.L. Kotkin and V.G. Serbo,*Collection of Problems in Classical Mechanics* (Pergamon Press,1971).

2.5 With $x = r\ \cos\theta$, $y = r\ \sin\theta$, equation (2.21) gives

$$s_2x^2 + s_1y^2 = 1 \quad .$$

This is the equation of an ellipse

$$(x/A)^2 + (y/B)^2 = 1 \text{ with } A = 1/\sqrt{s_2},\ B = 1/\sqrt{s_1}.$$

How is this related to the equation

$$(x'^2/a^2) + (y'^2/b^2) - (2\cos\delta/ab)x'y' = \sin^2\delta$$

found in problem 1.7?

2.6 Use equation (2.52) to evaluate the angle of precession per revolution when $\delta f(r)$ is small.

2.7 (i) Show that a particle with energy $E < 1/6$ and potential energy $U_{\rm HH}(x,y)$ given by (2.60) is trapped in the triangle of vertices
($x = 1, y = 0$), ($x = -1/2, y = \sqrt{3}/2$), ($x = -1/2, y = -\sqrt{3}/2$).

(ii) These vertices are saddle points. Show this for the $(1,0)$ vertex.

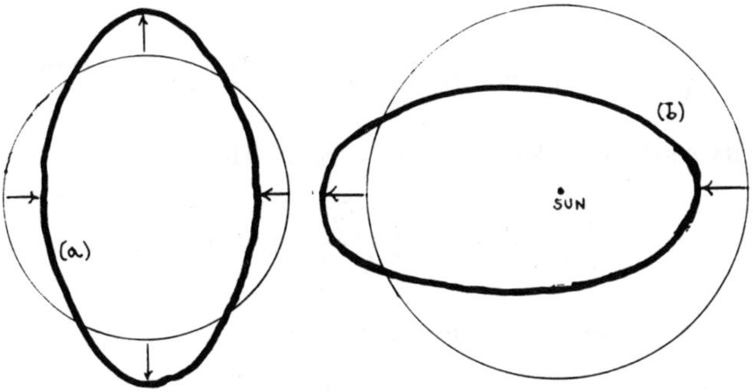

Figure 2.5: Quasi-circular orbits

Solutions to ch. 2 problems

S2.1 The force (τ, tension of the string) is constant if the motion of the puck is circular with center at the hole ($\tau = Mg$), the period being $T = 2\pi\sqrt{mR/Mg}$. However, if m executes small oscillations about the circular orbit, one has

$$m\ddot{r} = -\tau + l^2/mr^3,\ M\ddot{r} = \tau - Mg,\ (m+M)\ddot{r} = -Mg + l^2/mr^3,$$

$$\mathrm{d}^2(r-R)/\mathrm{d}t^2 \simeq -\omega_{\rm osc}^2\,(r-R) \text{ with } \omega_{\rm osc}^2 = 3l^2/mR^4(m+M),$$

so that $T_{\rm osc} = 2\pi\sqrt{(m+M)R/3Mg} = T\sqrt{(m+M)/3m}$. The force acting on m (the tension τ) is given by $\tau = M(\ddot{r}+g)$, and is clearly not constant.

S2.2 (i) The force must be attractive, and so $-\alpha r^{\alpha-1}C < 0$, $\alpha C > 0$. Stability requires $U_e'' = U'' + 3U'/R > 0$, $\alpha(\alpha-1)C + 3\alpha C > 0$, $C\alpha(\alpha+2) > 0$. Since $\alpha C > 0$, the condition for stability is $\alpha > -2$ ($\alpha \neq 0$).
(ii) For $U = C\ln(r/r_0)$ ($C > 0$), $f(r) = -C/r$,
$U_e'' = -(C/R^2) + (3C/R^2) = 2C/R^2 > 0$, indicating stability of all circular orbits.

S2.3 For the isotropic oscillator one finds $T_{\rm osc} = \pi\sqrt{m/k} = T/2$, where $T = 2\pi\sqrt{m/k}$ is the period of the circular orbit.

This can be understood analytically by considering the equations of motion $x = a\cos(\omega t)$, $y = b\sin(\omega t)$ with $a = R+\epsilon$ and $b = R-\epsilon$. Since $r = \sqrt{x^2+y^2} \simeq \sqrt{R^2 + 2R\epsilon\cos(2\omega t)}$, we see that $\omega_{\rm osc} = 2\omega$. For a graphic illustration, see figure 2.5(a).

For the Newton-Coulomb force we have $T_{\rm osc} = 2\pi\sqrt{mR^3/k} = T$, which is evident from figure 2.5(b).

S2.4 Inserting $U = -h/r^2$ in (2.8) with the square root taken with a minus sign, we find

$$\theta = \theta_0 + \frac{l}{\sqrt{2mh - l^2}}[\cosh^{-1}(r_{\rm max}/r) - \cosh^{-1}(r_{\rm max}/r_0)] \quad ,$$

$$\frac{1}{r} = \frac{1}{r_0}\cosh\left(\sqrt{\frac{2mh}{l^2}-1}\ (\theta-\theta_0)\right) - \frac{\sqrt{r_{\max}^2 - r_0^2}}{r_{\max} r_0}\sinh\left(\sqrt{\frac{2mh}{l^2}-1}\ (\theta-\theta_0)\right) \ .$$

For simplicity's sake we take $\theta_0 = 0$, $r_0 = r_{\max}$. Then

$$\frac{1}{r} = \frac{1}{r_{\max}}\cosh\left(\sqrt{\frac{2mh}{l^2}-1}\ \theta\right) \ .$$

It is clear that $r \to 0$ as $\theta \to \infty$, in which case

$$\theta \simeq \ln(r_{\max}/r)/\sqrt{\frac{2mh}{l^2}-1} \ .$$

The particle falls into the center of force after an infinite number of revolutions.

On the other hand, from equation (2.7) we have

$$t = t_0 - \int_{r_0}^{r} \mathrm{d}r/\sqrt{\frac{2}{m}\left(E + \frac{h}{r^2} - \frac{l^2}{2mr^2}\right)} = t_0 - \sqrt{\frac{m}{2|E|}}\int_{r_0}^{r} r\ \mathrm{d}r/\sqrt{r_{\max}^2 - r^2}$$

$$= t_0 + \sqrt{m/2|E|}(\sqrt{r_{\max}^2 - r^2} - \sqrt{r_{\max}^2 - {r_0}^2}\ ) \ .$$

Taking $t_0 = 0$ and $r_0 = r_{\max}$, we see that the fall into the center of force takes only the finite time $t_{\text{fall}} = r_{\max}\sqrt{m/2|E|} = \sqrt{2mh - l^2}/2|E|$. The velocity tends to infinity as r tends to zero because the potential energy tends to $-\infty$ so that the kinetic energy tends to $+\infty$.

Circular orbits are $E = 0$ orbits with the additional requirement $h = l^2/2m$. In fact $mv^2/R = 2h/R^3$ gives $l^2 = 2hm$ and $E = mv^2/2 - h/R^2 = 0$.

S2.5 The equation for (x, y) reduces to that for (x', y') by a rotation

$$x = x'\cos(\gamma/2) + y'\sin(\gamma/2)\ ,\ \ y = -x'\sin(\gamma/2) + y'\cos(\gamma/2)$$

if

$$[s_1 + s_2 + (s_1 - s_2)\cos\gamma]/2 = 1/(a^2\sin^2\delta),$$
$$[s_1 + s_2 - (s_1 - s_2)\cos\gamma]/2 = 1/(b^2\sin^2\delta),$$

and

$$(s_1 - s_2)\sin\gamma = 2\cos\delta/ab\ \sin^2\delta \ .$$

By working on the previous three equations we find

$a^2 + b^2 = (s_1 + s_2)/s_1 s_2 = 2E/m\omega^2$ and $a^2 b^2 \sin^2\delta = 1/s_1 s_2 = l^2/m^2\omega^2$.

This makes sense. From $x = a\ \cos(\omega t - \alpha)$ and $y = b\ \cos(\omega t - \beta)$, we have $E = [m(\dot{x}^2 + \dot{y}^2) + (k(x^2 + y^2)]/2 = m\omega^2(a^2 + b^2)/2$, $l = m(x\dot{y} - y\dot{x}) = m\omega ab\ \sin\delta$.

S2.6 Define $\mathrm{d}\theta_{\text{pr}}/\mathrm{d}t = (\hat{\mathbf{l}} \times \hat{\mathbf{A}})\cdot(\mathrm{d}\hat{\mathbf{A}}/\mathrm{d}t)$, where $\hat{\mathbf{l}}$ and $\hat{\mathbf{A}}$ are unit vectors in the directions of $\mathbf{l}$ and $\mathbf{A}$, respectively. Since $\mathrm{d}\hat{\mathbf{A}}/\mathrm{d}t = (\mathrm{d}\mathbf{A}/\mathrm{d}t)/A +$ term parallel to $\mathbf{A}$, $A = |\mathbf{A}|$, we have $(\hat{\mathbf{l}} \times \hat{\mathbf{A}})\cdot \mathrm{d}\hat{\mathbf{A}}/\mathrm{d}t = l^{-1}A^{-2}(\mathbf{l} \times \mathbf{A})\cdot \mathrm{d}\mathbf{A}/\mathrm{d}t$. Therefore

$$\frac{\mathrm{d}\theta_{\text{pr}}}{\mathrm{d}t} = \frac{(\mathbf{l} \times \mathbf{A})\cdot(\mathbf{r} \times \mathbf{l})\ \delta f(r)}{kmlrA^2} = \frac{l\ \delta f(r)}{kmrA^2}\left(r - \frac{l^2}{km}\right) \ ,$$

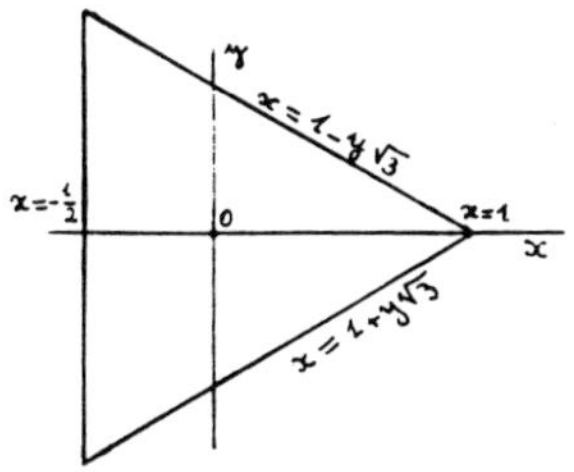

Figure 2.6: Hénon-Heiles potential energy

$$\Delta\theta_{\mathrm{pr}} \simeq \frac{l}{km\epsilon^2}\int_0^{2\pi}\left(1-\frac{l^2}{kmr}\right)\delta f(r)\frac{\mathrm{d}t}{\mathrm{d}\theta}\mathrm{d}\theta = \frac{1}{k\epsilon^2}\int_0^{2\pi}\left(1-\frac{l^2}{kmr}\right)\delta f(r)\, r^2\,\mathrm{d}\theta \ ,$$

where we have replaced A^2 by its value ϵ^2 for $\delta f = 0$.

For $\delta f(r) = -2h/r^3$, using (2.33) we have

$$\Delta\theta_{\mathrm{pr}} \simeq \frac{2h}{k\epsilon^2}\int_0^{2\pi}\left(\frac{l^2}{kmr^2}-\frac{1}{r}\right)\mathrm{d}\theta = \frac{2\pi mh}{l^2}$$

in agreement with (2.16).

S2.7 (i) Expressing U_{HH} in the form

$$U_{\mathrm{HH}} = \frac{1}{6} - \frac{1}{3}\left(x+\frac{1}{2}\right)\left(x+y\sqrt{3}-1\right)\left(x-y\sqrt{3}-1\right)$$

one sees at once that U_{HH} has the value 1/6 on the sides of the triangle shown in the figure,

$$U_{\mathrm{HH}}(x=-1/2,y) = U_{\mathrm{HH}}(x=1-y\sqrt{3},y) = U_{\mathrm{HH}}(x=1+y\sqrt{3},y) = 1/6 \ .$$

On the other hand, $U_{\mathrm{HH}}(0,0) = 0$, and $U_{\mathrm{HH}} < 1/6$ inside the triangle (see figure 2.6).

(ii) Near the vertex $(1,0)$, $(x = 1+\delta x, y = \delta y)$, the force components

$$F_x = -(x - x^2 + y^2),\ F_y = -y(1+2x)$$

are given approximately by $F_x \simeq \delta x$, $F_y \simeq -3\delta y$.

For $\delta y = 0$, we have $\mathrm{d}^2\delta x/\mathrm{d}t^2 \simeq +\delta x$, so that $(1,0)$ is a point of unstable equilibrium for motion along the x-axis.

For $\delta x = 0$, we have $\mathrm{d}^2\delta y/\mathrm{d}t^2 \simeq -3\delta y$, so that $(1,0)$ is a point of stable equilibrium for motion perpendicular to the x-axis.

Chapter 3

FIXED POINTS, OSCILLATIONS, CHAOS

The fixed points of a system of first-order differential equations of motion are defined, and the behavior of solutions in their neighborhood is examined.
Oscillations about equilibrium configurations are considered, as well as forced oscillations and parametric resonance.
The chapter ends with an outline of chaotic motion.

3.1 Fixed points

We shall be concerned with systems of differential equations of first order in the time

$$\dot{x}_i = f_i(x_1, \ldots x_N) \quad (i = 1, \ldots N = 2n) \quad , \tag{3.1}$$

where n of the x_i's are coordinates and the other n generalized momenta. Such equations replace n second-order differential equations, as outlined in Chapter 1.

Suppose all the f_i's vanish for $x_i = x_{0i}$ $(i = 1, \ldots N)$, where the x_{0i}'s are constants. Then $x_i = x_{0i}$ is a solution of the system. The point $\mathbf{x}_0$ is a "fixed point".

Example: Defining $x_1 = \theta$, $x_2 = \dot{\theta}$, the damped pendulum equation $\ddot{\theta} + \gamma\dot{\theta} + (g/l)\sin\theta = 0$ can be replaced by the system of equations $\dot{x}_1 = x_2$, $\dot{x}_2 = -(g/l)\sin x_1 - \gamma x_2$. There are two fixed points, $(x_1 = 0, x_2 = 0)$ and $(x_1 = \pi, x_2 = 0)$, or, more precisely,

$$(x_1 = 2n\pi, x_2 = 0) \text{ and } (x_1 = (2n+1)\pi, x_2 = 0).$$

The fixed-point solution is stable if any solution $x_i(t)$ such that $|x_i(0) - x_{0i}| < \epsilon$ (ϵ infinitesimal) differs infinitesimally from x_{0i} at all times.

The following is a criterion for stability. Linearize equation (3.1) about $(x_{01}, \ldots x_{0N})$. With $x'_i = x_i - x_{0i}$ one has

$$\dot{x}'_i = a_{ij} x'_j \quad \text{where} \quad a_{ij} = (\partial f_i / \partial x_j)_0 \quad , \tag{3.2}$$

or, more concisely,

$$\dot{\mathsf{x}}'(t) = \mathsf{A}\mathsf{x}'(t) \quad , \tag{3.3}$$

where

$$\mathsf{x}' = \begin{pmatrix} x'_1 \\ \vdots \\ x'_N \end{pmatrix} \quad . \tag{3.4}$$

and A is the $N \times N$ matrix with elements a_{ij}.

Example: For the damped pendulum, about the fixed point $(0, 0)$ one has $a_{11} = 0$, $a_{12} = 1$, $a_{21} = -g/l$, $a_{22} = -\gamma$, and about the fixed point $(\pi, 0)$ one has $a_{11} = 0$, $a_{12} = 1$, $a_{21} = +g/l$, $a_{22} = -\gamma$.

In order to study the stability of the fixed point solution $\mathsf{x}' = 0$, we consider solutions of the form

$$\mathsf{x}'(t) = \mathsf{X}\, \mathrm{e}^{\lambda t} \quad , \tag{3.5}$$

from which the general solution of equation (3.3) can be constructed.

Substituting in (3.3), we find

$$\mathsf{AX} = \lambda \mathsf{X} \quad .$$

The problem is reduced to finding the eigenvalues λ_i and the eigenvectors X_i of A. If the eigenvalues λ_i are all different, then the general solution can be expressed in the form

$$\mathsf{x}'(t) = \sum_{i=1}^{N} c_i \mathsf{X}_i \mathrm{e}^{\lambda_i t} \quad . \tag{3.6}$$

Let us consider the simple case $N = 2$,

$$\mathsf{A} = \begin{pmatrix} a_{11} & a_{12} \\ a_{21} & a_{22} \end{pmatrix} \quad . \tag{3.7}$$

The eigenvalues λ_1 and λ_2 are the solutions of the quadratic equation

$$\lambda^2 - S\,\lambda + \Delta = 0 \quad , \tag{3.8}$$

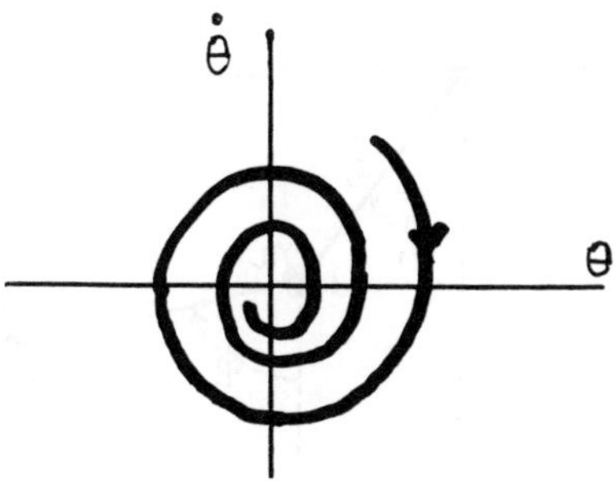

Figure 3.1: Spiral attractor

where $S \equiv \text{trace}(\mathsf{A}) = a_{11} + a_{22}$ and $\Delta = a_{11}a_{22} - a_{12}a_{21}$. They are

$$\begin{pmatrix} \lambda_1 \\ \lambda_2 \end{pmatrix} = \frac{1}{2}(S \pm \sqrt{S^2 - 4\Delta}\,) \tag{3.9}$$

corresponding to the eigenvectors

$$\mathsf{X}_i = \begin{pmatrix} a_{12} \\ \lambda_i - a_{11} \end{pmatrix} \quad (i = 1, 2) \quad . \tag{3.10}$$

We confine ourselves to the case $S^2 < 4\Delta$ (λ_1 and λ_2 complex numbers, $|\exp(\lambda_i t)| = \exp(St/2)$).

If $S < 0$ the fixed point is a "spiral attractor", if $S > 0$ a "spiral repellor".

Example of spiral attractor (figure 3.1): Damped pendulum about lower equilibrium
position, $S = -\gamma$, $\Delta = g/l$, while $S^2 < 4\Delta$ means $\gamma < 2\sqrt{g/l}$.
The eigenvalues are $\lambda_1 = (-\gamma + \mathrm{i}\sqrt{4gl^{-1} - \gamma^2})/2$, $\lambda_2 = \bar{\lambda}_1$. Note the solution
$\theta(t) = \theta_0(\lambda_2 \exp(\lambda_1 t) - \lambda_1 \exp(\lambda_2 t))/(\lambda_2 - \lambda_1)$ for $\theta(0) = \theta_0$, $\dot{\theta}(0) = 0$.
One finds

$$\theta(t) = \Theta\ \cos(\omega t + \alpha)\exp(-\gamma t/2)$$

with

$$\omega = \tfrac{1}{2}\sqrt{4gl^{-1} - \gamma^2},\ \Theta = \theta_0\sqrt{\gamma^2 + 4\omega^2}\,/2\omega,\ \tan\alpha = -\gamma/2\omega.$$

Other cases will be studied as problems.

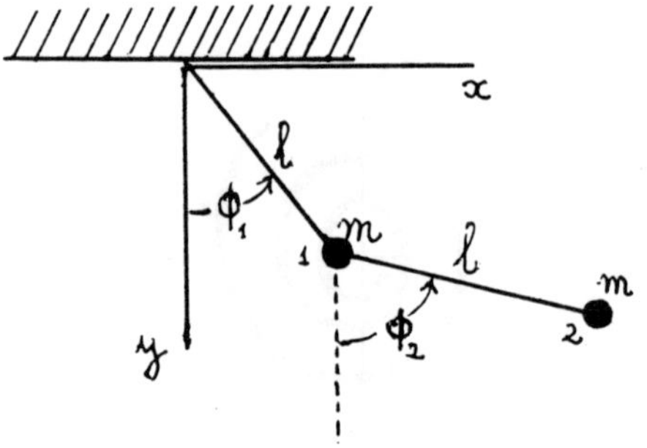

Figure 3.2: Double pendulum

3.2 Small oscillations

Let the matrix A in (3.3) be of the form

$$\mathsf{A} = \begin{pmatrix} 0 & -\mathsf{K} \\ \mathsf{M}^{-1} & 0 \end{pmatrix} \quad , \tag{3.11}$$

where K and M are $n \times n$ real symmetric matrices with constant coefficients, and x' in (3.4) be

$$\mathsf{x}' = \begin{pmatrix} \mathsf{p} \\ \mathsf{q} \end{pmatrix} \quad , \quad \mathsf{p} = \begin{pmatrix} p_1 \\ p_2 \end{pmatrix} \quad , \quad \mathsf{q} = \begin{pmatrix} q_1 \\ q_2 \end{pmatrix} \quad . \tag{3.12}$$

Equation (3.3) then gives

$$\dot{\mathsf{q}} = \mathsf{M}^{-1}\mathsf{p} \quad , \quad \dot{\mathsf{p}} = -\mathsf{K}\mathsf{q} \quad , \tag{3.13}$$

corresponding to the second order equation

$$\mathsf{M}\ddot{\mathsf{q}} = -\mathsf{K}\mathsf{q} \quad . \tag{3.14}$$

Restrictions on K and M will be imposed as we go along.

As an example, we consider the small amplitude oscillations of a double pendulum (figure 3.2) about the stable fixed point $\phi_1 = \phi_2 = 0$, $\dot{\phi}_1 = \dot{\phi}_2 = 0$. We leave it to the reader to establish the equations

$$\dot{\phi}_1 = (p_1 - p_2)/ml^2 \quad , \quad \dot{\phi}_2 = (-p_1 + 2p_2)/ml^2 \quad ,$$

$$\dot{p}_1 = -2mgl\phi_1 \quad , \quad \dot{p}_2 = -mgl\phi_2 \quad .$$

These can be written more concisely in the form (3.13) with

$$\mathsf{q} = \begin{pmatrix} \phi_1 \\ \phi_2 \end{pmatrix} \quad , \quad \mathsf{p} = \begin{pmatrix} p_1 \\ p_2 \end{pmatrix} \quad , \quad \mathsf{M} = ml^2 \begin{pmatrix} 2 & 1 \\ 1 & 1 \end{pmatrix} \quad ,$$

$$\mathsf{M}^{-1} = \frac{1}{ml^2}\begin{pmatrix} 1 & -1 \\ -1 & 2 \end{pmatrix} \quad , \quad \mathsf{K} = mgl\begin{pmatrix} 2 & 0 \\ 0 & 1 \end{pmatrix} \quad .$$

For a solution of the form

$$\mathsf{q}(t) = \mathsf{q}(0)\exp(\lambda t) \quad ,$$

substituting in equation (3.14) we find a system of homogeneous equations for the components $(\mathsf{q}_1(0), \mathsf{q}_2(0))$. This has a non-trivial solution if λ is a solution of the equation $\lambda^4 + (4g/l)\lambda^2 + 2g^2/l^2 = 0$. The solutions are pure imaginary, $\lambda_i = \pm i\omega_i$ $(i = 1, 2)$,

$$\omega_1 = \sqrt{g(2+\sqrt{2})/l} \quad , \quad \omega_2 = \sqrt{g(2-\sqrt{2})/l} \quad .$$

These are the eigenfrequencies of the system.

The eigenvalues of M are

$$\begin{pmatrix} \mu_1 \\ \mu_2 \end{pmatrix} = \frac{ml^2}{2}(3 \pm \sqrt{5}) > 0 \quad ,$$

while K is already in diagonal form with elements $\kappa_1 = 2mgl$ and $\kappa_2 = mgl$, both positive. Notice that the quadratic forms $m_{ij}\xi_i\xi_j = ml^2[\xi_1^2 + (\xi_1 + \xi_2)^2]$ and $k_{ij}\xi_i\xi_j = mgl(2\xi_1^2 + \xi_2^2)$ are positive definite.

Returning to the general case, one finds at once that the "energy"

$$E = (\dot{\mathsf{q}}^{\mathrm{T}}\mathsf{M}\dot{\mathsf{q}} + \mathsf{q}^{\mathrm{T}}\mathsf{K}\mathsf{q})/2 \tag{3.15}$$

is an integral of motion. In fact

$$\begin{aligned} \mathrm{d}E/\mathrm{d}t &= (\ddot{\mathsf{q}}^{\mathrm{T}}\mathsf{M}\dot{\mathsf{q}} + \dot{\mathsf{q}}^{\mathrm{T}}\mathsf{M}\ddot{\mathsf{q}} + \dot{\mathsf{q}}^{\mathrm{T}}\mathsf{K}\mathsf{q} + \mathsf{q}^{\mathrm{T}}\mathsf{K}\dot{\mathsf{q}})/2 \\ &= \{[\dot{\mathsf{q}}^{\mathrm{T}}(\mathsf{M}\ddot{\mathsf{q}} + \mathsf{K}\mathsf{q})] + [\ldots]^{\mathrm{T}}\}/2 = 0 \quad , \end{aligned}$$

where we have used the symmetry of the matrices and equation (3.14). Since

$$\dot{\mathsf{q}}^{\mathrm{T}}\mathsf{M}\dot{\mathsf{q}} = \mathsf{p}^{\mathrm{T}}\mathsf{M}^{-1\,\mathrm{T}}\mathsf{p} = \mathsf{p}^{\mathrm{T}}\mathsf{M}^{-1}\mathsf{p} \quad ,$$

the energy can also be expressed in the form

$$E = (\mathsf{p}^{\mathrm{T}}\mathsf{M}^{-1}\mathsf{p} + \mathsf{q}^{\mathrm{T}}\mathsf{K}\mathsf{q})/2 \quad . \tag{3.16}$$

We assume that the quadratic form $m_{ij}\xi^i\xi^j$ $((\xi^i,\ \xi^j)$ real) is positive definite. This is natural since the first term in equation (3.15) is the kinetic energy. We assume also that $k_{ij}\xi^i\xi^j$ is positive definite. This is also natural if equations (3.14) are linearizations about a stable fixed point. Then if the potential energy U has a minimum at the fixed point, $(\partial U/\partial q_i)_0 = 0$, $k_{ij} = (\partial^2 U(q)/\partial q_i \partial q_j)_0$ (symmetric), and $k_{ij}\xi^i\xi^j > 0$ (condition for minimum).

An immediate consequence of these assumptions is that the frequencies are real. In fact, for a solution of the form

$$\mathsf{q}(t) = \mathsf{Q}e^{i\omega t} \tag{3.17}$$

one finds

$$(\mathsf{K} - \omega^2 \mathsf{M})\mathsf{Q} = 0 \quad , \tag{3.18}$$

$$\omega^2 = \frac{\mathsf{Q}^T \mathsf{K} \mathsf{Q}}{\mathsf{Q}^T \mathsf{M} \mathsf{Q}} > 0 \tag{3.19}$$

since the numerator and the denominator are both positive.

Equations (3.18) have a nontrivial solution if

$$\det(\mathsf{K} - \omega^2 \mathsf{M}) = 0$$

(secular equation).

For the double pendulum

$$\begin{vmatrix} 2g - 2l\omega^2 & -l\omega^2 \\ -l\omega^2 & g - l\omega^2 \end{vmatrix} = 0 \quad , \quad l^2\omega^4 - 4lg\omega^2 + 2g^2 = 0 \quad ,$$

$$\omega_1 = \sqrt{g(2+\sqrt{2})/l} \quad , \quad \omega_2 = \sqrt{g(2-\sqrt{2})/l} \quad ,$$

as we have already found.

Our equations can be cast into a canonical form. Let u_i be the eigenvectors of M,

$$\mathsf{M}\mathsf{u}_i = \mu_i \mathsf{u}_i \quad (\text{no sum over } i) \quad . \tag{3.20}$$

The eigenvalues μ_i are all positive because of the assumption $m_{ij}\xi^i\xi^j > 0$. We normalize the eigenvectors so that $\mathsf{u}_j^T \mathsf{u}_i = (1/\mu_i)\delta_{ij}$. Representing q as a linear combination of these eigenvectors, $\mathsf{q} = Q_i \mathsf{u}_i$, we have the kinetic energy

$$K = \dot{\mathsf{q}}^T \mathsf{M} \dot{\mathsf{q}}/2 = \dot{Q}_i \dot{Q}_j \mathsf{u}_i^T \mathsf{M} \mathsf{u}_j/2 \quad ,$$

$$K = \dot{Q}_i \dot{Q}_j \mu_j \mathsf{u}_i^T \mathsf{u}_j/2 = \dot{Q}_i \dot{Q}_i/2 = \dot{\mathsf{Q}}^T \dot{\mathsf{Q}}/2 \quad , \tag{3.21}$$

invariant under rotations in the n-dimensional space of the Q_i's. By a suitable rotation the potential energy can be brought to the form

$$U = \omega_i^2 Q_i'^2/2 \quad . \tag{3.22}$$

Equations (3.14) are replaced by

$$\ddot{Q}_i' + \omega_i^2 Q_i' \quad (\text{no sum over } i) \tag{3.23}$$

or by the first-order equations

$$\dot{P}_i' = -\omega_i^2 Q_i' \quad , \quad \dot{Q}_i' = P_i' \quad . \tag{3.24}$$

In some cases it may be convenient to interchange the roles of M and K in the above prescription.
For example, $U = mgl(2q_1^2 + q_2^2)/2$, the potential energy of the double pendulum, takes the form $U = mgl(Q_1^2 + Q_2^2)/2$ by putting $q_1 = Q_1/\sqrt{2}$, $q_2 = Q_2$.
The kinetic energy is then $K = (\dot{Q}_1^2 + \dot{Q}_2^2 + \sqrt{2}\dot{Q}_1\dot{Q}_2)/2$.
Performing the rotation $Q_1 = (Q_1' + Q_2')/\sqrt{2}$, $Q_2 = (-Q_1' + Q_2')/\sqrt{2}$, we find

$$U = mgl(Q_1'^2 + Q_2'^2)/2 \quad , \quad K = ml^2[(1 - 1/\sqrt{2})\dot{Q}_1'^2 + (1 + 1/\sqrt{2})\dot{Q}_2'^2]/2$$

and the corresponding equations of motion are

$$\ddot{Q}_1' = -gl^{-1}(2 + \sqrt{2})Q_1' \quad , \quad \ddot{Q}_2' = -gl^{-1}(2 - \sqrt{2})Q_2' \quad .$$

The frequency $\omega_1 = \sqrt{g(2 + \sqrt{2})/l}$ corresponds to the mode $Q_2' = 0$, $Q_1 + Q_2 = 0$, $\sqrt{2}q_1 + q_2 = 0$, while $\omega_2 = \sqrt{g(2 - \sqrt{2})/l}$ corresponds to $Q_1' = 0$, $Q_1 - Q_2 = 0$, $\sqrt{2}q_1 - q_2 = 0$.

3.3 Parametric resonance

We study the solutions of the equation

$$\ddot{q} + \omega^2(t)\, q = 0 \tag{3.25}$$

in which the time-dependent frequency $\omega(t)$ is periodic.

Example: By retracting and extending the legs, a child sitting on a swing changes the parameters of the system, and so its $\omega(t)$, setting it in motion.

Equation (3.25) is equivalent to

$$\dot{\mathsf{x}} = \begin{pmatrix} 0 & 1 \\ -\omega^2 & 0 \end{pmatrix} \mathsf{x} \quad \text{where} \quad \mathsf{x} = \begin{pmatrix} q \\ \dot{q} \end{pmatrix} \quad . \tag{3.26}$$

Two solutions q_1 and q_2 of equation (3.25) are linearly independent if their Wronskian $W = q_1\dot{q}_2 - q_2\dot{q}_1$,

$$W = \mathsf{x}_1^{\mathrm{T}} \mathsf{J} \mathsf{x}_2 \quad \text{with} \quad \mathsf{J} = \begin{pmatrix} 0 & 1 \\ -1 & 0 \end{pmatrix} \quad , \tag{3.27}$$

is different from zero.

It is easy to show that $\mathrm{d}W/\mathrm{d}t = 0$, $W =$ constant. Therefore two solutions $q_1(t)$ and $q_2(t)$ are independent if their W is different from zero at some arbitrary time.

We choose as a fundamental system set of independent solutions $\mathsf{x}_1(t)$ and $\mathsf{x}_2(t)$ such that

$$\mathsf{x}_1(0) = \begin{pmatrix} 1 \\ 0 \end{pmatrix} \quad \text{and} \quad \mathsf{x}_2(0) = \begin{pmatrix} 0 \\ 1 \end{pmatrix} \quad , \tag{3.28}$$

namely $(q_1(0) = 1, \dot{q}_2(0) = 0)$ and $(q_2(0) = 0, \dot{q}_2(0) = 1)$.

At $t = T$, where T is the period of $\omega(t)$,

$$\mathsf{x}_1(T) = \begin{pmatrix} a_{11} \\ a_{21} \end{pmatrix} \quad \text{and} \quad \mathsf{x}_2(T) = \begin{pmatrix} a_{12} \\ a_{22} \end{pmatrix} \quad , \tag{3.29}$$

where $a_{11} = q_1(T)$, $a_{12} = q_2(T)$, $a_{21} = \dot{q}_1(T)$, and $a_{22} = \dot{q}_2(T)$. These can be written in the form

$$\mathsf{x}_1(T) = \begin{pmatrix} a_{11} & a_{12} \\ a_{21} & a_{22} \end{pmatrix} \begin{pmatrix} 1 \\ 0 \end{pmatrix} = \mathsf{A}\mathsf{x}_1(0) \quad \text{and} \quad \mathsf{x}_2(T) = \mathsf{A}\mathsf{x}_2(0) \quad , \tag{3.30}$$

with

$$\mathsf{A} = \begin{pmatrix} a_{11} & a_{12} \\ a_{21} & a_{22} \end{pmatrix} \quad . \tag{3.31}$$

There will be no occasion to confuse this A with that of equation (3.3). Since any solution can be expressed in the form $\mathsf{x}(t) = c_1\mathsf{x}_1(t) + c_2\mathsf{x}_2(t)$, for any solution we have

$$\mathsf{x}(T) = \mathsf{A}\mathsf{x}(0) \quad . \tag{3.32}$$

We want to find solutions such that

$$\mathsf{x}(t+T) = \lambda\mathsf{x}(t) \quad . \tag{3.33}$$

Putting

$$\mathsf{x}(t) = \alpha\mathsf{x}_1(t) + \beta\mathsf{x}_2(t) \quad , \tag{3.34}$$

for $t = 0$ we have
$\mathsf{x}(T) = \mathsf{A}\mathsf{x}(0) = \alpha\mathsf{A}\mathsf{x}_1(0) + \beta A x_2(0)$, and $\mathsf{x}(T) = \lambda\mathsf{x}(0) = \lambda[\alpha\mathsf{x}_1(0) + \beta\mathsf{x}_2(0)]$. Hence

$$\begin{cases} a_{11}\alpha + a_{12}\beta = \lambda\alpha \quad , \\ a_{21}\alpha + a_{22}\beta = \lambda\beta \quad , \end{cases}$$

or

$$(\mathsf{A} - \lambda) \begin{pmatrix} \alpha \\ \beta \end{pmatrix} = 0 \quad . \tag{3.35}$$

Note that $\det \mathsf{A} = W = 1$,

$$\det \mathsf{A} = \det \begin{pmatrix} q_1(0) & q_2(0) \\ \dot{q}_1(0) & \dot{q}_2(0) \end{pmatrix} = \begin{pmatrix} 1 & 0 \\ 0 & 1 \end{pmatrix} = 1 \quad .$$

The secular equation $\det(\mathsf{A} - \lambda\mathsf{I}) = 0$ gives the equation for λ

$$\lambda^2 - S\lambda + 1 = 0 \quad , \tag{3.36}$$

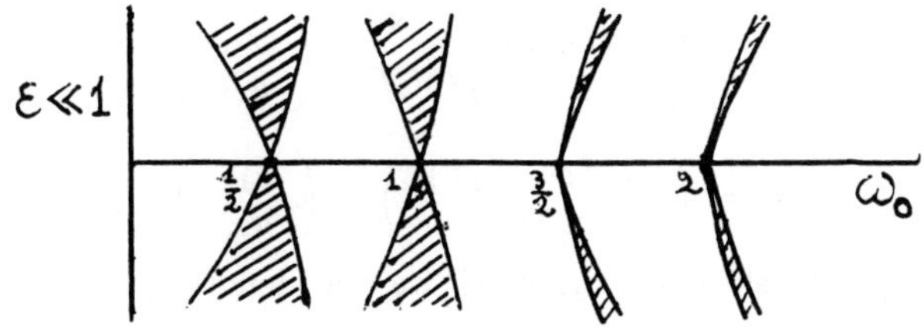

Figure 3.3: Instability zones

where $S = a_{11} + a_{22}$. The eigenvalues are

$$\begin{pmatrix} \lambda_1 \\ \lambda_2 \end{pmatrix} = \left(S \pm \sqrt{S^2 - 4}\right)/2 \quad , \tag{3.37}$$

$\lambda_1 + \lambda_2 = S$, $\lambda_1 \lambda_2 = 1$.

If $|S| > 2$, the eigenvalues are real and of the same sign ($\lambda_1 \lambda_2 = 1$). If they are both positive, we might take $\lambda_1 > 1$, $\lambda_2 < 1$, and write $\lambda_1 = \exp(\alpha)$, $\lambda_2 = \exp(-\alpha)$ (α real and positive). The solutions are clearly unbounded, $\mathsf{x}(nT) = \exp(n\alpha)\mathsf{x}(0)$.

If $|S| < 2$, the eigenvalues are complex ($\lambda_2 = \bar{\lambda}_1$). Since their product equals unity, they lie on the unit circle, $\lambda_1 = \exp(\mathrm{i}\gamma)$, $\lambda_2 = \exp(-\mathrm{i}\gamma)$ (γ real). Then $q(\lambda_1; t+T) = \exp(\mathrm{i}\gamma)\, q(\lambda_1; t)$. The function

$$\varphi_1(t) = q(\lambda_1; t) \exp(-\mathrm{i}\gamma t/T) \tag{3.38}$$

is periodic with period T,

$$\varphi_1(t+T) = \varphi_1(t) \quad .$$

Hence

$$q(\lambda_1; t) = \mathrm{e}^{\mathrm{i}\gamma t/T} \varphi_1(t) \quad \text{and} \quad q(\lambda_2; t) = \mathrm{e}^{-\mathrm{i}\gamma t/T} \varphi_2(t) \quad , \tag{3.39}$$

where φ_1 and φ_2 are periodic with period T. This result is known to physicists as Bloch's theorem.

S is a function of T. The roots in T of the equations $S = \pm 2$ separate the zones of stability from those of instability. For $\ddot{q} = -\omega_0^2(1 + \epsilon \cos t)q$ ($0 < \epsilon \ll 1$) the instability zones are shown in figure 3.3 .

It may be useful to go step by step through the above mathematics for the harmonic oscillator equation $\ddot{q} + \omega_0^2 q = 0$ (ω_0= constant). Here $q_1(t) = \cos(\omega_0 t)$ ($q_1(0) = 1$, $\dot{q}_1(0) = 0$), $q_2(t) = \omega_0^{-1} \sin(\omega_0 t)$ ($q_2(0) = 0$, $\dot{q}_2(0) = 1$),

$$\mathsf{x}_1(T) = \begin{pmatrix} \cos(\omega_0 T) \\ -\omega_0 \sin(\omega_0 T) \end{pmatrix} = \begin{pmatrix} a_{11} \\ a_{21} \end{pmatrix} \quad ,$$

$$\mathsf{x}_2(T) = \begin{pmatrix} \omega_0^{-1} \sin(\omega_0 T) \\ \cos(\omega_0 T) \end{pmatrix} = \begin{pmatrix} a_{12} \\ a_{22} \end{pmatrix} \quad ,$$

while the matrix with elements a_{ij}, which we now denote by A_0, is

$$\mathsf{A}_0 = \begin{pmatrix} \cos(\omega_0 T) & \omega_0^{-1} \sin(\omega_0 T) \\ -\omega_0 \sin(\omega_0 T) & \cos(\omega_0 T) \end{pmatrix} \quad .$$

Note that here T can be anything, it must not be confused with $T_0 = 2\pi/\omega_0$.

The eigenvalues of A_0 are $\lambda_1 = \exp(\mathrm{i}\omega_0 T)$ and $\lambda_2 = \exp(-\mathrm{i}\omega_0 T)$. The equation $\mathsf{x}(t+T) = \lambda \mathsf{x}(t)$, with $\mathsf{x}(t) = \alpha \mathsf{x}_1(t) + \beta \mathsf{x}_2(t)$, gives $\beta = (\lambda - a_{11})\alpha/a_{12}$. Hence $q(\lambda_1) = \alpha \exp(\mathrm{i}\omega_0 t)$ and $q(\lambda_2) = \alpha \exp(-\mathrm{i}\omega_0 t)$, because in this case $\gamma = \mathrm{i}^{-1} \ln \lambda_1 = \omega_0 T$.

Referring to figure 3.3, for $\epsilon = 0$ and $T = 2\pi$ we have $S_0 = |\mathrm{tr}\mathsf{A}_0| = |2\cos(2\pi\omega_0)|$, and so $S_0 = 2$ for $\omega_0 = k/2$ (k integer), intersection of instability zones with ω_0 axis.

3.4 Periodically jerked oscillator

The equation

$$\ddot{q} + [\omega_0^2 + f \sum_{n=-\infty}^{\infty} \delta(t - nT)]q = 0 \tag{3.40}$$

provides an exactly solvable model for equation (3.25).

Take $q_1(0^-) = 1$, $\dot{q}_1(0^-) = 0$, $q_2(0^-) = 0$, $\dot{q}_2(0^-) = 1$. At 0^+ we must have $q_1(0^+) = q_1(0^-)$, $q_2(0^+) = q_2(0^-)$, and, by integrating equation (3.40) from $-\epsilon$ to ϵ ($\epsilon \to 0$),

$\dot{q}_1(0^+) - \dot{q}_1(0^-) = -f q_1(0) = -f$, $\dot{q}_2(0^+) - \dot{q}_2(0^-) = -f q_2(0) = 0$.

In the interval $0 < t < T$ we take

$$q_1(t) = \cos(\omega_0 t) - f\omega_0^{-1} \sin(\omega_0 t) \quad , \quad q_2(t) = \omega_0^{-1} \sin(\omega_0 t) \quad . \tag{3.41}$$

The matrix A, defined by $\mathsf{x}_1(T^-) = \mathsf{A}\mathsf{x}_1(0^-)$ and $\mathsf{x}_2(T^-) = \mathsf{A}\mathsf{x}_2(0^-)$, is

$$\mathsf{A} = \begin{pmatrix} \cos(\omega_0 T) - f\omega_0^{-1} \sin(\omega_0 T) & \omega_0^{-1} \sin(\omega_0 T) \\ -\omega_0 \sin(\omega_0 T) - f \cos(\omega_0 T) & \cos(\omega_0 T) \end{pmatrix} \quad . \tag{3.42}$$

We have $\det(\mathsf{A}) = 1$ and $S = \mathrm{trace}(\mathsf{A}) = 2\cos(\omega_0 T) - f\omega_0^{-1} \sin(\omega_0 T)$.
If $|S| < 2$, we find $\lambda_1 = \exp(\mathrm{i}\gamma)$, $\lambda_2 = \bar{\lambda}_1$,

$\cos\gamma = S/2 = \cos(\omega_0 T) - f \sin(\omega_0 T)/2\omega_0$.

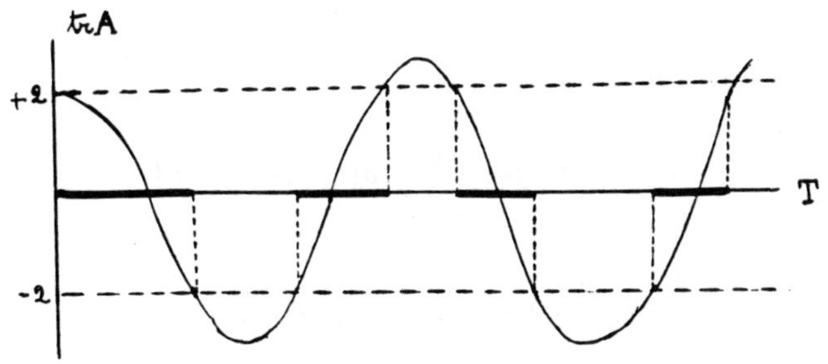

Figure 3.4: Stability zones for (3.40)

In figure 3.4 (S *versus* T for given ω_0), the thick lines show the stability intervals.
In the interval $0 < t < T$, the functions $q(\lambda; t)$ of equation (3.39) are given by

$$q_1(\lambda; t) = a_{12} q_1(t) + (\lambda - a_{11}) q_2(t)$$
$$= a_{12} \cos(\omega_0 t) + (-a_{12} f + \lambda - a_{11}) \omega_0^{-1} \sin \omega_0 t) \quad ,$$

and in the interval $-T < t < 0$ by

$$q_0(\lambda; t) = a_{12} \cos(\omega_0 t) + (\lambda - a_{11}) \omega_0^{-1} \sin(\omega_0 t) \quad .$$

As an exercise, verify that $q_1(\lambda; T/2) = \lambda q_0(\lambda; -T/2)$.

The graph in figure 3.4 is familiar to physicists. In fact, the present example is presented for comparison with the Kronig-Penney model of wave mechanics. Our equation corresponds to the Schrödinger equation

$$-(\hbar^2/2m) u'' + v_0 \sum_{n=-\infty}^{\infty} \delta(x - n\alpha) u = Eu \tag{3.43}$$

in G. Baym, *Lectures on Quantum Mechanics* (Addison Wesley, 1969), eq. 4-121 ($\cos(k\alpha) = \cos(q\alpha) + (mv_0\alpha/\hbar^2) \sin(q\alpha)/(q\alpha)$): $\gamma/T \leftrightarrow k$, $T \leftrightarrow \alpha$, $\omega_0^2 \leftrightarrow 2mE/\hbar^2$, $f \leftrightarrow -2mv_0\hbar^2$.

Let us return to equation (3.30). Putting $q_n = q(nT^-)$, $\dot{q}_n = \dot{q}(nT^-)$ (value of q and $\dot{q}$ just before the impulse at time nT), we have

$$\mathsf{x}_{n+1} = \mathsf{A}\mathsf{x}_n \quad , \tag{3.44}$$

where x_n has components $(q_n, \dot{q}_n)$ and similarly for x_{n+1}. Since f in the expression (3.42) for A is a constant, the map $(q_n, \dot{q}_n) \to (q_{n+1}, \dot{q}_{n+1})$ is manifestly linear.

However, this is equivalent to

$$\begin{pmatrix} q_{n+1} \\ \dot{q}_{n+1} \end{pmatrix} = \mathsf{A}_{f=0} \begin{pmatrix} q_n \\ \dot{q}_n - f q_n \end{pmatrix} \tag{3.45}$$

as is easy to verify.

Consider now the equation

$$\ddot{q} + \omega_0^2 q + f(q) \sum_{n=-\infty}^{\infty} \delta(t - nT)\, q = 0 \quad , \tag{3.46}$$

in general non-linear in q.

By a procedure identical to that used for f = constant, one finds the non-linear map

$$\begin{pmatrix} q_{n+1} \\ \dot{q}_{n+1} \end{pmatrix} = \mathsf{A}_{f=0} \begin{pmatrix} q_n \\ \dot{q}_n - f(q_n)q_n \end{pmatrix} \quad . \tag{3.47}$$

Consider finally the periodically jerked damped rotator

$$\ddot{q} + \gamma\dot{q} + f(q) \sum_{n=-\infty}^{\infty} \delta(t - nT) q = 0 \quad , \tag{3.48}$$

where $q(\mathrm{mod} 2\pi)$ is the angle of the rotator.

From (3.46) for the oscillator with $\omega_0^2 q$ replaced by $\gamma\dot{q}$, one finds easily

$$\begin{aligned} q_{n+1} &= q_n + \gamma^{-1}[\dot{q}_n - f(q_n)q_n][1 - \exp(-\gamma T)] \quad , \\ \dot{q}_{n+1} &= [\dot{q}_n - f(q_n)q_n] \exp(-\gamma T) \quad . \end{aligned} \tag{3.49}$$

For $f(q) = \delta[rq + (1 - r)]$, $\gamma \to \infty$, $\delta/\gamma \to 1$, one obtains the "logistic map"

$$q_{n+1} = rq_n(1 - q_n) \quad . \tag{3.50}$$

This derives its name from the equation $\mathrm{d}x/\mathrm{d}t = rx(1 - x)$ describing population growth.

3.5 Discrete maps, bifurcation, chaos

Consider a discrete one-dimensional map $x_{n+1} = f(x_n)$, such as the map $q_{n+1} = rq_n(1 - q_n)$ derived from the overdamped-overjerked rotator, equation (3.50).

A "fixed point" of the map is a solution of $x^\star = f(x^\star)$. The fixed point is stable if, starting from $x^\star + \epsilon$,

$$|f(x^\star + \epsilon) - f(x^\star)| < |(x^\star + \epsilon) - x^\star| = |\epsilon| \quad . \tag{3.51}$$

This requires

$$\left| \left(\frac{\mathrm{d}f(x)}{\mathrm{d}x} \right)_{x=x^\star} \right| < 1 \quad . \tag{3.52}$$

For the logistic map $x_{n+1} = rx_n(1 - x_n)$ $(0 < r < 4)$ in the interval $0 \leq x \leq 1$, there is always the fixed point $x^\star = 0$. Since the derivative of $rx(1-x)$ with respect to x for $x = 0$ equals r, $x^\star = 0$ is stable if $r < 1$. In this case $(r < 1)$, starting from any x_0 in the interval $0 < x < 1$, one has $\lim_{n\to\infty} x_n = 0$. Of course if $x_0 = 1$, $x_n = 0$ without having to go to the limit.

The other fixed point $x^\star = (r-1)/r$ exists only if $r > 1$, for the simple reason that $x^\star$ is assumed to be positive. Since the derivative of $rx(1-x)$ with respect to x for $x = (r-1)/r$ equals $2 - r$, $x^\star = (r-1)/r$ is stable for $1 < r < 3$. Note that the fixed point $x^\star = (r-1)/r$ is born (at $r = 1$) when the other, $x^\star = 0$, becomes unstable. For $1 < r < 3$, starting from an arbitrary $0 < x_0 < 1$ one has $\lim_{n=\infty} x_n = (r-1)/r$.

What happens at $r = 3$, when $x^\star = (r-1)/r$ becomes unstable? A limit cycle is born consisting of only two points since we are in one dimension. The disappearance of a stable fixed point accompanied by the appearance of a "limit cycle" is known as a "Hopf bifurcation".

For $3 < r < 1 + \sqrt{6} \simeq 3.45$, starting from any value $0 < x_0 < 1$ $(x_0 \neq (r-1)/r)$ we have

$$\lim_{n\to\infty} x_n = \left(r + 1 \pm \sqrt{r^2 - 2r - 3}\right)/2r \quad . \tag{3.53}$$

This means that for large values of n, x_n keeps flipping to and fro between two limit values. These are fixed points of the square of the map, defined as $f^2(x) = f(f(x))$ $(f^2(x) \neq (f(x))^2!)$. If we start from x, one application of the map gives $x' = rx(1-x)$, and a second application gives $x'' = rx'(1-x') = r^2x(1-x)(1-rx+rx^2)$. To search for a fixed point of f^2, put $x'' = x \to x^\star$. We get the equation

$$x^\star = r^2x^\star[1 - (1+r)x^\star + 2rx^{\star 2} - rx^{\star 3}] \quad . \tag{3.54}$$

Now $x^\star = 0$ is a solution. It is a fixed point of f, and so also of f^2, but it is unstable for $r > 1$ and, *a fortiori*, for $r > 3$. Knowing this we first reduce (3.54) to $(x^\star - (r-1)/r)[rx^{\star 2} - (r+1)x^\star + (1+r)/r] = 0$, and, since also $x^\star = (r-1)/r$ is not acceptable, to the quadratic equation

$$rx^{\star 2} - (r+1)x^\star + (1+r)/r = 0 \quad . \tag{3.55}$$

This has the roots

$$\begin{pmatrix} x_1^\star \\ x_2^\star \end{pmatrix} = \frac{1}{2r}\left(r + 1 \pm \sqrt{r^2 - 2r - 3}\right) \tag{3.56}$$

$(r^2 - 2r - 3 > 0$ for $r > 3)$. Note that $f(x_1^\star) = x_2^\star$ and $f(x_2^\star) = x_1^\star$ or, more explicitly, $x_2^\star = rx_1^\star(1 - x_1^\star)$, $x_1^\star = rx_2^\star(1 - x_2^\star)$.

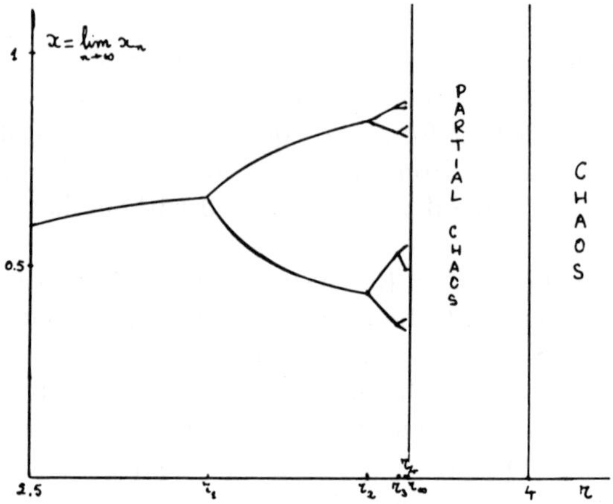

Figure 3.5: $x = \lim_{n\to\infty} x_n$ for logistic map

Is each of the above two values a stable fixed point of the f^2 map? In $1 > |dx''/dx| = |dx''/dx'||dx'/dx| = r^2|(1-2x')(1-2x)|$ put $x = x_1^*$, $x' = x_2^*$, obtaining $|4 + 2r - r^2| < 1$. This is satisfied for $3 < r < 1 + \sqrt{6}$.

After the first bifurcation for the value $r_1 = 3$ of the control parameter, another occurs at $r_2 = 1 + \sqrt{6} \simeq 3.45$, when each of the two fixed points splits into two, another at $r_3 \simeq 3.54$, another at $r_4 \simeq 3.56$ etc.

The values of r at the onsets of successive bifurcations are given by Feigenbaum's empirical formula

$$r_k = r_\infty - c/\delta^k \quad (k = 1, 2, 3...) \tag{3.57}$$

where $\delta \simeq 4.67$, $c \simeq 2.64$, $r_\infty \simeq 3.57$. This seems to apply also to other quadratic maps ("universality"). Some authors are so enthusiastic about this formula as if they believed that one day it will be equal in importance to the Balmer formula for the hydrogen spectrum.

From Feigenbaum's formula one sees that

$$\frac{r_k - r_{k-1}}{r_{k+1} - r_k} = \delta > 0 \quad , \tag{3.58}$$

showing that the interval between the onsets of bifurcations decreases.

We have reached the value $r_\infty \simeq 3.57$. What happens after that? For r between r_∞ and 4 there are chaotic intervals interrupted by odd periodic cycles, until chaos prevails at $r = 4$. Figure 3.5 is the result of a rough calculation performed with the use of a pocket calculator.

When we speak of chaos, we do not only mean that the iterates are distributed at random. Sensitivity to initial conditions is an essential element of chaotic behavior.

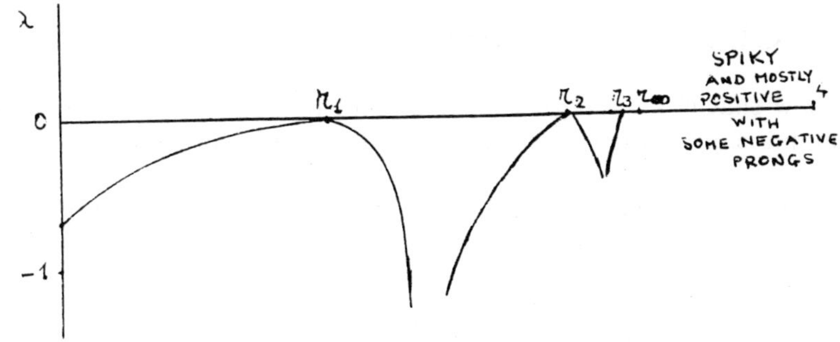

Figure 3.6: Lyapunov exponent for logistic map

Suppose we start from two different values x_0 and $x_0' = x_0 + \epsilon$, with ϵ arbitrarily small. In a chaotic regime, the iterates $f^n(x_0)$ and $f^n(x_0 + \epsilon)$ ($f^0(x_0) = x_0$ and $f^0(x_0 + \epsilon) = x_0 + \epsilon$) will diverge from each other as n increases.

If we assume the divergence to be exponential and write

$$|f^n(x_0 + \epsilon) - f^n(x_0)| = \epsilon \, \mathrm{e}^{n\lambda(x_0)} \quad , \tag{3.59}$$

for $|\epsilon| \ll x_0$ we find

$$\mathrm{e}^{n\lambda(x_0)} = \left| \frac{\mathrm{d}f^n(x)}{\mathrm{d}x} \right|_{x_0} \quad . \tag{3.60}$$

We define the Lyapunov exponent $\lambda(x_0)$ as

$$\lambda(x_0) = \lim_{n \to \infty} \frac{1}{n} \ln \left| \frac{\mathrm{d}f^n(x)}{\mathrm{d}x} \right|_{x_0} \quad . \tag{3.61}$$

Note that $f^n(x_0) \equiv f(f(\ldots f(x_0)))$,

$$\left(\frac{\mathrm{d}f^n(x)}{\mathrm{d}x} \right)_{x_0} = \left(\frac{\mathrm{d}f}{\mathrm{d}x} \right)_{x_{n-1}} \left(\frac{\mathrm{d}f}{\mathrm{d}x} \right)_{x_{n-2}} \cdots \left(\frac{\mathrm{d}f}{\mathrm{d}x} \right)_{x_0} \quad ,$$

and so

$$\lambda(x_0) = \lim_{n \to \infty} \frac{1}{n} \sum_{i=0}^{n-1} \ln |f'(x_i)| \quad . \tag{3.62}$$

Values of λ for the logistic map are shown in figure 3.6 (rough rendition from R. Shaw, *Z. Naturforsch.*, **36a**(1981)80). Positive values of λ are indicative of chaotic behavior.

Using a pocket calculator, it is easy to verify that equation (3.59) works perfectly for the tent map of problem 3.11. Why are we not surprised?

In chapter 8 we shall briefly consider chaos for systems with time-independent Hamiltonians.

3.6 Chapter 3 problems

3.1 Show that if $\gamma > 2\sqrt{g/l}$ the fixed point ($\theta = 0$, $\dot{\theta} = 0$) of the damped pendulum is an attractor ($S < 0, \Delta > 0$) (overdamped pendulum).

3.2 For the fixed point ($\theta = \pi$,$\dot{\theta} = 0$) of the damped pendulum one has $S = -\gamma$, $\Delta = -g/l$. The eigenvalues are ($\lambda_1 = (-\gamma + \sqrt{\gamma^2 + 4g/l}\ )/2$), positive, and

($\lambda_2 = (-\gamma - \sqrt{\gamma^2 + 4g/l}\ )/2$), negative.

The fixed point is a "saddle point" or "hyperbolic point".

Study the behavior of the solutions $\mathsf{X}_1 \exp(\lambda_1 t)$ and $\mathsf{X}_2 \exp(\lambda_2 t)$ by drawing graphs in the $(\theta, \dot{\theta})$ plane. Consider also the case $\gamma = 0$.

3.3 Consider the case $S^2 = 4\Delta$ in which $\lambda = \lambda_1 = \lambda_2 = S/2$. Show that the general solution can be expressed in the for

$$\mathsf{x}'(t) = [c_1\mathsf{X}_1 + c_2(\mathsf{X}_1 t + \mathsf{X}_2)] \exp(\lambda t) \quad ,$$

with X_1 and X_2 satisfying $\mathsf{A}\mathsf{X}_1 = \lambda \mathsf{X}_1$ and $(\mathsf{A} - \lambda)\mathsf{X}_2 = \mathsf{X}_1$. Find X_1 and X_2.

3.4 Apply the result of Problem 3.3 to the damped pendulum with $\gamma = 2\sqrt{g/l}$ at the fixed point $(0, 0)$.

3.5 A wire hoop of radius R rotates counterclockwise about the vertical z axis with angular velocity ω. Let $(x^2 + z^2 = R^2, y = 0)$ be the configuration of the hoop at $t = 0$. A bead slides on the hoop with coefficient of kinetic friction γ.

(i) Show that the angle θ of the position vector $\mathbf{R}$ of the bead with the negative z axis obeys the equation

$$\ddot{\theta} + \gamma\dot{\theta} + (g/R)\sin\theta - \omega^2 \sin\theta\ \cos\theta = 0 \quad .$$

(ii) Find the fixed points, linearize the equations about them etc.

3.6 The equations $\dot{p} = -(U_0/a)\cos(q/a)$, $\dot{q} = p/m$ for a particle of mass m with potential energy $U = U_0 \sin(q/a)$ ($U_0 > 0$) have the fixed points $(p = 0, q = -\pi a/2)$ and $(p = 0, q = \pi a/2)$.

(i) Linearizing about $(p = 0, q = -\pi a/2)$, show that the solution can be expressed in the form

$$\begin{pmatrix} p'(t) \\ q'(t) \end{pmatrix} = \mathrm{e}^{\mathrm{i}\hat{n}\cdot\sigma\omega t} \begin{pmatrix} p'(0) \\ q'(0) \end{pmatrix} \quad ,$$

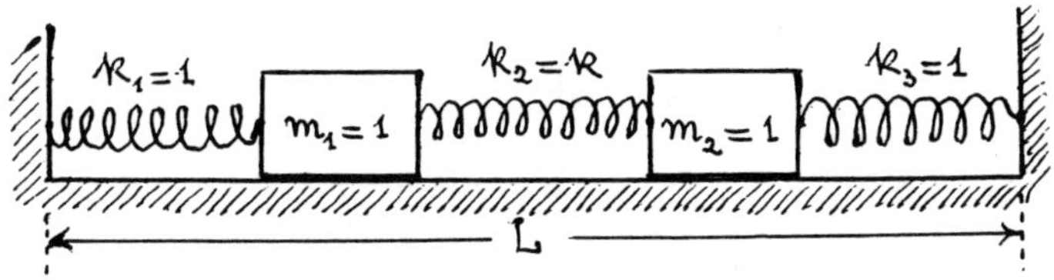

Figure 3.7: Problem 3.8

where

$$\sigma_x = \begin{pmatrix} 0 & 1 \\ 1 & 0 \end{pmatrix} \;, \quad \sigma_y = \begin{pmatrix} 0 & -\mathrm{i} \\ \mathrm{i} & 0 \end{pmatrix} \;, \quad \sigma_z = \begin{pmatrix} 1 & 0 \\ 0 & -1 \end{pmatrix} \;,$$

and (n_x, n_y, n_z), not necessarily real, are components of a unit vector. Use the formula $\exp(\mathrm{i}\hat{\mathbf{n}} \cdot \sigma\alpha) = \cos\alpha + \mathrm{i}\,\hat{\mathbf{n}} \cdot \sigma \sin\alpha$ $(\hat{\mathbf{n}}^2 = 1)$ to express $p'(t)$ and $q'(t)$ as linear combinations of $p'(0)$ and $q'(0)$.

(ii) Follow the same procedure for the fixed point $(p = 0, q = \pi a/2)$.

3.7 The system of equations

$$p_1 = m\dot{q}_1,\; p_2 = m\dot{q}_2,\; \dot{p}_1 = -k_1 q_1 - f q_2,\; \dot{p}_2 = -k_2 q_2 - f q_1$$

is of the type (3.13). Find the normal modes.

3.8 Two blocks, each of unit mass, are connected to each other by a spring of strength k and natural length l, and by springs of unit strength and equal length l_1 to two walls. They can move on a frictionless horizontal floor as shown in figure 3.7. Denote by $L = l + 2l_1$ the distance of the walls. Find the normal modes.

3.9 Three particles (a, b, c) of equal mass m, connected by springs of equal elastic constant k and natural length l, are in equilibrium at the vertexes of an equilateral triangle, $\mathbf{a}_0 = (a_{0x}, a_{0y}) = (-d\sqrt{3}/2, -d/2)$, $\mathbf{b}_0 = (b_{0x}, b_{0y}) = (d\sqrt{3}/2, -d/2)$, $\mathbf{c}_0 = (c_{0x}, c_{0y}) = (0, d)$, $d = l/\sqrt{3}$, as shown in figure 3.8. Study the small amplitude oscillations.

3.10 Consider the two systems of axes shown in figure 3.9. The unprimed system of axes is at rest in the laboratory, with the y axis vertical and pointing downwards. The origin O' of the primed system oscillates with angular frequency Ω and amplitude a along the y-axis. O' is the point of

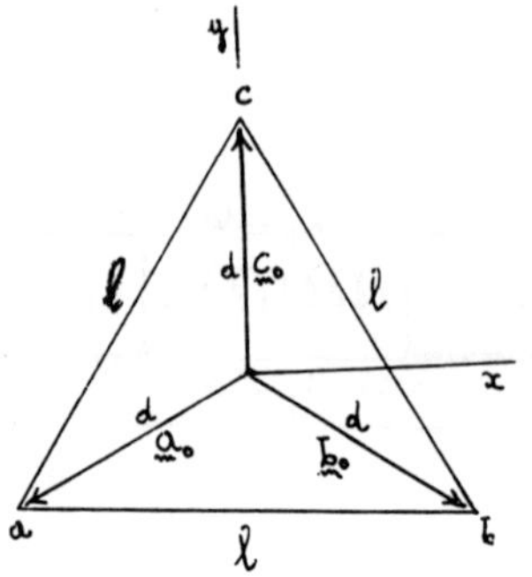

Figure 3.8: Triatomic molecule

suspension of a pendulum of length l. We have

$$m\ddot{x}' = -\tau \sin\phi \quad , \quad m\ddot{y}' = mg - \tau \cos\phi + ma\Omega^2 \cos(\Omega t) \quad .$$

Putting $x' = l \sin\phi$ and $y' = l \cos\phi$, we have

$$\ddot{x}' = l\ddot{\phi}\cos\phi - l\dot{\phi}^2 \sin\phi, \; \ddot{y}' = -l\ddot{\phi}\sin\phi - l\dot{\phi}^2 \cos\phi.$$

Hence

$$\tau = -m\ddot{x}'/\sin\phi = -ml(\ddot{\phi}\cos\phi - \dot{\phi}^2 \sin\phi)/\sin\phi,$$

$$\ddot{\phi} + l^{-1}[g + a\Omega^2 \cos(\Omega t)] \sin\phi = 0 \quad .$$

Try putting $\phi = \Phi + \xi$, where Φ is a constant and ξ is small. We obtain

$$\ddot{\xi} + l^{-1}(g + a\Omega^2 \cos(\Omega t))(\sin\Phi + \cos\Phi\, \xi) \simeq 0 \quad ,$$

$$\ddot{q} + l^{-1} \cos\Phi (g + a\Omega^2 \cos(\Omega t))q \simeq 0 \quad ,$$

where $q = \sin\Phi + \cos\Phi\, \xi$.

Simulate the above situation by solving the equation

$$\ddot{q} + l^{-1} \cos\Phi \left[g + f \sum_{n=-\infty}^{\infty} (-)^n \delta(t - nT/2) \right] q = 0 \quad ,$$

where $T = 2\pi/\Omega$.

3.11 For the "tent map" (also "delta map")

$$x' = r(1 - 2|x - 1/2|) = 2r(x \text{ for } 0 \le x \le 1/2 \;,\; 1 - x \text{ for } 1/2 \le x \le 1)$$

show that the chaotic regime can be expected to begin with $r > r_\infty = 1/2$.

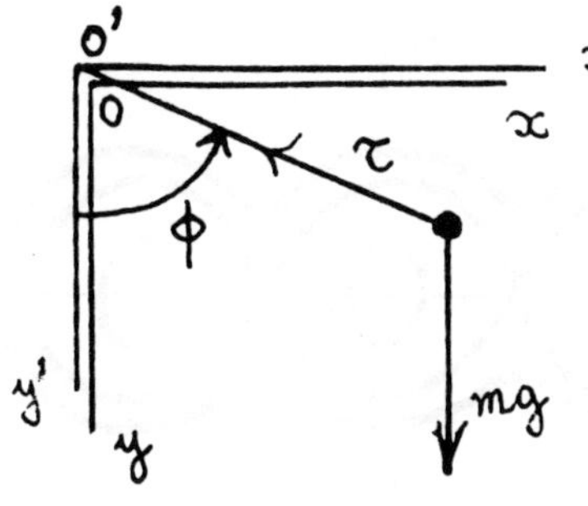

Figure 3.9: Problem 3.10

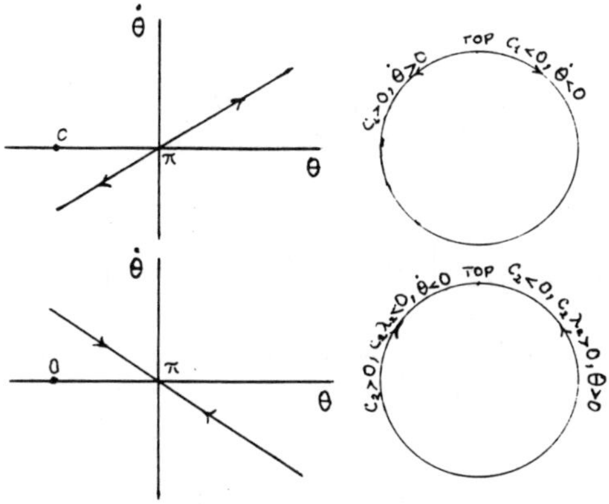

Figure 3.10: Escape from or approach to top position

Solutions to ch. 3 problems

S3.1 The eigenvalues $\lambda_1 = (-\gamma + \sqrt{\gamma^2 - 4g/l})/2$ and $\lambda_2 = (-\gamma - \sqrt{\gamma^2 - 4g/l})/2$ are real and negative.

S3.2 $\mathsf{X}_i \exp(\lambda_i t)$ $(i = 1, 2)$ correspond to

$$\begin{pmatrix} \theta - \pi \\ \dot{\theta} \end{pmatrix} = c_i \begin{pmatrix} 1 \\ \lambda_i \end{pmatrix} e^{\lambda_i t} \quad , \quad \dot{\theta} = \lambda_i(\theta - \pi) \quad .$$

For $\gamma = 0$, $\lambda_1 = \sqrt{g/l}$ the straight lines $\dot{\theta} = \pm\sqrt{g/l}(\theta - \pi)$ are tangent at $(\theta = \pi, \dot{\theta} = 0)$ to the curve $E = g/l$. In fact

$$0 = E - g/l = (\dot{\theta}^2/2) - g(1 + \cos\theta)/l = (\dot{\theta}^2/2) - g[1 - \cos(\theta - \pi)]/l$$
$$\simeq (\dot{\theta}^2/2) - g(\theta - \pi)^2/2l = [\dot{\theta} + \sqrt{g/l}(\theta - \pi)][\dot{\theta} - \sqrt{g/l}(\theta - \pi)]/2 \text{ for } \theta \text{ near } \pi.$$

The curves $E = g/l$ separate regions of librational motion $(\theta(t + T) = \theta(t))$ from regions of rotational motion $(\theta(t + T) = \theta(t) + 2\pi)$ (see figure 3.11).

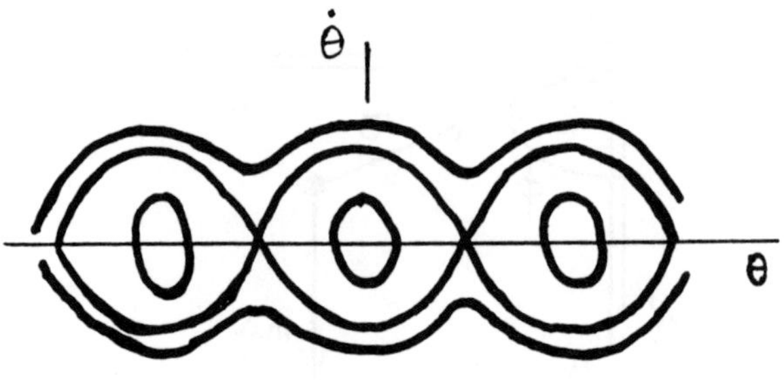

Figure 3.11: Librational and rotational motion

S3.3

$$\mathrm{d}[(\mathsf{X}_1 t + \mathsf{X}_2)\exp(\lambda t)]/\mathrm{d}t = \lambda \mathsf{X}_1 t \exp(\lambda t) + (\mathsf{X}_1 + \lambda \mathsf{X}_2)\exp(\lambda t)$$
$$= \mathsf{A}(\mathsf{X}_1 t + \mathsf{X}_2)\exp(\lambda t).$$

While equation (3.10) gives

$$\mathsf{X}_1 = \begin{pmatrix} a_{12} \\ (a_{22} - a_{11})/2 \end{pmatrix} \text{ one finds } \mathsf{X}_2 = \begin{pmatrix} (a_{22} - a_{11})/2 \\ 1 - a_{21} \end{pmatrix} .$$

In checking this latter, note that $S^2 = 4\Delta$, $(a_{11} + a_{22})^2 = 4(a_{11}a_{22} - a_{12}a_{21})$ can be written in the form $(a_{11} - a_{22})^2 = -4a_{12}a_{21}$.

S3.4

$$\begin{pmatrix} \theta(t) \\ \dot{\theta}(t) \end{pmatrix} = \begin{pmatrix} c_1 + c_2(t - \gamma/2) \\ -\gamma c_1/2 + c_2[-\gamma t/2 + (1 + g/l)] \end{pmatrix} \mathrm{e}^{-\gamma t/2} ,$$

$$c_1 = (1 + \gamma^2/4)\theta_0 + \gamma\dot{\theta}_0/2,\ c_2 = (\gamma\theta_0/2) + \dot{\theta}_0.$$

Eliminating c_1 we find $\dot{\theta} + \gamma\theta/2 = c_2 \exp(-\gamma t/2)$.
For brevity's sake we do not discuss some pathological cases which occur for $\lambda_2 = \lambda_1$.

S3.5 (i) $\mathrm{d}(mR^2\dot{\theta})/\mathrm{d}t = m(\omega^2 R^2 \sin\theta)\cos\theta - (\gamma m R\dot{\theta})R - mgR\sin\theta$
(ii) Putting $x_1 = \theta$ and $x_2 = \dot{\theta}$ we have

$$\dot{x}_1 = x_2,\ \dot{x}_2 = -\gamma x_2 - (g/R)\sin x_1 + \omega^2 \sin x_1 \cos x_1.$$

There are two fixed points, $(x_1 = 0, x_2 = 0)$ and $(x_1 = \cos^{-1}(g/R\omega^2), x_2 = 0)$, the latter only for $\omega^2 > g/R$.
Linearize about $(0,0)$:

$$a_{11} = 0,\ a_{12} = 1,\ a_{21} = \omega^2 - g/R,\ a_{22} = -\gamma,\ \Delta = (g/R) - \omega^2,$$
$$\lambda = (-\gamma \pm \sqrt{\gamma^2 + 4\omega^2 - 4g/R}\,)/2.$$

If $\omega^2 < (g/R) - \gamma^2/4$, λ_1 and λ_2 are complex conjugate with negative real part, $(0,0)$ is a spiral attractor.
If $\omega^2 = (g/R) - \gamma^2/4$, $\lambda_1 = \lambda_2 = -\gamma/2$.
If $\omega^2 = g/R$, $\lambda_1 = 0$, $\lambda_2 < 0$. In this case the equation linearized about $(0,0)$ is

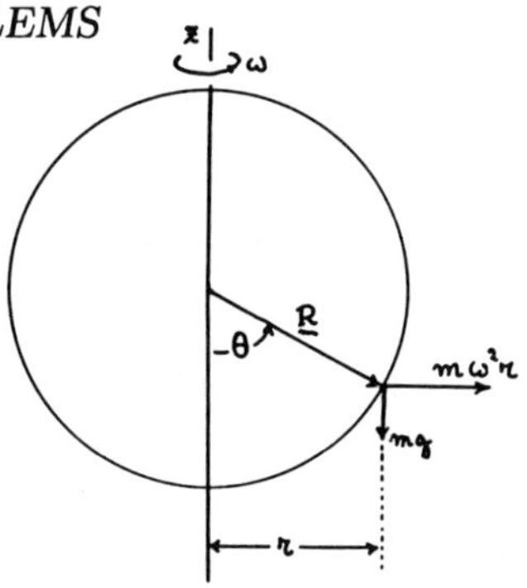

Figure 3.12: Bead on rotating hoop

$\ddot{\theta} = -\gamma\dot{\theta}$, with solution $\theta = \theta_0 + \dot{\theta}_0[1 - \exp(-\gamma t)]/\gamma$, $\dot{\theta} = \dot{\theta}_0 \exp(-\gamma t)$.
If $\omega^2 > g/R$, $\lambda_1 > 0$, $\lambda_2 < 0$, $(0,0)$ is unstable.
Linearize about $(\cos^{-1}(g/R\omega^2), 0)$:

$$a_{11} = 0,\ a_{12} = 1,\ a_{21} = (g^2/R^2\omega^2) - \omega^2,\ a_{22} = -\gamma,\ \Delta = \omega^2 - g^2/R^2\omega^2,$$

$$\lambda = (-\gamma \pm \sqrt{\gamma^2 + (4g^2/R^2\omega^2) - 4\omega^2}\,)/2.$$

The square root is imaginary, and the fixed point is a spiral attractor, if

$$\omega^2 > \sqrt{(g^2/R^2) + (\gamma^2/64)} + \gamma^2/8.$$

Note that for $\gamma = 0$, $(0,0)$ becomes unstable and $(\cos^{-1}(g/R\omega^2), 0)$ comes into existence as a stable fixed point for $\omega = \sqrt{g/R}$.

S3.6 (i) With $p' = p$ and $q' = q + \pi a/2$, the linearized equations

$$\begin{pmatrix} \dot{p}'(t) \\ \dot{q}'(t) \end{pmatrix} = \mathsf{A} \begin{pmatrix} p'(t) \\ q'(t) \end{pmatrix}$$

with

$$\mathsf{A} = \begin{pmatrix} 0 & -U_0/a^2 \\ 1/m & 0 \end{pmatrix}$$

have the formal solution

$$\begin{pmatrix} p'(t) \\ q'(t) \end{pmatrix} = e^{\mathsf{A}t} \begin{pmatrix} p'(0) \\ q'(0) \end{pmatrix} .$$

The matrix A can be expressed in the form $\mathsf{A} = \mathrm{i}(n_x\sigma_x + n_y\sigma_y)\omega$, $n_x = (a^2 - mU_0)/2\mathrm{i}a\sqrt{mU_0}$, $n_y = -(a^2 + mU_0)/2a\sqrt{mU_0}$ $(n_x^2 + n_y^2 = 1)$, $\omega = a^{-1}\sqrt{U_0/m}$. Then

$$\begin{pmatrix} p'(t) \\ q'(t) \end{pmatrix} = [\cos(\omega t) + \mathrm{i}(n_x\sigma_x + n_y\sigma_y)\sin(\omega t)] \begin{pmatrix} p'(0) \\ q'(0) \end{pmatrix} ,$$

$$\begin{cases} p'(t) = \cos(\omega t)p'(0) - a^{-1}\sqrt{mU_0}\sin(\omega t)q'(0) , \\ q'(t) = (a/\sqrt{mU_0}\,)\sin(\omega t)p'(0) + \cos(\omega t)q'(0) . \end{cases}$$

(ii) Linearizing about $(p = 0, q = \pi a/2)$, we have

$$\mathsf{A} = \begin{pmatrix} 0 & U_0/a^2 \\ 1/m & 0 \end{pmatrix} ,$$

$\mathsf{A} = (n_x\sigma_x + n_y\sigma_y)\omega$, $n_x = (a^2 + mU_0)/2a\sqrt{mU_0}$, $n_y = \mathrm{i}(-a^2 + mU_0)/2a\sqrt{mU_0}$ $(n_x^2 + n_y^2 = 1)$, $\omega = a^{-1}\sqrt{U_0/m}$. Then

$$\begin{pmatrix} p'(t) \\ q'(t) \end{pmatrix} = [\cosh(\omega t) + (n_x\sigma_x + n_y\sigma_y)\sinh(\omega t)] \begin{pmatrix} p'(0) \\ q'(0) \end{pmatrix} ,$$

$$\begin{cases} p'(t) = \cosh(\omega t)p'(0) + a^{-1}\sqrt{mU_0}\sinh(\omega t)q'(0) , \\ q'(t) = (a/\sqrt{mU_0}\,)\sinh(\omega t)p'(0) + \cosh(\omega t)q'(0) . \end{cases}$$

S3.7

$$\mathsf{M} = \begin{pmatrix} m & 0 \\ 0 & m \end{pmatrix} , \quad \mathsf{K} = \begin{pmatrix} k_1 & f \\ f & k_2 \end{pmatrix}$$

The secular equation $\det(\mathsf{K} - \omega^2\mathsf{M}) = 0$ gives

$$m^2(\omega^2)^2 - m(k_1 + k_2)\omega^2 + (k_1k_2 - f^2) = 0$$

with solutions

$$\omega_1^2 = [k_1 + k_2 + \sqrt{(k_1 - k_2)^2 + 4f^2}]/2m$$

and ω_2^2 with a minus before the square root.
The kinetic and potential energies are

$$K = m(\dot{q}_1^2 + \dot{q}_2^2)/2,\ U = (k_1q_1^2 + k_2q_2^2 + 2fq_1q_2)/2.$$

With $Q_1 = \sqrt{m}q_1$ and $Q_2 = \sqrt{m}q_2$ we have

$$K = (\dot{Q}_1^2 + \dot{Q}_2^2)/2,\ U = (k_1Q_1^2 + k_2Q_2^2 + 2fQ_1Q_2)/2m.$$

The rotation $Q_1 = Q_1{}'\cos\alpha + Q_2{}'\sin\alpha$, $Q_2 = -Q_1'\sin\alpha + Q_2'\cos\alpha$ gives $K = (\dot{Q}_1'^2 + \dot{Q}_2'^2)/2$ and an expression for U which, by choosing $\tan(2\alpha) = -2f/(k_1 - k_2)$, reduces to $U = (\omega_1^2Q_1'^2 + \omega_2^2Q_2'^2)/2$.

The normal modes correspond to $Q_2{}' = 0$ and $Q_1{}' = 0$.

S3.8 With $q_1 = x_1 - l_1$ and $q_2 = x_2 - l - l_1$, the equations

$$\ddot{x}_1 = k(x_2 - x_1 - l) - (x_1 - l_1) \text{ and } \ddot{x}_2 = -k(x_2 - x_1 - l) - (x_2 - l - l_1)$$

take the form

$$\ddot{q}_1 = -(k+1)q_1 + kq_2 \text{ and } \ddot{q}_2 = kq_1 - (k+1)q_2.$$

With $\mathsf{M} = \mathsf{I}$, the unit matrix, and K having diagonal elements $k + 1$ and off diagonal elements $-k$, the equation $\det(\mathsf{K} - \omega^2\mathsf{M}) = 0$ yields the eigenfrequencies $\omega_1 = 1$ and $\omega_2 = \sqrt{2k+1}$. With $Q_1 = q_1$ and $Q_2 = q_2$ we have $K = (\dot{Q}_1^2 + \dot{Q}_2^2)/2$ and $U = [(k+1)(Q_1^2 + Q_2^2) - 2kQ_1Q_2]/2$. The $\pi/4$ rotation $Q_1 = (Q_1' - Q_2')/\sqrt{2}$ and $Q_2 = (Q_1' + Q_2')/\sqrt{2}$ yields $K = (\dot{Q}_1'^2 + \dot{Q}_2'^2)/2$ and $U = (\omega_1^2Q_1'^2 + \omega_2^2Q_2'^2)/2$ with $\omega_1 = 1$ and $\omega_2 = \sqrt{2k+1}$.
$\omega_1 = 1$ corresponds to the mode $Q_2{}' = 0$, $Q_2 - Q_1 = 0$, $q_2 - q_1 = 0$, $x_2 = l + x_1$. The two blocks, at a fixed distance from each other, oscillate together with period

2π, $x_1 = l_1 + A\ \cos(t+\alpha)$, $x_2 = l + l_1 + A\ \cos(t+\alpha)$.
$\omega_2 = \sqrt{2k+1}$ corresponds to

$$Q_1' = 0,\ Q_2 + Q_1 = 0,\ q_2 = -q_1,$$
$$x_1 = l_1 + A\ \cos(\sqrt{2k+1}t+\alpha),\ x_2 = l + l_1 - A\ \cos(\sqrt{2k+1}t+\alpha).$$

The center of mass of the blocks remains fixed halfway between the walls, $(x_1 + x_2)/2 = L/2$, the blocks oscillate towards and away from each other.

S3.9 Denoting by $\mathbf{a} = \mathbf{a}_0 + \mathbf{a}'$, $\mathbf{b} = \mathbf{b}_0 + \mathbf{b}'$, and $\mathbf{c} = \mathbf{c}_0 + \mathbf{c}'$ the positions of the particles, Newton's second law yields for particle a the (exact) equations

$$\begin{cases} m\ddot{a}_x = -k[x_{ab}(1 - l/r_{ab}) + x_{ac}(1 - l/r_{ac})] \ , \\ m\ddot{a}_y = -k[y_{ab}(1 - l/r_{ab}) + y_{ac}(1 - l/r_{ac})] \ , \end{cases}$$

where $r_{ab} = |\mathbf{a} - \mathbf{b}|$, $x_{ab} = a_x - b_x$, $y_{ab} = a_y - b_y$ etc. Similar equations hold for $\mathbf{b}$ and $\mathbf{c}$.

If $\mathbf{a}'$, $\mathbf{b}'$, and $\mathbf{c}'$ are small, then, with $x'_{ab} = a'_x - b'_x$ etc. one has $1 - l/r_{ab} \simeq (x_{0ab}x'_{ab} + y_{0ab}y'_{ab})/l^2$ etc. One finds

$$\begin{cases} m\ddot{a}'_x = -k[x_{0ab}(x_{0ab}x'_{ab} + y_{0ab}y'_{ab}) + x_{0ac}(x_{0ac}x'_{ac} + y_{0ac}y'_{ac})]/l^2 \ , \\ m\ddot{a}'_y = -k[y_{0ab}(\ldots + \ldots) + y_{0ac}(\ldots + \ldots)]/l^2 \ , \end{cases}$$

and similar equations for $\mathbf{b}'$ and $\mathbf{c}'$.

Denoting by q' a column matrix of which, to save space, we write the transpose

$$\mathsf{q}'^{\mathrm{T}} = (a'_x\ b'_x\ c'_x\ a'_y\ b'_y\ c'_y) \ ,$$

we find $m\ddot{\mathsf{q}}' = -\mathsf{K}\mathsf{q}'$, where

$$\mathsf{K} = \frac{k}{4}\begin{pmatrix} 5 & -4 & -1 & \sqrt{3} & 0 & -\sqrt{3} \\ -4 & 5 & -1 & 0 & -\sqrt{3} & \sqrt{3} \\ -1 & -1 & 2 & -\sqrt{3} & \sqrt{3} & 0 \\ \sqrt{3} & 0 & -\sqrt{3} & 3 & 0 & -3 \\ 0 & -\sqrt{3} & \sqrt{3} & 0 & 3 & -3 \\ -\sqrt{3} & \sqrt{3} & 0 & -3 & -3 & 6 \end{pmatrix} .$$

The determinant of this matrix is zero. This can be seen by summing the elements of the second and third rows to those of the first, thereby getting a row of zeros. Therefore one of the eigenvalues of M is zero. We shall see that zero is a triple root of the secular equation, and that there are three independent eigenvectors belonging to the eigenvalue zero.

Note at once that $\mathrm{d}^2(\mathbf{a}' + \mathbf{b}' + \mathbf{c}')/\mathrm{d}t^2 = 0$. Therefore our equations allow for a uniform motion of the center of mass, initially at $\mathbf{a} + \mathbf{b} + \mathbf{c} = 0$. The two

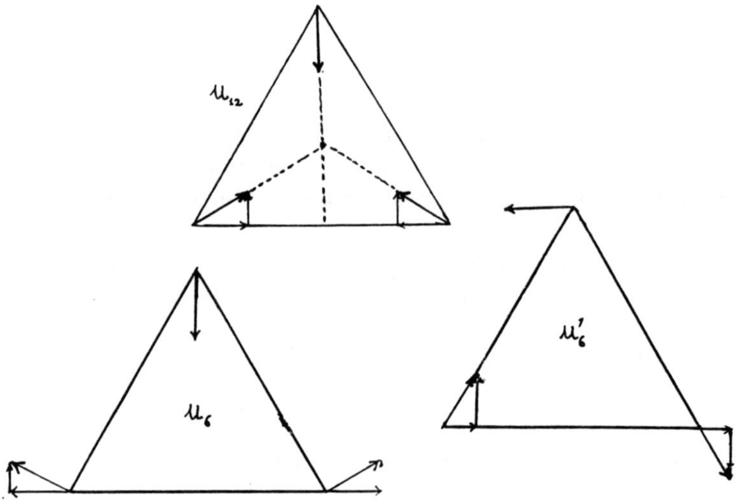

Figure 3.13: Triatomic molecule ($\lambda = 6$ and $\lambda = 12$)

mutually orthogonal eigenvectors of K,

$$\mathsf{u}_0^{(x)} = \begin{pmatrix} 1 \\ 1 \\ 1 \\ 0 \\ 0 \\ 0 \end{pmatrix} \text{ and } \mathsf{u}_0^{(y)} = \begin{pmatrix} 0 \\ 0 \\ 0 \\ 1 \\ 1 \\ 1 \end{pmatrix} ,$$

both belonging to the eigenvalue zero, provide for this: $\mathsf{q}' = vt\mathsf{u}_0^{(x)}$ and $\mathsf{q}' = vt\mathsf{u}_0^{(y)}$ (v =constant) represent uniform motion in the x and y direction, respectively.

K is rather formidable. We can reduce the eigenvalue problem to a simpler one in four dimensions by assuming that the center of mass is at rest, so that $\mathbf{c}' = -\mathbf{a}' - \mathbf{b}'$. Using this condition we reduce the problem to $m\ddot{\mathsf{q}}' = -\mathsf{K}\mathsf{q}'$, with $\mathsf{q}'^{\mathrm{T}} = (a'_x\ b'_x\ a'_y\ b'_y)$ and

$$\mathsf{K} = \frac{k}{4} \begin{pmatrix} 6 & -3 & 2\sqrt{3} & \sqrt{3} \\ -3 & 6 & -\sqrt{3} & -2\sqrt{3} \\ 2\sqrt{3} & \sqrt{3} & 6 & 3 \\ -\sqrt{3} & -2\sqrt{3} & 3 & 6 \end{pmatrix} .$$

From $m\omega^2\mathsf{u} = \mathsf{K}\mathsf{u}$ we obtain the equation $(6-\lambda)^4 - 36(6-\lambda)^2 = 0$ for $\lambda = 4m\omega^2/k$. We see at once that $\lambda = 0$ and $\lambda = 12$ are simple roots, while $\lambda = 6$ is a double root.

The eigenvector corresponding to $\lambda = 0$ is

$$\mathsf{u}_0^{(\mathrm{rot})} = \begin{pmatrix} 1 \\ 1 \\ -2 \\ -\sqrt{3} \\ \sqrt{3} \\ 0 \end{pmatrix} .$$

Since $\mathsf{q} = \mathsf{q}_0 - (d\epsilon/2)\mathsf{u}_0^{(\mathrm{rot})}$ gives $a_x = a_{0x} + \epsilon a_{0y}$, $a_y = -\epsilon a_{0x} + a_{0y}$, and similar equations for the components of **b** and **c**, it clearly describes an infinitesimal rotation about the center of mass. Note that the angular momentum about the center of mass is a conserved quantity.

The eigenvector corresponding to the eigenvalue $\lambda = 12$ is

$$\mathsf{u}_{12} = \begin{pmatrix} \sqrt{3} \\ -\sqrt{3} \\ 0 \\ 1 \\ 1 \\ -2 \end{pmatrix} .$$

It describes oscillations as in figure 3.13a with $\omega = \sqrt{3k/m}$.

There are two orthogonal eigenvectors for $\lambda = 6$, $\omega = \sqrt{3k/2m}$,

$$\mathsf{u}_6 = \begin{pmatrix} -\sqrt{3} \\ \sqrt{3} \\ 0 \\ 1 \\ 1 \\ -2 \end{pmatrix} \quad \text{and} \quad \mathsf{u}_6{}' = \begin{pmatrix} 1 \\ 1 \\ -2 \\ \sqrt{3} \\ -\sqrt{3} \\ 0 \end{pmatrix} .$$

They describe oscillations as in figures 3.13b and 3.13c , respectively.

S3.10 Starting with $q_1(0^-) = 1$, $\dot{q}_1(0^-) = 0$, $q_2(0^-) = 0$, $\dot{q}_2(0^-) = 1$, with the notation of section 3.4 we have

$$\mathsf{x}(T^-/2) = \mathsf{A}_+ \mathsf{x}(0^-) \quad ,$$

where

$$\mathsf{A}_+ = \begin{pmatrix} \cos(\omega_0' T/2) - f' \omega_0'^{-1} \sin(\omega_0' T/2) & \omega_0'^{-1} \sin(\omega_0' T/2) \\ -\omega_0' \sin(\omega_0' T/2) - f' \cos(\omega_0' T/2) & \cos(\omega_0' T/2) \end{pmatrix} ,$$

$\omega_0' = \sqrt{(g \cos\Phi)/l}$, $f' = (f \cos\Phi)/l$.
Then

$$\mathsf{x}(T^-) = \mathsf{A}\mathsf{x}(0^-) \quad ,$$

with

$$\mathsf{A} = \mathsf{A}_-\mathsf{A}_+ \quad , \quad \mathsf{A}_- = [\mathsf{A}_+]_{f'\to -f'} \quad .$$

For our purpose we need

$$\mathrm{trace}(\mathsf{A}) = \left(2 + \frac{f'^2}{2{\omega'_0}^2}\right)\cos(\omega'_0 T) - \frac{f'^2}{2{\omega'_0}^2} \quad .$$

We consider the case $\cos\Phi < 0$ (pendulum rod above horizontal position) for which $\omega'_0 = \mathrm{i}\sqrt{g|\cos\Phi|/l}$. For small T we then have

$$\mathrm{trace}(\mathsf{A}) = [2 + (f^2\cos\Phi)/2gl]\cosh(\sqrt{g|\cos\Phi|/l}\,T) - (f^2\cos\Phi)/2gl$$

$$\simeq 2 + (g/l)|\cos\Phi|T^2 + (f^2\cos^2\Phi\, T^2)/4l^2 \quad .$$

At equilibrium must be $\gamma = 0$, $S = \mathrm{trace}(\mathsf{A}) = 2$, $\cos\Phi = -4lg/f^2$.

For the frequency of small oscillations about $\Phi = \pi$, we must have

$$\exp(\pm\mathrm{i}\gamma) = \lambda = (S \pm \mathrm{i}\sqrt{4 - S^2}\,)/2$$

with $S = 2 + (g/l)T^2 - (f^2T^2)/(4l^2)$. Then

$$\sin^2(\gamma/2) = (1 - \cos\gamma)/2 = (2 - S)/4 = (f^2T^2/16l^2) - (gT^2/4l),$$

$$\gamma^2 \simeq (f^2T^2/4l^2) - (gT^2/l),\ \omega^2 = \gamma^2/T^2 \simeq (f^2/4l^2) - (g/l).$$

S3.11

$$|\mathrm{d}x'/\mathrm{d}x| = 2r,\ \lambda = \lim_{n\to\infty} n^{-1}\ln(2r)^n = \ln(2r).$$

Chapter 4

COORDINATE SYSTEMS

A detailed treatment of rotations in three dimensions, which will be widely used in the book, is followed by a discussion of the fictitious forces which must be included if the motion of a body is referred to a non-inertial frame of reference.

4.1 Translations and rotations

Let P be a point with coordinates x_i ($i = 1, 2, 3$) with respect to a system of orthogonal axes S with origin at O. The coordinates x'_i of P with respect to another system of orthogonal axes S' with origin at O' are related to the x_i's by equations of the form

$$x_i = R_{ij}x'_j + x_{0i} \quad , \tag{4.1}$$

where x_{0i} are the coordinates of O' with respect to S (dummy indices convention). If the axes of S' are parallel to, and equally oriented as those of S, then $R_{ij} = \delta_{ij}$ ($= 1$ if $i = j, = 0$ if $i \neq j$).

Introducing the notation

$$\mathsf{r} = \begin{pmatrix} x_1 \\ x_2 \\ x_3 \end{pmatrix} \quad , \quad \mathsf{r}' = \begin{pmatrix} x'_1 \\ x'_2 \\ x'_3 \end{pmatrix} \quad , \quad \mathsf{r}_0 = \begin{pmatrix} x_{01} \\ x_{02} \\ x_{03} \end{pmatrix} \quad , \tag{4.2}$$

equation (4.1) can be written in the form

$$\mathsf{r} = \mathsf{R}\mathsf{r}' + \mathsf{r}_0 \quad , \tag{4.3}$$

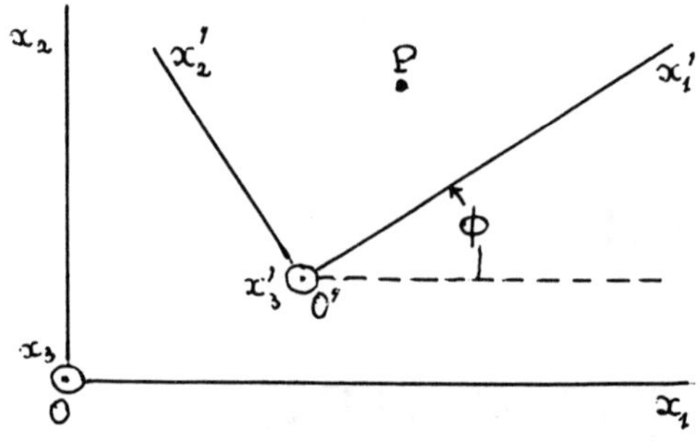

Figure 4.1: Example of translation and rotation

where R is the matrix with elements R_{ij}.

Let $\hat{\mathbf{e}}_i$ denote the unit vector along the i-th axis of S. The distinction between $\mathbf{r} = \vec{OP} = x_i\hat{\mathbf{e}}_i$ and the one-column matrix r, though formal, must be remembered.

Example (figure 4.1):

$$\left\{\begin{array}{l} x_1 = x_1' \cos\phi - x_2' \sin\phi + x_{01} \quad , \\ x_2 = x_1' \sin\phi + x_2' \cos\phi + x_{02} \quad , \\ x_3 = x_3' \quad , \end{array}\right.$$

$$\mathsf{R} = \begin{pmatrix} \cos\phi & -\sin\phi & 0 \\ \sin\phi & \cos\phi & 0 \\ 0 & 0 & 1 \end{pmatrix} \Rightarrow \mathsf{R}_3(\phi) \quad . \tag{4.4}$$

The S' system is obtained by first translating S so that O coincides with O', thus obtaining a system S'' with origin $O'' = O'$ and axes parallel to those of S, and then by rotating S'' counterclockwise through ϕ around x_3''. The matrix for this rotation is $\mathsf{R}_3(\phi)$.

Similarly we would have

$$\mathsf{R}_1(\phi) = \begin{pmatrix} 1 & 0 & 0 \\ 0 & \cos\phi & -\sin\phi \\ 0 & \sin\phi & \cos\phi \end{pmatrix} \tag{4.5}$$

and

$$\mathsf{R}_2(\phi) = \begin{pmatrix} \cos\phi & 0 & \sin\phi \\ 0 & 1 & 0 \\ -\sin\phi & 0 & \cos\phi \end{pmatrix} \quad . \tag{4.6}$$

If $\phi \to \delta\phi$ infinitesimal, we have

$$\mathsf{R}_i(\delta\phi) = \mathsf{I} - \mathsf{J}_i\, \delta\phi \quad , \tag{4.7}$$

where I is the unit matrix and

$$\mathsf{J}_1 = \begin{pmatrix} 0 & 0 & 0 \\ 0 & 0 & 1 \\ 0 & -1 & 0 \end{pmatrix}, \mathsf{J}_2 = \begin{pmatrix} 0 & 0 & -1 \\ 0 & 0 & 0 \\ 1 & 0 & 0 \end{pmatrix}, \mathsf{J}_3 = \begin{pmatrix} 0 & 1 & 0 \\ -1 & 0 & 0 \\ 0 & 0 & 0 \end{pmatrix}. \tag{4.8}$$

These skew-symmetric matrices are the "generators of infinitesimal rotations". Note that their matrix elements are given by $(\mathsf{J}_i)_{jk} = \epsilon_{ijk}$, where ϵ_{ijk} is the Ricci-Levi Civita symbol, skew-symmetric in all indices (e.g. $\epsilon_{kji} = -\epsilon_{ijk}$) and $\epsilon_{123} = 1$.

The generators satisfy the relations

$$[\mathsf{J}_i, \mathsf{J}_j] = -\epsilon_{ijk}\mathsf{J}_k \quad , \tag{4.9}$$

where the "commutator" of two matrices A and B is denoted by $[\mathsf{A}, \mathsf{B}] \equiv \mathsf{AB} - \mathsf{BA}$.

Equation (4.9) can be verified by matrix multiplication obtaining $[\mathsf{J}_1, \mathsf{J}_2] = -\mathsf{J}_3$ and cyclic permutations, or by comparing the matrix elements of the two sides with the help of $\epsilon_{abc}\epsilon_{cde} = \delta_{ad}\delta_{be} - \delta_{ae}\delta_{bd}$.

For an infinitesimal rotation about an arbitrary direction specified by a unit vector $\hat{\mathbf{n}}$ the matrix R is given by (see problem 4.1)

$$\mathsf{R} = \mathsf{I} - \mathsf{J}(\hat{\mathbf{n}})\,\delta\phi \quad , \tag{4.10}$$

where

$$\mathsf{J}(\hat{\mathbf{n}}) = n_i\mathsf{J}_i = \begin{pmatrix} 0 & n_3 & -n_2 \\ -n_3 & 0 & n_1 \\ n_2 & -n_1 & 0 \end{pmatrix} \quad . \tag{4.11}$$

To obtain a formula for finite rotations, we write

$$\mathsf{r} - \mathsf{r}_0 = \lim_{N\to\infty} \left(\mathsf{I} - \mathsf{J}(\hat{\mathbf{n}})\phi/N\right)^N \mathsf{r}' = \mathrm{e}^{-\mathsf{J}(\hat{\mathbf{n}})\phi}\, \mathsf{r}' \quad , \tag{4.12}$$

where, we repeat, $\mathsf{J}(\hat{\mathbf{n}}) = n_i\mathsf{J}_i$, n_i being the components of $\hat{\mathbf{n}}$ with respect to S.

This formal expression can be reduced to a practical formula by using the following properties of the J_i's.

First of all, one has

$$\mathsf{J}_i\mathsf{J}_j\mathsf{J}_k + \mathsf{J}_k\mathsf{J}_j\mathsf{J}_i = -\mathsf{J}_i\delta_{jk} - \delta_{ij}\mathsf{J}_k \quad . \tag{4.13}$$

Multiplying by $n_i n_j n_k$ this gives

$$\mathsf{J}(\hat{\mathbf{n}})^3 = -\mathsf{J}(\hat{\mathbf{n}}) \quad , \tag{4.14}$$

so that ($n = 1, 2 \ldots$)

$$\mathsf{J}(\hat{\mathbf{n}})^{2n+1} = (-)^n \mathsf{J}(\hat{\mathbf{n}}) \quad \text{and} \quad \mathsf{J}(\hat{\mathbf{n}})^{2n} = (-)^{n+1} \mathsf{J}(\hat{\mathbf{n}})^2 \quad .$$

Furthermore, it is easy to show that

$$(\mathsf{J}(\hat{\mathbf{n}})^2)_{ij} = n_a n_b (\mathsf{J}_a \mathsf{J}_b)_{ij} = -\delta_{ij} + n_i n_j \quad .$$

Now for $\mathsf{R} = \exp(-\mathsf{J}(\hat{\mathbf{n}})\phi)$ (equation (4.12)) we have

$$\mathsf{R} = \mathsf{I} + \sum_{n=0}^{\infty} [-\mathsf{J}(\hat{\mathbf{n}})\phi]^{2n}/(2n)! + \sum_{n=0}^{\infty} [-\mathsf{J}(\hat{\mathbf{n}})\phi]^{2n+1}/(2n+1)! \ ,$$

$$\mathsf{R} = \mathsf{I} + \mathsf{J}(\hat{\mathbf{n}})^2 (1 - \cos\phi) - \mathsf{J}(\hat{\mathbf{n}}) \sin\phi \quad , \tag{4.15}$$

and so

$$R_{ij} = \delta_{ij} \cos\phi + (1 - \cos\phi) n_i n_j - \epsilon_{ijk} n_k \sin\phi \quad . \tag{4.16}$$

Note that changing the sign of ϕ in (4.15) is equivalent to transposing the skew-symmetric matrices J_i. Therefore

$$\mathsf{R}^{-1} = \mathsf{R}^{\mathrm{T}} \quad . \tag{4.17}$$

The important formula

$$\mathsf{R}^{-1} \mathsf{J}_i \mathsf{R} = R_{ij} \mathsf{J}_j \tag{4.18}$$

will be proved in problem 4.2 .

The informed reader may expect some grouptheoretical references. The above rotations are the natural representation of the rotation group $R(3)$. This is a subgroup of $O(3)$, the orthogonal group in three dimensions, defined by $\mathsf{r}' = \mathsf{O}\mathsf{r}$, $\mathsf{r}'^{\mathrm{T}}\mathsf{r}' = \mathsf{r}^{\mathrm{T}}\mathsf{r}$, $\mathsf{O}^{\mathrm{T}}\mathsf{O} = \mathsf{I}$, $\mathsf{O}^{\mathrm{T}} = \mathsf{O}^{-1}$. Besides rotations, $O(3)$ contains the inversion $\mathsf{O} = -\mathsf{I}$. Since $\det \mathsf{O}^{\mathrm{T}} = \det \mathsf{O}$, taking the determinant of $\mathsf{O}^{\mathrm{T}}\mathsf{O} = \mathsf{I}$ one finds $\det \mathsf{O}^2 = 1$, $\det \mathsf{O} = \pm 1$.

The rotation group $R(3)$ is restricted to have $\det \mathsf{O} = 1$. Thus $R(3) = SO(3) \subset O(3)$. In $SO(3)$, the S stands for "special".

Expressing a rotation matrix in the form $\mathsf{R} = \exp \mathsf{A}$, since $\det(\exp \mathsf{A}) = \exp(\mathrm{trace}\mathsf{A})$, $\det \mathsf{R} = 1$ implies $\mathrm{trace}(\mathsf{A}) = 0$. This is less restrictive than the condition $\mathsf{A}^{\mathrm{T}} = -\mathsf{A}$, following from (4.17). The tracelessness of our J_i's is due to their skewsymmetry.

The connection between $R(3)$ and the special unitary group in two dimensions $SU(2)$ will be discussed in chapter 10 as a special case of that of the Lorentz group with $SL(2)$, the special linear group in two dimensions.

4.2 Some kinematics

If S' is in motion with respect to S, R and r_0 are functions of t. Let P be a moving point. Differentiating $\mathsf{r} = \mathsf{R}\mathsf{r}' + \mathsf{r}_0$ with respect to t, we have

$$\dot{\mathsf{r}} = \dot{\mathsf{R}}\mathsf{r}' + \mathsf{R}\dot{\mathsf{r}}' + \dot{\mathsf{r}}_0 \quad . \tag{4.19}$$

Case 1: S' is in translational motion with respect to S ($\dot{\mathsf{R}} = 0$). The components of the velocity with respect to S, $v_i = \dot{x}_i$, are related to those with respect to S', $v_i' = \dot{x}_i'$, by the equation $v_i = R_{ij}v_j' + v_{0i}$, where $v_{0i} = \dot{x}_{0i}$. More concisely $\mathsf{v} = \mathsf{R}\mathsf{v}' + \mathsf{v}_0$.

If the axes of S' are parallel to those of S ($\mathsf{R} = \mathsf{I}$), then $\mathsf{v} = \mathsf{v}' + \mathsf{v}_0$. Carefully distinguish v and $\mathbf{v}$. Even if $\mathsf{R} \neq \mathsf{I}$, one has $\mathbf{v} = \mathbf{v}' + \mathbf{v}_0$, which is nothing but $\mathrm{d}\vec{OP}/\mathrm{d}t = \mathrm{d}\vec{O'P}/\mathrm{d}t + \mathrm{d}\vec{OO'}/\mathrm{d}t$.

Case 2: $\mathsf{r}_0 = 0$, $\dot{\mathsf{r}}' = 0$, P at rest with respect to S', whose origin coincides with that of S, S' obtained from S by rotation with constant angular velocity ω around $\hat{\mathbf{n}}$. Then, with $\mathsf{J}(\omega) = \omega_i \mathsf{J}_i$,

$$\dot{\mathsf{r}} = \frac{\mathrm{d}}{\mathrm{d}t}\left(\mathrm{e}^{-\mathsf{J}(\omega)t}\mathsf{r}'\right) = -\mathsf{J}(\omega)\mathsf{r} \quad , \tag{4.20}$$

$\dot{x}_i = -\omega_k(\mathsf{J}_k)_{ij}x_j = \epsilon_{ikj}\omega_k x_j$ ($\mathbf{v} = \omega \times \mathbf{r}$ in the usual notation).

Case 3: The point P is at rest with respect to S', $\mathsf{r}_0 = 0$, but R is a function of t.

At each time t one can find an instantaneous angular velocity $\omega(t)$ such that

$$\dot{\mathsf{r}} = -\mathsf{J}(\omega(t))\mathsf{r} \quad . \tag{4.21}$$

Proof: $\dot{\mathsf{r}} = \dot{\mathsf{R}}\mathsf{r}' = \mathsf{A}\mathsf{r}$ with $\mathsf{A} = \dot{\mathsf{R}}\mathsf{R}^{-1} = \dot{\mathsf{R}}\mathsf{R}^{\mathrm{T}}$, $\mathrm{d}(\mathsf{R}\mathsf{R}^{\mathrm{T}})/\mathrm{d}t = 0$, $\dot{\mathsf{R}}\mathsf{R}^{\mathrm{T}} + \mathsf{R}\dot{\mathsf{R}}^{\mathrm{T}} = 0$, $\mathsf{A} + \mathsf{A}^{\mathrm{T}} = 0$, A is skewsymmetric.

The J_i's form a complete basis for all 3×3 skewsymmetric matrices. Hence $\mathsf{A} = -\mathsf{J}(\omega(t))$, where $-\omega_i(t)$ are the components of A in that basis.

4.3 Fictitious forces

Case 1: S' in translational motion with respect to the inertial frame S, S and S' axes parallel, $\ddot{\mathsf{r}} = \ddot{\mathsf{r}}' + \ddot{\mathsf{r}}_0$. The applied force f in Newton's second law in S, $m\ddot{\mathsf{r}} = \mathsf{f}$, is supplemented by an "inertial force" when the motion is referred to S', $m\ddot{\mathsf{r}}' = \mathsf{f} - m\ddot{\mathsf{r}}_0$.

Case 2: $\mathsf{r}_0 = 0$, S' in uniform rotational motion with respect to the inertial frame S.

Differentiating $\mathsf{r} = \mathsf{R}\mathsf{r}'$ twice with respect to t we have
$\ddot{\mathsf{r}} = \ddot{\mathsf{R}}\mathsf{r}' + 2\dot{\mathsf{R}}\dot{\mathsf{r}}' + \mathsf{R}\ddot{\mathsf{r}}'$. Multiplying by $m\mathsf{R}^{-1}$ and rearranging this gives

$m\ddot{\mathsf{r}}' = \mathsf{R}^{-1}(m\ddot{\mathsf{r}}) - m\mathsf{R}^{-1}\ddot{\mathsf{R}}\mathsf{r}' - 2m\mathsf{R}^{-1}\dot{\mathsf{R}}\dot{\mathsf{r}}'$. Using $\mathsf{R} = \exp(-\mathsf{J}(\omega)t)$ (ω constant) we find

$$m\ddot{\mathsf{r}}' = \mathsf{f}' - m\mathsf{J}(\omega)^2\mathsf{r}' + 2m\mathsf{J}(\omega)\dot{\mathsf{r}}' \quad , \tag{4.22}$$

where

$$\mathsf{f}' = \mathsf{R}^{-1}\mathsf{f} = \begin{pmatrix} f_1' \\ f_2' \\ f_3' \end{pmatrix} \quad , \tag{4.23}$$

and f_i' are the components of the applied force with respect to S'.

In problem 4.4 it will be shown that

$$\mathsf{J}(\omega) = \omega_i \mathsf{J}_i = \omega n_i \mathsf{J}_i = \omega n_i' \mathsf{J}_i = \omega_i' \mathsf{J}_i .$$

The second term on the right side of equation (4.22) is the "centrifugal force"

$$\mathsf{f}'_{\text{ctf}} = -m\mathsf{J}(\omega)^2\mathsf{r}' \tag{4.24}$$

with components $m\omega^2[x_i' - (n_j'x_j')n_i']$ with respect to S'. These are the S' components of $m\omega^2(\mathbf{r} - (\hat{\mathbf{n}} \cdot \mathbf{r})\hat{\mathbf{n}}) = -m\omega \times (\omega \times \mathbf{r})$, where $\mathbf{r} - (\hat{\mathbf{n}} \cdot \mathbf{r})\hat{\mathbf{n}}$ is the component or $\mathbf{r}$ normal to the rotation axis.

The last term in equation (4.22) is the "Coriolis force"

$$\mathsf{f}'_{\text{Cor}} = 2m\mathsf{J}(\omega)\dot{\mathsf{r}}' = 2m\omega_j \mathsf{J}_j \dot{\mathsf{r}}' = 2m\omega_j' \mathsf{J}_j \dot{\mathsf{r}}' \quad . \tag{4.25}$$

The i-th component of the Coriolis force with respect to S' is then

$$-2m\epsilon_{ijk}\omega_j'\dot{x}_k' = \hat{\mathbf{e}}_i' \cdot (-2m\omega \times [\mathbf{v} - \omega \times \mathbf{r}]) \quad .$$

Note that $\mathbf{v} - \omega \times \mathbf{r}$ is precisely the velocity of P as seen by S'.

4.4 Chapter 4 problems

4.1 Let the system of axes S' be obtained from S by a counterclockwise infinitesimal rotation through the angle $\delta\phi$ about the unit vector $\hat{\mathbf{n}}$. Show that the coordinates x_i and x_i' of a fixed point with respect to S and S', respectively, are related by the formula

$$x_i \simeq x_i' - \epsilon_{ijk} x_j' n_k \, \delta\phi \quad .$$

4.2 Show that $\mathsf{R}^{-1}\mathsf{J}_i\mathsf{R} = R_{ij}\mathsf{J}_j$.

4.3 Find eigenvalues and eigenvectors of $\mathsf{J}(\hat{\mathbf{n}}$ (equation (4.11)).

4.4 In equation (4.10) we have written $\mathsf{J}(\hat{\mathbf{n}})$ rather than $\mathsf{J}(\hat{\mathsf{n}})$. Why?

4.5 A thin massless pipe is tilted from the vertical direction by an angle θ and forced to rotate about the vertical axis with angular velocity ω. A bug crawls inside the pipe at constant speed v heading towards the upper end of the pipe.

(a) Find the magnitude and direction of each force acting on the bug in the frame of the rotating pipe.

(b) Find the power input of friction (P_{f}), of gravity (P_{g}), and of the centrifugal force (P_{c}).

(c) Find the power P_{m} that must be developed by the motor in order to keep the system rotating.

(d) Verify that

$$P_{\mathrm{m}} + P_{\mathrm{f}} + P_{\mathrm{g}} = \mathrm{d}K/\mathrm{d}t \quad ,$$

where K is the kinetic energy of the bug in the lab frame.

(e) Verify that $P_{\mathrm{m}} = \tau\omega$, where τ is the torque exerted by the motor on the bug.

Solutions to ch. 4 problems

S4.1 The vector $\mathbf{b}$ obtained by rotating the vector $\mathbf{a}$ counterclockwise about $\hat{\mathbf{n}}$ through $\delta\phi$ is $\mathbf{b} \simeq \mathbf{a} + \hat{\mathbf{n}} \times \mathbf{a}\,\delta\phi$. Then the unit vectors $\hat{\mathbf{e}}_i$ and $\hat{\mathbf{e}}'_i$ along the axes of S and S' are related by $\hat{\mathbf{e}}'_i \simeq \hat{\mathbf{e}}_i + \hat{\mathbf{n}} \times \hat{\mathbf{e}}_i\,\delta\phi$. Therefore

$$x_i = \vec{OP}\cdot\hat{\mathbf{e}}_i \simeq \vec{OP}\cdot(\hat{\mathbf{e}}'_i - \hat{\mathbf{n}} \times \hat{\mathbf{e}}'_i\,\delta\phi) \text{ etc.}$$

S4.2 With $\mathsf{R} = \exp(-\mathsf{J}(\hat{\mathbf{n}})\phi)$, we have

$$\mathrm{d}(\mathsf{R}^{-1}\mathsf{J}_i\mathsf{R})/\mathrm{d}\phi = \mathsf{R}^{-1}[\mathsf{J}(\hat{\mathbf{n}}), \mathsf{J}_i]\mathsf{R}$$

$$= n_j\mathsf{R}^{-1}[\mathsf{J}_j, \mathsf{J}_i]\mathsf{R} = \epsilon_{ijk}n_j\mathsf{R}^{-1}\mathsf{J}_k\mathsf{R} \quad .$$

It is easy to verify that the solution of this equation is $\mathsf{R}^{-1}\mathsf{J}_i\mathsf{R} = R_{ij}\mathsf{J}_j$.

S4.3 $\det(\mathsf{J}(\hat{\mathbf{n}}) - \lambda\mathsf{I}) = 0$ gives $\lambda^3 + \lambda = 0$, $\lambda = (0, \pm\mathrm{i})$. The eigenvalue $\lambda = 0$ corresponds to the eigenvector $\mathsf{v}_0 = \hat{\mathbf{n}}$. The eigenvalues $\pm\mathrm{i}$ correspond to the eigenvectors

$$\mathsf{v}_{\pm\mathrm{i}} = \begin{pmatrix} n_1n_3 \mp \mathrm{i}n_2 \\ n_2n_3 \pm \mathrm{i}n_1 \\ n_3^2 - 1 \end{pmatrix} \quad .$$

The trace of $\mathsf{J}(\hat{\mathbf{n}})$ equals the sum of the eigenvalues, $0 + \mathrm{i} - \mathrm{i} = 0$, as we expect.

Since the $R(3)$ transformations are defined as real transformations in three-dimensional real space, the two complex eigenvalues are not acceptable.

S4.4 The components of $\hat{\mathbf{n}}$ with respect to S and S' are equal:

$$n'_i = R^{-1}_{ij}n_j = [\delta_{ij}\cos\phi + (1-\cos\phi)n_in_j + \epsilon_{ijk}n_k\sin\phi]n_j$$
$$= n_i\cos\phi + (1-\cos\phi)n_i + \text{zero} = n_i .$$

S4.5 As shown in figure 4.2, $\hat{\mathbf{e}}'_2 = \hat{\mathbf{e}}_2\cos\theta - \hat{\mathbf{e}}_3\sin\theta$, $\hat{\mathbf{e}}'_3 = \hat{\mathbf{e}}_2\sin\theta + \hat{\mathbf{e}}_3\cos\theta$.

(a) $\mathbf{F}'_{\mathrm{g}} = mg(\hat{\mathbf{e}}'_2\sin\theta - \hat{\mathbf{e}}'_3\cos\theta)$, $\mathbf{F}'_{\mathrm{ctf}} = m\omega^2 r\sin\theta(\hat{\mathbf{e}}'_2\cos\theta + \hat{\mathbf{e}}'_3\sin\theta)$,
$\mathbf{F}'_{\mathrm{Cor}} = 2m\omega v\sin\theta\,\hat{\mathbf{e}}'_1$, $\mathbf{F}'_{\mathrm{f}} = m(g\cos\theta - \omega^2 r\sin^2\theta)\hat{\mathbf{e}}'_3$,
and the normal reactions of the constraints
$\mathbf{N}'_1 = -2m\omega v\sin\theta\,\hat{\mathbf{e}}'_1$, $\mathbf{N}'_2 = -m\sin\theta(g + \omega^2 r\cos\theta)\hat{\mathbf{e}}'_2$
Verify that the sum of these six forces is zero.

(b) $P_{\mathrm{f}} = \mathbf{F}'_{\mathrm{f}}\cdot\mathbf{v}' = mv(g\cos\theta - \omega^2 r\sin^2\theta)$, $P_{\mathrm{g}} = \mathbf{F}'_{\mathrm{g}}\cdot\mathbf{v}' = -mgv\cos\theta$,
$P_{\mathrm{ctf}} = \mathbf{F}'_{\mathrm{ctf}}\cdot\mathbf{v}' = mv\omega^2 r\sin^2\theta$.
One has $P_{\mathrm{f}} + P_{\mathrm{g}} + P_{\mathrm{ctf}} = 0$.

(c) $P_{\mathrm{m}} = \mathbf{N}'_1\cdot(-\omega r\sin\theta\,\hat{\mathbf{e}}_1) = 2m\omega^2 vr\sin^2\theta$

(d) From (b) we have $P_{\mathrm{m}} + P_{\mathrm{f}} + P_{\mathrm{g}} = P_{\mathrm{m}} - P_{\mathrm{ctf}}$.

The kinetic energy of the system is $K = (mv^2 + I\omega^2)/2$, where I is the moment of inertia with respect to the rotation axis. Since v =constant, ω =constant, and the change of the moment of inertia is due only to the motion of the bug, we have $\mathrm{d}K/\mathrm{d}t = (1/2)\omega^2\mathrm{d}(mr^2\sin^2\theta)/\mathrm{d}t = m\omega^2 rv\sin^2\theta = P_{\mathrm{m}} - P_{\mathrm{ctf}}$. Hence $P_{\mathrm{m}} = 2m\omega^2 rv\sin^2\theta$.

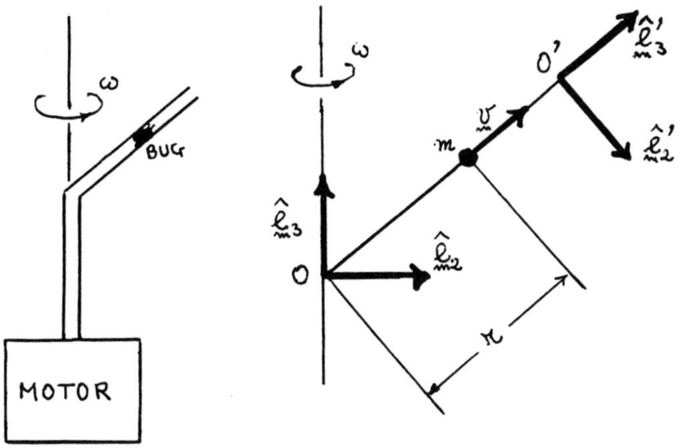

Figure 4.2: Bug in tilted rotating pipe

It is sometimes stated that the power delivered by the motor is half as much, an obvious confusion between P_{m} and $P_{\mathrm{m}} - P_{\mathrm{ctf}}$.

That we have found the correct expression for P_{m} is shown by the following:
(e) $\tau = \omega\ \mathrm{d}I/\mathrm{d}t = 2m\omega vr\ \sin^2\theta,\ \tau\omega = 2m\omega^2 vr\ \sin^2\theta = P_{\mathrm{m}}$

Chapter 5

RIGID BODIES

In this chapter we study the dynamics of rigid bodies intentionally abstaining from the use of Lagrangian methods. Some problems requiring a fair amount of work will be proposed again in the following chapter on Lagrangians, where it will be easier to establish the equations of motion and to find conserved quantities.

5.1 Angular momentum

Let us consider two systems of axes, S fixed in the laboratory and S' attached to the body. We assume that the origins O and O' of the systems coincide. Therefore the point O' of the body is fixed.

Let $\mathbf{L}$ denote the angular momentum with respect to $O' = O$. When translated into the notation of Chapter 4, the familiar expression

$$\mathbf{L} = \int \mathbf{r} \times \mathbf{v} \, \mathrm{d}m \tag{5.1}$$

reads

$$\mathsf{L} = -\int \mathsf{J}(\mathsf{r})\mathsf{v} \, \mathrm{d}m \quad . \tag{5.2}$$

Here $\mathsf{J}(\mathsf{r}) = x_i \mathsf{J}_i$ (see (4.8)),

$$\mathsf{L} = \begin{pmatrix} L_1 \\ L_2 \\ L_3 \end{pmatrix} \quad , \quad \mathsf{r} = \begin{pmatrix} x_1 \\ x_2 \\ x_3 \end{pmatrix} \quad , \quad \mathsf{v} = \begin{pmatrix} v_1 \\ v_2 \\ v_3 \end{pmatrix} \quad . \tag{5.3}$$

The distinction between bold-faced roman and bold-faced sans-serif should be remembered. Although $\mathbf{r}$ and r denote the same vector, $\mathbf{r}$ indicates its intrinsic geometrical meaning, while r stands for the coordinates with respect to S. Therefore we shall have r', the coordinates with respect to another system S', but not $\mathbf{r}'$.

Below ω stands for a column matrix with elements ω_i. There is no sans-serif omega! Since

$$\mathsf{v} = -\mathsf{J}(\omega)\mathsf{r} \quad , \tag{5.4}$$

where ω is the instantaneous angular velocity (see 4.21)), we find

$$\mathsf{L} = \int \mathsf{J}(\mathsf{r})\mathsf{J}(\omega)\mathsf{r}\ dm = -\left[\int \mathsf{J}(\mathsf{r})^2 dm\right]\omega = \mathsf{I}\,\omega \quad , \tag{5.5}$$

where

$$\mathsf{I} = -\int \mathsf{J}(\mathsf{r})^2 dm \quad , \tag{5.6}$$

is the inertia matrix with elements

$$\mathsf{I}_{ij} = \int \left(r^2\delta_{ij} - x_i x_j\right) dm \quad . \tag{5.7}$$

We have used the formula

$$\mathsf{J}(\omega)\mathsf{r} = -\mathsf{J}(\mathsf{r})\omega \quad , \tag{5.8}$$

which is easy to prove:

$$\begin{aligned} (\mathsf{J}(\omega)\mathsf{r})_i &= (\mathsf{J}_k\omega_k)_{ij} x_j = \epsilon_{kij} x_j \omega_k \\ &= -x_j \epsilon_{jik}\omega_k = -(\mathsf{J}_j x_j)_{ik}\omega_k = -(\mathsf{J}(\mathsf{r})\omega)_i \quad . \end{aligned}$$

Note that the inertia matrix (5.6) depends on the spatial relation of the rigid body with S, and therefore changes during the motion. For this reason the argument of $\mathsf{J}(\bullet)$ in (5.6) is r, rather than $\mathbf{r}$.

5.2 Euler's equations

In the laboratory system

$$\dot{\mathsf{L}} = \tau \quad , \tag{5.9}$$

the rate of change of the angular momentum equals the applied torque.

Let R be the rotation matrix that transforms a', the components of a vector with respect to the body system S', into a. Differentiating $\mathsf{RL}' = \mathsf{L}$, we have $\dot{\mathsf{R}}\mathsf{L}' + \mathsf{R}\dot{\mathsf{L}}' = \tau = \mathsf{R}\tau'$. Proceeding as in chapter 4, we find

$$\dot{\mathsf{L}}' = \tau' + \mathsf{R}^{-1}\mathsf{J}(\omega)\mathsf{L} = \tau' + \mathsf{J}(\omega')\mathsf{R}^{-1}\mathsf{L},$$

$$\dot{\mathsf{L}}' = \tau' + \mathsf{J}(\omega')\mathsf{L}' \quad , \tag{5.10}$$

in the usual language $\dot{\mathbf{L}} = \mathbf{L}' \times \omega' + \tau'$.
Here $\mathsf{J}(\omega) = \omega_i \mathsf{J}_i$, $\mathsf{J}(\omega') = \omega_i' \mathsf{J}_i$, and

$$\mathsf{L}' = \mathsf{R}^{-1}\mathsf{L} = \mathsf{R}^{-1} \| \, \omega = \|' \, \omega' \tag{5.11}$$

with

$$\|' = -\int \mathsf{J}(\mathsf{r}')^2 \mathrm{d}m \quad , \tag{5.12}$$

$$\|'_{ij} = \int (r'^2 \delta_{ij} - x_i' x_j') \mathrm{d}m \qquad (r'^2 = r^2) \quad . \tag{5.13}$$

Note that $\|$ and its elements do <u>not</u> change during the motion.

If the S' axes are principal inertia axes of the body, then $\|'$ is diagonal, and

$$L_i' = I_i \omega_i' \quad (\text{no sum over } i) \quad , \tag{5.14}$$

where I_i are the "principal moments of inertia".

The kinetic energy is

$$K = \omega^{\mathrm{T}} \| \omega/2 = \omega^{\mathrm{T}} \mathsf{L} \omega/2 \quad . \tag{5.15}$$

Expressing this in terms of primed quantities, we have

$$K = \omega'^{\mathrm{T}} \|' \omega'/2 \quad , \tag{5.16}$$

and, if the inertia matrix is diagonal,

$$K = I_i \omega_i'^2/2 \quad . \tag{5.17}$$

We now assume that the origin $O' = O$ coincides with the center of mass of a body acted upon by neither forces nor torques. Then

$$\dot{L}_i' = \epsilon_{ijk} L_j' \omega_k' \quad . \tag{5.18}$$

If the S' axes are principal inertia axes, this equation reduces to Euler's equations

$$\left\{ \begin{array}{l} I_1 \dot{\omega}_1' = (I_2 - I_3) \omega_2' \omega_3' \quad , \\ I_2 \dot{\omega}_2' = (I_3 - I_1) \omega_3' \omega_1' \quad , \\ I_3 \dot{\omega}_3' = (I_1 - I_2) \omega_1' \omega_2' \quad . \end{array} \right. \tag{5.19}$$

From these equations one easily verifies that $K = I_i {\omega_i'}^2/2$ is a constant of motion.

If $I_1 = I_2$ (rotational symmetry about the x_3'-axis) the third of (5.19) yields ω_3' =constant. Then, putting $\alpha = [(I_3 - I_1)/I_1]\omega_3'$. the first two can be written in the form

$$\dot{\omega}_1' = -\alpha \omega_2' \quad \text{and} \quad \dot{\omega}_2' = \alpha \omega_1' \quad . \tag{5.20}$$

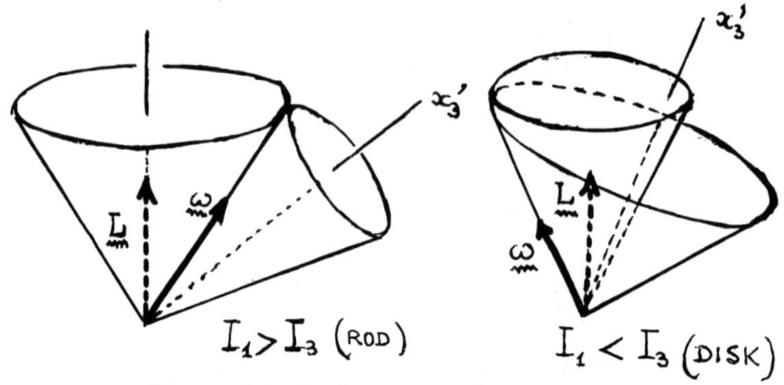

Figure 5.1: Body-cone and space-cone

Multiplying the first by ω_1' and the second by ω_2', and summing, we find that $\omega_1'^2 + \omega_2'^2$ is a constant of motion, which might have been inferred from K and $\omega_3{}'$ being constants of motion.

For $I_1 = I_2$ the general solution of equations (5.19) is

$$\omega_1' = c\ \cos(\alpha t + \beta) \quad \text{and} \quad \omega_2' = c\ \sin(\alpha t + \beta) \tag{5.21}$$

showing that, as seen by S', the angular velocity rotates around $\hat{\mathbf{e}}_3'$ with period $2\pi/|\alpha|$. For the earth $(I_3 - I_1)/I_1 \simeq 1/300$, $\omega_3' \simeq \omega = 2\pi/\text{day}$, so that the period $2\pi/\alpha$ is 300 days $=$ 10 months. This is shorter than the observed value of 14 months (Chandler's period). The discrepancy is due to earth deformation caused by the polar fluctuations themselves.

The angle formed by the angular velocity with $\hat{\mathbf{e}}_3'$ is
$$\theta' = \tan^{-1}(\sqrt{\omega_1'^2 + \omega_2'^2}/\omega_3').$$

We wish to discuss more carefully the matter of the rotation of the angular velocity of a rigid body free from torques with $I_1 = I_2 \neq I_3$. For such a body the angular momentum $\mathbf{L}$ is a fixed vector. On the other hand, the angular velocity ω is not.

If θ is the angle formed by $\mathbf{L}$ and ω, then $\cos\theta = \omega \cdot \mathbf{L}/\omega L = 2K/\omega L$. The kinetic energy K is a constant of motion, and $\omega = \sqrt{\omega_1'^2 + \omega_2'^2 + \omega_3'^2}$ is also a constant of motion because the sum of the first two terms and the last term under square root are both constants of motion.

Therefore ω is a vector of constant magnitude rotating arounf $\mathbf{L}$ on a cone of semiaperture θ ("space cone" or, better, "lab cone").

We have also seen that the angular velocity rotates around $\hat{\mathbf{e}}_3'$ on a cone of semiaperture θ' ("body cone"). Furthermore, at each instant
$\mathbf{L} = I_1(\omega_1'\hat{\mathbf{e}}_1' + \omega_2'\hat{\mathbf{e}}_2') + I_3\omega_3'\hat{\mathbf{e}}_3'$, $\omega = \omega_1'\hat{\mathbf{e}}_1' + \omega_2'\hat{\mathbf{e}}_2' + \omega_3'\hat{\mathbf{e}}_3'$, and $\hat{\mathbf{e}}_3'$ lie on the same plane $((\mathbf{L} \times \omega) \cdot \hat{\mathbf{e}}_3' = 0)$.

Finally the angle θ'' of $\mathbf{L}$ with $\hat{\mathbf{e}}_3'$ is given by
$\tan\theta'' = \sqrt{L_1'^2 + L_2'^2}/L_3' = I_1\sqrt{\omega_1'^2 + \omega_2'^2}/I_3\omega_3' = (I_1/I_3)\tan\theta'$.

For $I_1 > I_3$ (rod-like body) one has $\tan\theta'' > \tan\theta'$, whereas for $I_1 < I_3$ (earth, flat disk) $\tan\theta'' < \tan\theta'$. In the former case, ω is to be found between $\mathbf{L}$ and $\hat{\mathbf{e}}_3'$, in the latter $\mathbf{L}$ is between ω and $\hat{\mathbf{e}}_3'$.

The body-cone rolls on the space-cone. At each instant the angular velocity is along their line of contact (see figure 5.1).

One must not confuse the angular velocity with which ω is seen to rotate around $\hat{\mathbf{e}}_3'$ in the primed system attached to the $I_1 = I_2$ body, with the precessional angular velocity of $\hat{\mathbf{e}}_3'$ (the symmetry axis of the body) around the fixed angular momentum $\mathbf{L}$.

We can write $\mathbf{L} = I_1(\omega_1'\hat{\mathbf{e}}_1' + \omega_2'\hat{\mathbf{e}}_2') + I_3\omega_3'\hat{\mathbf{e}}_3' = I_1\omega + (I_3 - I_1)\omega_3'\hat{\mathbf{e}}_3'$,

$$\omega = (\mathbf{L}/I_1) - \alpha\hat{\mathbf{e}}_3' \quad \text{with} \quad \alpha = [(I_3 - I_1)/I_1]\omega_3' \quad . \tag{5.22}$$

Then $\mathrm{d}\hat{\mathbf{e}}_3'/\mathrm{d}t = \omega \times \hat{\mathbf{e}}_3' = (\mathbf{L} \times \hat{\mathbf{e}}_3')/I_1$, showing that $\hat{\mathbf{e}}_3'$ rotates around $\mathbf{L}$ with angular velocity

$$\omega_{\mathbf{pr}} = |\mathbf{L}|/I_1 \quad . \tag{5.23}$$

On the other hand we have seen that with respect to the primed system attached to the body, ω appears to rotate around $\hat{\mathbf{e}}_3'$ with angular velocity $\alpha = [(I_3 - I_1)/I_1]\omega_3'$.

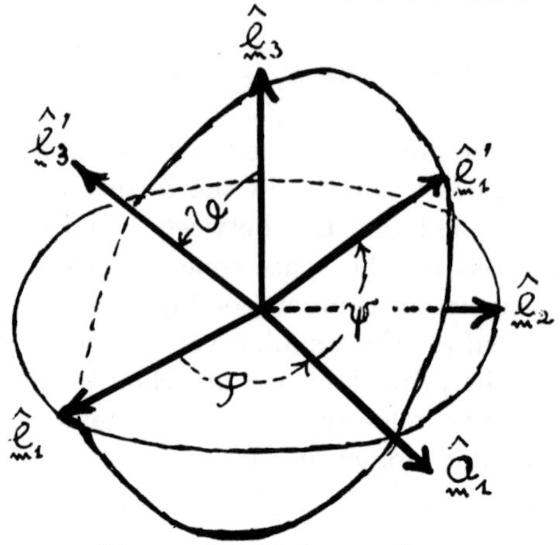

Figure 5.2: Euler angles

5.3 Euler angles

Think of the S' system as being obtained as follows (see figure 5.2).

By a counterclockwise rotation through φ around $\hat{\mathbf{e}}_3$, S' becomes the system S_a with basic vectors $\hat{\mathbf{a}}_1$, $\hat{\mathbf{a}}_2$, and $\hat{\mathbf{a}}_3 = \hat{\mathbf{e}}_3$.

By a counterclockwise rotation through ϑ about $\hat{\mathbf{a}}_1$, S_a becomes the system S_b with basic vectors $\hat{\mathbf{b}}_1 = \hat{\mathbf{a}}_1$, $\hat{\mathbf{b}}_2$, $\hat{\mathbf{b}}_3$.

By a counterclockwise rotation through ψ about $\hat{\mathbf{b}}_3$, S_b becomes the system S' with basic vectors $\hat{\mathbf{e}}'_1$, $\hat{\mathbf{e}}'_2$, $\hat{\mathbf{e}}'_3 = \hat{\mathbf{b}}_3$.

Thus

$$\hat{\mathbf{e}}'_i = \mathsf{R}(\psi, \vartheta, \varphi)\hat{\mathbf{e}}_i \tag{5.24}$$

with

$$\mathsf{R}(\psi, \vartheta, \varphi) = \mathrm{e}^{-\mathsf{J}(\hat{\mathbf{b}}_3)\psi}\mathrm{e}^{-\mathsf{J}(\hat{\mathbf{a}}_1)\vartheta}\mathrm{e}^{-\mathsf{J}(\hat{\mathbf{e}}_3)\varphi} \quad . \tag{5.25}$$

Using the formula $\mathsf{R}^{-1}\mathsf{J}(\mathsf{a})\mathsf{R} = \mathsf{J}(\mathsf{R}^{-1}\mathsf{a})$ we now find

$$\mathsf{R}(\psi, \vartheta, \varphi) = \mathsf{R}_3(\varphi)\mathsf{R}_1(\vartheta)\mathsf{R}_3(\psi) \quad , \tag{5.26}$$

where $\mathsf{R}_3(\bullet)$ and $\mathsf{R}_1(\bullet)$ are the matrices for rotations around the S axes given by equations (4.4) and (4.5).

Proof: Putting $\mathsf{R}(\hat{\mathbf{x}}_i, \alpha) = \exp(-\mathsf{J}(\hat{\mathbf{x}}_i)\alpha)$, we have

$$\mathsf{R}(\hat{\mathbf{b}}_3, \psi)\mathsf{R}(\hat{\mathbf{a}}_1, \vartheta)\mathsf{R}(\hat{\mathbf{e}}_3, \varphi)$$

$$\begin{aligned} &= [\mathsf{R}(\hat{\mathbf{a}}_1, \vartheta)\mathsf{R}(\hat{\mathbf{a}}_3, \psi)\mathsf{R}(\hat{\mathbf{a}}_1, -\vartheta)]\ \mathsf{R}(\hat{\mathbf{a}}_1, \vartheta)\mathsf{R}(\hat{\mathbf{e}}_3, \varphi) \\ &= \mathsf{R}(\hat{\mathbf{a}}_1, \vartheta)\mathsf{R}(\hat{\mathbf{e}}_3, \varphi)\ [\mathsf{R}(\hat{\mathbf{e}}_3, -\varphi)\mathsf{R}(\hat{\mathbf{a}}_3, \psi)\mathsf{R}(\hat{\mathbf{e}}_3, \varphi)] \\ &= \mathsf{R}(\hat{\mathbf{a}}_1, \vartheta)\mathsf{R}(\hat{\mathbf{e}}_3, \varphi)\mathsf{R}(\hat{\mathbf{e}}_3, \psi) \end{aligned}$$

$$= \mathsf{R}(\hat{\mathbf{e}}_3,\varphi)\ [\mathsf{R}(\hat{\mathbf{e}}_3,-\varphi)\mathsf{R}(\hat{\mathbf{a}}_1,\vartheta)\mathsf{R}(\hat{\mathbf{e}}_3,\varphi)]\ \mathsf{R}(\hat{\mathbf{e}}_3,\psi)$$
$$= \mathsf{R}(\hat{\mathbf{e}}_3,\varphi)\mathsf{R}(\hat{\mathbf{e}}_1,\vartheta)\mathsf{R}(\hat{\mathbf{e}}_3,\psi)$$
$$= \mathsf{R}_3(\varphi)\mathsf{R}_1(\vartheta)\mathsf{R}_3(\psi)$$

By equation (5.26)

$$\mathsf{R}(\psi,\vartheta,\varphi)\begin{pmatrix}1\\0\\0\end{pmatrix} = \begin{pmatrix}\cos\varphi\ \cos\psi - \cos\vartheta\ \sin\varphi\ \sin\psi\\ \sin\varphi\ \cos\psi + \cos\vartheta\ \cos\varphi\ \sin\psi\\ \sin\vartheta\ \sin\psi\end{pmatrix} \quad . \tag{5.27}$$

From this we read out that

$$\hat{\mathbf{e}}_1' = (\cos\varphi\ \cos\psi - \cos\vartheta\ \sin\varphi\ \sin\psi)\hat{\mathbf{e}}_1$$
$$+(\sin\varphi\ \cos\psi + \cos\vartheta\ \cos\varphi\ \sin\psi)\hat{\mathbf{e}}_2 + \sin\vartheta\ \sin\psi\ \hat{\mathbf{e}}_3 \quad . \tag{5.28}$$

Similarly we find

$$\hat{\mathbf{e}}_2' = -(\cos\varphi\ \sin\psi + \cos\vartheta\ \sin\varphi\ \cos\psi)\hat{\mathbf{e}}_1$$
$$+(-\sin\varphi\ \sin\psi + \cos\vartheta\ \cos\varphi\ \cos\psi)\hat{\mathbf{e}}_2 + \sin\vartheta\ \cos\psi\ \hat{\mathbf{e}}_3 \tag{5.29}$$

and

$$\hat{\mathbf{e}}_3' = \sin\varphi\ \sin\vartheta\ \hat{\mathbf{e}}_1 - \cos\varphi\ \sin\vartheta\ \hat{\mathbf{e}}_2 + \cos\vartheta\ \hat{\mathbf{e}}_3 \quad . \tag{5.30}$$

We shall frequently refer to the following formulae

$$\hat{\mathbf{a}}_1 = \hat{\mathbf{e}}_1\cos\varphi + \hat{\mathbf{e}}_2\sin\varphi\ ,\ \hat{\mathbf{a}}_2 = -\hat{\mathbf{e}}_1\sin\varphi + \hat{\mathbf{e}}_2\cos\varphi\ ,\ \hat{\mathbf{a}}_3 = \hat{\mathbf{e}}_3 \tag{5.31}$$

$$\hat{\mathbf{b}}_1 = \hat{\mathbf{a}}_1\ ,\ \hat{\mathbf{b}}_2 = \hat{\mathbf{a}}_2\cos\vartheta + \hat{\mathbf{a}}_3\sin\vartheta\ ,\ \hat{\mathbf{b}}_3 = -\hat{\mathbf{a}}_2\sin\vartheta + \hat{\mathbf{a}}_3\cos\vartheta \tag{5.32}$$

$$\hat{\mathbf{e}}_1' = \hat{\mathbf{b}}_1\cos\psi + \hat{\mathbf{b}}_2\sin\psi\ ,\ \hat{\mathbf{e}}_2' = -\hat{\mathbf{b}}_1\sin\psi + \hat{\mathbf{b}}_2\cos\psi\ ,\ \hat{\mathbf{e}}_3' = \hat{\mathbf{b}}_3\ . \tag{5.33}$$

These can be used to check equations (5.28,29,30).

Since infinitesimal rotations around different axes are additive in the first order, the angular velocity can be expressed in the form

$$\omega = \dot{\varphi}\hat{\mathbf{e}}_3 + \dot{\vartheta}\hat{\mathbf{a}}_1 + \dot{\psi}\hat{\mathbf{b}}_3 \quad . \tag{5.34}$$

Using repeatedly equations (5.31,32,33), we have

$$\begin{aligned}\omega &= (\sin\vartheta\ \sin\psi\ \dot{\varphi} + \cos\psi\ \dot{\vartheta})\hat{\mathbf{e}}_1' + (\sin\vartheta\ \cos\psi\ \dot{\varphi} - \sin\psi\ \dot{\vartheta})\hat{\mathbf{e}}_2' + (\cos\vartheta\ \dot{\varphi} + \dot{\psi})\hat{\mathbf{e}}_3'\\ &= (\dot{\vartheta}\cos\varphi + \dot{\psi}\sin\vartheta\ \sin\varphi)\hat{\mathbf{e}}_1 + (\dot{\vartheta}\sin\varphi - \dot{\psi}\sin\vartheta\cos\varphi)\hat{\mathbf{e}}_2 + (\dot{\varphi} + \dot{\psi}\cos\vartheta)\hat{\mathbf{e}}_3\ .\end{aligned} \tag{5.35}$$

The kinetic energy of a rigid body with $I_1 = I_2$ can be written as

$$K = [I_1(\dot{\varphi}^2\sin^2\vartheta + \dot{\vartheta}^2) + I_3(\dot{\psi} + \dot{\varphi}\cos\vartheta)^2]/2 \quad . \tag{5.36}$$

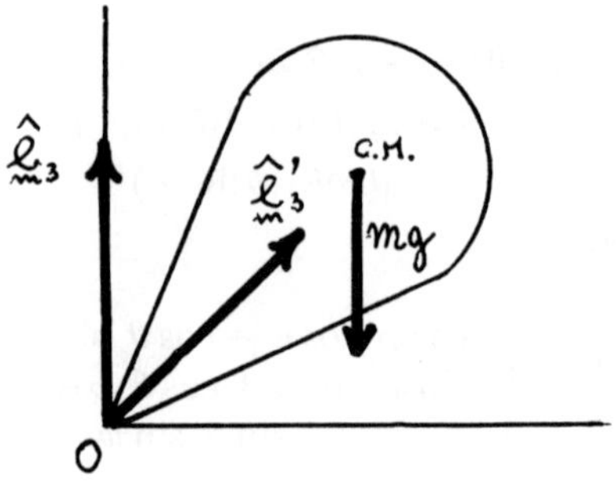

Figure 5.3: Spinning top

5.4 Spinning top

The spinning top ($I_1 = I_2$) shown in figure 5.3 has a fixed point O other than the center of mass. Denoting by ℓ the distance of the center of mass and $O = O'$, the torque due to gravity is $\tau = \ell \hat{\mathbf{e}}_3' \times (-mg\hat{\mathbf{e}}_3)$. Since

$$\hat{\mathbf{e}}_3 = (\hat{\mathbf{e}}_1' \sin\psi + \hat{\mathbf{e}}_2' \cos\psi)\sin\vartheta + \hat{\mathbf{e}}_3' \cos\vartheta \ ,$$

we have

$$\tau = \ell mg(\hat{\mathbf{e}}_1' \cos\psi - \hat{\mathbf{e}}_2' \sin\psi)\sin\vartheta \quad . \tag{5.37}$$

By a simple extension of Euler's equations (5.19) to include the torque, we have

$$\begin{cases} I_1\dot{\omega}_1' = (I_1 - I_3)\omega_2'\omega_3' + \ell mg\ \sin\vartheta\ \cos\psi \ , \\ I_1\dot{\omega}_2' = (I_3 - I_1)\omega_3'\omega_1' - \ell mg\ \sin\vartheta\ \sin\psi \ , \\ I_3\dot{\omega}_3' = 0 \ . \end{cases} \tag{5.38}$$

Multiplying the first of (5.38) by ω_1' and the second by ω_2', summing, and using (5.35), we find

$$I_1(\omega_1'^2 + \omega_2'^2)/2 + \ell mg\cos\vartheta = K - I_3\omega_3'^2/2 + \ell mg\cos\vartheta = \text{ constant } . \tag{5.39}$$

Since ω_3' is also a constant of motion, so is the total energy $E = K + U$ with $U = \ell mg\cos\vartheta$.

There is another constant of motion, $L_3 = \mathbf{L}\cdot\hat{\mathbf{e}}_3$. In fact, $\dot{L}_3 = \dot{\mathbf{L}}\cdot\hat{\mathbf{e}}_3 = \tau\cdot\hat{\mathbf{e}}_3 = 0$. A simple calculation yields

$$L_3 = (I_1\sin^2\vartheta + I_3\cos^2\vartheta)\dot{\varphi} + I_3\cos\vartheta\ \dot{\psi} \quad .$$

It is convenient to express the constant of motion $\omega_3' = \cos\vartheta\ \dot{\varphi} + \dot{\psi}$ in the form $\omega_3' = aI_1/I_3$, and the constant of motion L_3 in the form $L_3 = I_1 b$. Then from the expression for L_3 we find

$$\sin\vartheta\ \dot{\varphi} = [I_1 b - I_3\cos\vartheta(\dot{\psi} + \cos\vartheta\ \dot{\varphi})]/I_1\sin\vartheta = (b - a\ \cos\vartheta)/\sin\vartheta.$$

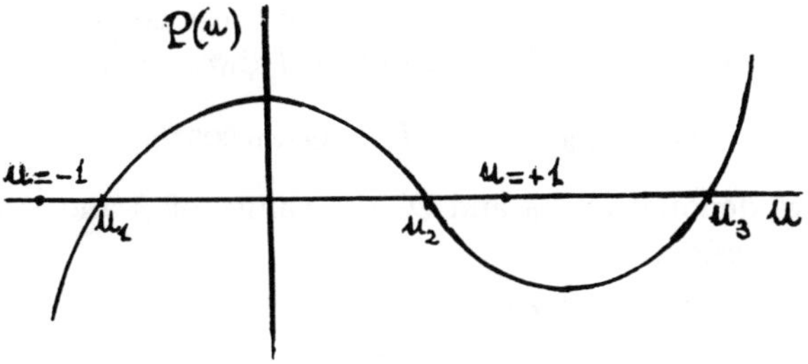

Figure 5.4: Graph of $P(u)$

The energy can be expressed in the form

$$E = (I_1/2)\dot{\vartheta}^2 + U_e(\vartheta) \quad , \tag{5.40}$$

where

$$U_e = \frac{I_1}{2}\left(\frac{b - a\,\cos\vartheta}{\sin\vartheta}\right)^2 + \frac{I_3}{2}\left(\frac{I_1 a}{I_3}\right)^2 + \ell mg\,\cos\vartheta \tag{5.41}$$

is an effective potential energy for the ϑ motion.

The angles φ and ψ have disappeared from this expression for the energy, in the same way as the angle θ did from the expression for the energy of a particle moving under the action of a central force.

Introducing the variable $u = \cos\vartheta$, one finds $\dot{u}^2 = P(u)$, where

$$P(u) = \left(\frac{2E}{I_1} - \frac{I_1 a^2}{I_3} - \frac{2\ell mgu}{I_1}\right)(1 - u^2) - (b - au)^2 \tag{5.42}$$

is a polynomial of third degree in u.

Since $P(\pm 1) < 0$ and $\lim_{u\to\pm\infty} P(u) = \pm\infty$, we expect the graph of $P(u)$ to be as in figure 5.4 .

5.4.1 Regular precession of top

The "regular precession" is the analogue of circular orbits of Chapter 2. We seek the conditions under which $\vartheta = \vartheta_0$(constant) during the motion. For $\vartheta = \vartheta_0$, $\dot{\vartheta} = 0$, $\ddot{\vartheta} = 0$, the first two of Euler's equations (5.38) read

$$I_1 \sin\vartheta_0 \mathrm{d}(\sin\psi\ \dot{\varphi})/\mathrm{d}t = (I_1 - I_3)\sin\vartheta_0 \cos\psi\ \dot{\varphi}\ \omega_3' + \ell mg \sin\vartheta_0 \cos\psi \tag{5.43}$$

and

$$I_1 \sin\vartheta_0 \mathrm{d}(\cos\psi\ \dot{\varphi})/\mathrm{d}t = (I_3 - I_1)\sin\vartheta_0 \sin\psi\ \dot{\varphi}\ \omega_3' - \ell mg \sin\vartheta_0 \sin\psi \quad . \tag{5.44}$$

Assuming $\vartheta_0 \neq 0$, multiplying the first of these equations by $\cos\psi$, the second by $\sin\psi$, and subtracting, we obtain $I_1\dot{\varphi}_0\dot{\psi}_0 = (I_1 - I_3)\dot{\varphi}_0\omega_3' + \ell mg$,

$$\ell mg = \dot{\varphi}_0[I_3\dot{\psi}_0 - (I_1 - I_3)\dot{\varphi}_0\cos\vartheta_0] \quad , \tag{5.45}$$

where we have denoted by $\dot{\varphi}_0$ and $\dot{\psi}_0$ the values of $\dot{\varphi}$ and $\dot{\psi}$ pertaining to the steady precession.

Neglecting the square of $\dot{\varphi}_0$, we find

$$\dot{\varphi}_0 \simeq \frac{\ell mg}{I_3\dot{\psi}_0} \tag{5.46}$$

in agreement with the elementary treatment: $\mathrm{d}\mathbf{L} = \tau\ \mathrm{d}t$ with $|\tau| = \ell mg\ \sin\vartheta$ and $|\mathrm{d}\mathbf{L}| \simeq (|\mathbf{L}|\sin\vartheta)\dot{\varphi}_0\mathrm{d}t$ yields $\dot{\varphi}_0 = \ell mg/|\mathbf{L}|$. If the top precesses slowly and spins fast, then $|\mathbf{L}| \simeq I_3\dot{\psi}_0$.

5.4.2 Sleeping top

Consider the special case in which the the axis of the top is vertical, $\vartheta = 0$, $\dot{\vartheta} = 0$, $\omega_3 = \omega_3' \Rightarrow \omega_0$. One has $b = a$,

$$E = (I_3\omega_0^2)/2 + \ell mg = I_1^2a^2/2I_3 + \ell mg.$$

We wish to study the stability of this vertical top. We start from the equation $I_1\mathrm{d}(\omega_1' + \mathrm{i}\omega_2')/\mathrm{d}t = \mathrm{i}(I_3 - I_1)\omega_3'(\omega_1' + \mathrm{i}\omega_2') + \ell mg\ \sin\vartheta\ \exp(-\mathrm{i}\psi)$. Putting $\omega_1' + \mathrm{i}\omega_2' = \delta\omega$, $\omega_3' = (I_1/I_3)a + \delta\omega_3'$, $\sin\vartheta\ \exp(-\mathrm{i}\psi) = \mathrm{i}\delta\chi$, we have linearly in the δ quantities

$$I_1\mathrm{d}\delta\omega/\mathrm{d}t = \mathrm{i}(I_3 - I_1)(I_1/I_3)a\ \delta\omega + \mathrm{i}\ell mg\ \delta\chi \tag{5.47}$$

and

$$\mathrm{i}\ \mathrm{d}\delta\chi/\mathrm{d}t = (I_1/I_3)a\ \delta\chi + \delta\omega \quad . \tag{5.48}$$

Putting $\delta\omega = \Omega\exp(\mathrm{i}\lambda t)$ and $\delta\chi = X\exp(\mathrm{i}\lambda t)$ we obtain the equations

$$\begin{cases} [I_1\lambda + (I_1 - I_3)(I_1/I_3)a]\Omega - \ell mgX = 0 \ , \\ \Omega + [(I_1/I_3)a + \lambda)]X = 0 \ . \end{cases} \tag{5.49}$$

This system of equation has a non-trivial solution if the determinant of the coefficients is zero,

$$I_1\lambda^2 + aI_1[2(I_1/I_3) - 1]\lambda + (I_1 - I_3)(I_1a/I_3)^2 + \ell mg = 0 \quad . \tag{5.50}$$

The vertical spinning is stable if λ is real,

$$[aI_1(2(I_1/I_3) - 1)]^2 - 4I_1[(I_1 - I_3)(I_1a/I_3)^2 + \ell mg] > 0 \quad , \tag{5.51}$$

namely

$$a^2 > 4\ell mg/I_1 \quad . \tag{5.52}$$

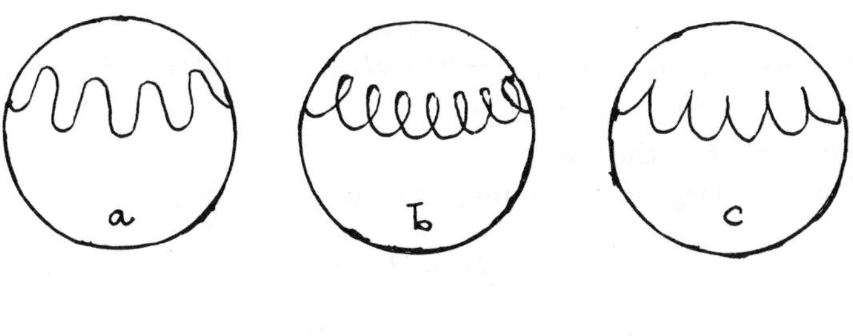

Figure 5.5: Irregular precessions of top

5.4.3 Irregular precessions of top

Returning to section 5.4 and figure 5.4 , $1 > u_2 > u_1 > -1$ $(\theta_2 < \theta_1)$, remember that

$$\dot{\varphi} = \frac{b - au}{1 - u^2} = \frac{a(a^{-1}b - u)}{1 - u^2} \quad . \tag{5.53}$$

If $a^{-1}b > u_2$, then $\dot{\varphi}$ never changes sign, and the motion of the axis of the top is as in figure 5.5(a) .

If $u_2 > a^{-1}b > u_1$, then $\dot{\varphi}$ changes sign at $u = a^{-1}b$, and the axis describes loops as in figure 5.5(b) .

If $u_2 = a^{-1}b$, then $[\dot{\varphi}]_{u=u_2} = 0$. The axis describes cusps as in figure 5.5(c) . Note that $u_2 = a^{-1}b$ means that $a^{-1}b$ is a root of $P(u)$. This gives $E = (I_1^2 a^2/2I_3) + \ell mgb/a$,

$$K = E - \ell mgu = (I_1^2 a^2/2I_3) + \ell mg(a^{-1}b - u)$$

$$= (I_3/2)\omega_3'^2 + \ell mg(a^{-1}b - u) \quad . \tag{5.54}$$

At $u = u_2$ $(\vartheta = \vartheta_{\min})$, the kinetic energy of the top reduces to the spin kinetic energy. If I hold the axis of a spinning top at a certain angle, and then I release it, the axis falls and rises again as in figure 5.5(c).

Suppose we do precisely this, releasing a top spinning with $\omega_3' = (I_1/I_3)a$ from the angle ϑ_0 of the axis with the vertical. The total energy will be $E = (I_1^2 a^2/2I_3) + \ell mg \cos\vartheta_0$. If we put $a^{-1}b = \cos\vartheta_0 = u_0$ in $P(u)$, this becomes

$$P(u) = (u_0 - u)[(2\ell mg/I_1)(1 - u^2) - a^2(u_0 - u)] \quad . \tag{5.55}$$

One root of $P(u)$ is u_0, the other two are the roots of the quadratic polynomial in the square brackets,

$$u = u_1 = \left[I_1 a^2 - \sqrt{I_1^2 a^4 - 8\ell mg I_1 a^2 u_0 + 16\ell^2 m^2 g^2} \right] / 4\ell mg$$

being the smaller of the two.

If a^2 is very large (top spinning very fast) one finds

$$u_1 \simeq u_0 - 2\ell mg(1 - u_0^2)/I_1 a^2 \quad ,$$

$$\cos\vartheta_1 \simeq \cos\vartheta_0 - (2\ell mg/I_1 a^2)\sin^2\vartheta_0 \quad ,$$

$$\vartheta_1 - \vartheta_0 \simeq (2\ell mg I_1/I_3^2 \omega_3'^2) \sin\vartheta_0 \quad .$$

Thus the axis to the top falls from the release inclination ϑ_0 to ϑ_1, to rise again to ϑ_0. The larger the spinning angular velocity, the smaller the difference between ϑ_1 and ϑ_0.

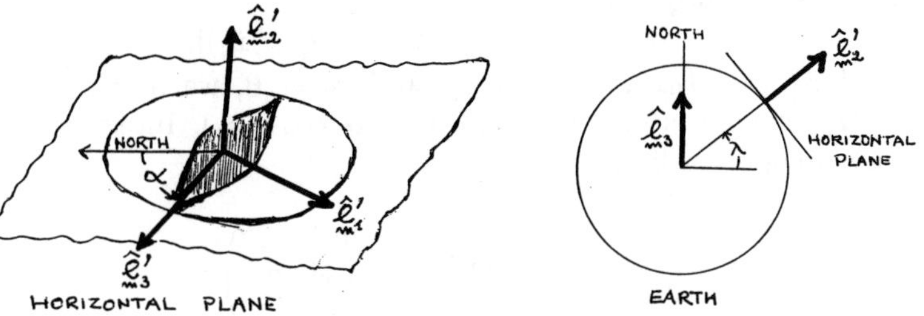

Figure 5.6: Gyrocompass

5.5 Gyrocompass

In its simplest form this consists of a gyroscope rotating about its axis which coincides with the diameter of a horizontal ring. The ring can rotate freely around its own axis ($\hat{\mathbf{e}}_2'$ in figure 5.6). Clearly $\omega_1' = 0$, $\omega_2' = \dot{\alpha}$. We anticipate that ω_3' is a constant of motion, $\omega_3' = \omega_0$ (constant). In the real thing, ω_3' is kept constant by a motor in order to overcome slowing down by friction.

The gyroscope is in the non-inertial frame of the earth rotating with angular velocity

$$\boldsymbol{\Omega} = \Omega\ \hat{\mathbf{e}}_3 = \Omega[\hat{\mathbf{e}}_2' \sin\lambda + \cos\lambda\ (-\hat{\mathbf{e}}_1' \sin\alpha + \hat{\mathbf{e}}_3' \cos\alpha)] \quad . \tag{5.56}$$

Therefore it will be subject to a centrifugal torque and a Coriolis torque.

The components of the centrifugal force acting on an element dm will be

$$\mathrm{d}f_i' = \mathrm{d}m\ [\Omega^2 x_i' - \Omega_i'(\Omega_j' x_j')]$$

and those of the torque will be

$$\mathrm{d}\tau_i' = \epsilon_{ijk} x_j' \mathrm{d}f_k' = -\mathrm{d}m(\Omega_l' x_l')\epsilon_{ijk} x_j' \Omega_k'.$$

Integrating over the whole gyroscope, and using $\int x_i' x_j' \mathrm{d}m = 0$ if $i \neq j$, $\int x_1'^2 \mathrm{d}m = \int x_2'^2 \mathrm{d}m = I_3/2$, $\int x_3'^2 \mathrm{d}m = I_1 - I_3/2$, we find

$$\tau_{\mathrm{ctf}}' = \Omega^2 \cos\alpha\ \cos\lambda\ (I_3 - I_1) \begin{pmatrix} \sin\lambda \\ -\cos\lambda\ \sin\alpha \\ 0 \end{pmatrix} \quad . \tag{5.57}$$

The components of the Coriolis force acting on dm are

$$\mathrm{d}f_i' = -2\mathrm{d}m\epsilon_{ijk}\Omega_j' \dot{x}_k'$$

and those of the torque are

$$d\tau_i' = 2dm[\dot{x}_i'(x_j'\Omega_j') - \Omega_i'(x_j'\dot{x}_j')].$$

Since[1] $\dot{x}_1' = -\omega_0 x_2'$, $\dot{x}_2' = \omega_0 x_1'$, and $\dot{x}_3' = 0$, we find $x_j'\dot{x}_j' = 0$ and $x_j'\Omega_j' = \Omega(-\sin\alpha\ \cos\lambda\, x_1' + \sin\lambda\, x_2' + \cos\alpha\ \cos\lambda\, x_3')$. Integrating over the whole gyroscope, we find

$$\tau_{\rm Cor}' = -I_3\omega_0\Omega \begin{pmatrix} \sin\lambda \\ \cos\lambda\ \sin\alpha \\ 0 \end{pmatrix} \quad . \tag{5.58}$$

Euler's equations will read

$$I_1\dot{\omega}_1' = (I_1 - I_3)\omega_2'\omega_3' + \Omega^2(I_3 - I_1)\sin\lambda\ \cos\lambda\ \cos\alpha - I_3\omega_0\Omega\sin\lambda + \ldots \quad ,$$

$$I_1\dot{\omega}_2' = (I_3 - I_1)\omega_3'\omega_1' - \Omega^2(I_3 - I_1)\cos^2\lambda\ \sin\alpha\ \cos\alpha - I_3\omega_0\Omega\cos\lambda\ \sin\alpha \quad ,$$

$$I_3\dot{\omega}_3' = 0 \quad ,$$

where the dots denote a mechanical torque exerted by the ring.

With $\omega_1' = 0$, $\omega_2' = \dot{\alpha}$, and $\omega_3' = \omega_0$, we find that the first equation is satisfied only by virtue of the constraint reaction. The second equation yields

$$I_1\ddot{\alpha} = -\Omega^2(I_3 - I_1)\sin\alpha\ \cos\alpha\ \cos^2\lambda - I_3\omega_0\Omega\cos\lambda\ \sin\alpha \quad . \tag{5.59}$$

Neglecting the square of the earth angular velocity, the frequency of small oscillations about the northern direction (stable equilibrium position) is

$$\omega = \sqrt{(I_3/I_1)\omega_0\Omega\cos\lambda} \quad . \tag{5.60}$$

[1]The axes $\hat{e}_i'$ $(i = 1, 2, 3)$ are <u>not</u> Euler axes, the coordinates x_i' of dm are <u>not</u> fixed.

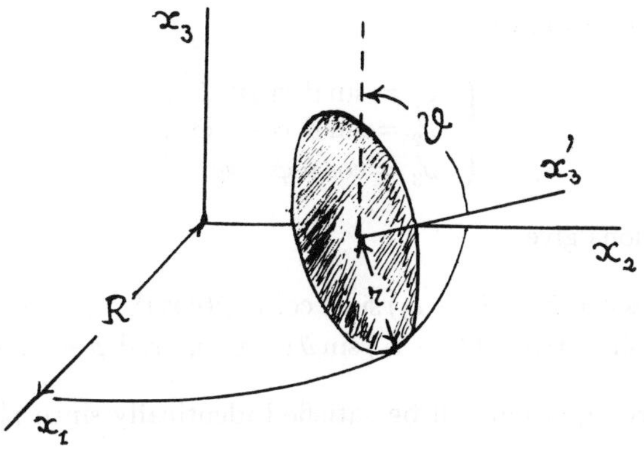

Figure 5.7: Rolling disk

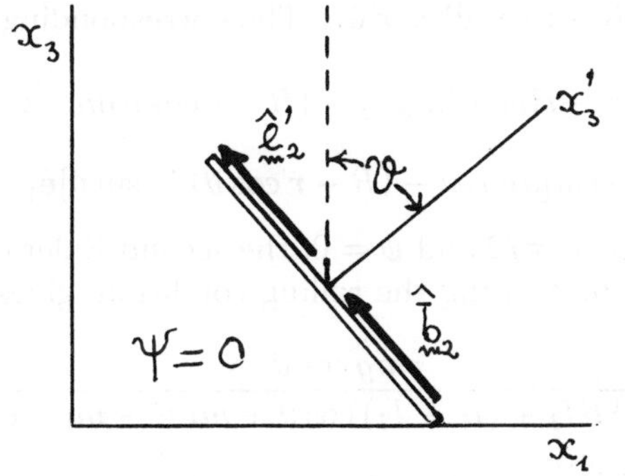

Figure 5.8: $\hat{\mathbf{e}}_1'$ into page, $\hat{\mathbf{e}}_2' = \hat{\mathbf{b}}_2$, $\hat{\mathbf{e}}_3' = \hat{\mathbf{b}}_3$

5.6 Tilted disk rolling in a circle

A uniform disk of radius r rolls on a perfectly rough horizontal plane, the point of contact describing a circle of radius R (figure 5.7). The normal to the disk forms a fixed angle ϑ with the vertical direction (see figure 5.8).

The Euler axes are attached to the disk, with $\hat{\mathbf{e}}_3'$ normal to it. At each instant the unit vector $\hat{\mathbf{b}}_2$ points from the point of contact to the center of the disk. Figure 5.8 shows the situation at a certain time, say $t = 0$, when $\varphi = \pi/2$ and $\psi = 0$.

Note the rolling condition $R\dot{\varphi} = -r\dot{\psi}$, with $\dot{\varphi}$ and $\dot{\psi}$ both constant.

Since $\dot{\vartheta} = 0$, we have

$$\begin{cases} \omega'_1 = \sin\vartheta \ \sin\psi \ \dot{\varphi} \ , \\ \omega'_2 = \sin\vartheta \ \cos\psi \ \dot{\varphi} \ , \\ \omega'_3 = \cos\vartheta \ \dot{\varphi} + \dot{\psi} \ . \end{cases}$$

Euler's equations give

$$\begin{cases} I_1 \sin\vartheta \cos\psi \ \dot{\psi}\dot{\varphi} = (I_1 - I_3) \sin\vartheta \cos\psi \ \dot{\varphi}(\cos\vartheta \ \dot{\varphi} + \dot{\psi}) + \tau'_1 \ , \\ -I_1 \sin\vartheta \sin\psi \ \dot{\psi}\dot{\varphi} = (I_3 - I_1) \sin\vartheta \sin\psi \ \dot{\varphi}(\cos\vartheta \ \dot{\varphi} + \dot{\psi}) + \tau'_2 \ , \end{cases} \tag{5.61}$$

while the third equation will be satisfied identically since τ'_3 will be found to vanish.

At $t = 0$ (figure 5.8) the forces acting on the disk at the point of contact with the plane are a vertical force $\mathbf{F}_{\mathrm{v}} = mg\hat{\mathbf{e}}_3$ and a horizontal force (friction) $\mathbf{F}_{\mathrm{h}} = -(R - r\cos\vartheta)m\dot{\varphi}^2\hat{\mathbf{e}}_1$. The corresponding total torque is

$$\begin{aligned} \tau &= -r\hat{\mathbf{b}}_2 \times [mg\hat{\mathbf{e}}_3 - (R - r\cos\vartheta)m\dot{\varphi}^2\hat{\mathbf{e}}_1] \\ &= -rm[g\cos\vartheta - (R - r\cos\vartheta)\dot{\varphi}^2 \sin\vartheta]\hat{\mathbf{e}}'_1 \ . \end{aligned} \tag{5.62}$$

At $t = 0$, when $\varphi = \pi/2$ and $\psi = 0$, the second Euler equation is clearly satisfied, while the first, using the rolling condition, gives

$$\dot{\varphi}^2 = \frac{rmg\cos\vartheta}{\sin\vartheta[r^{-1}RI_3 + (I_1 - I_3)\cos\vartheta + mrR - mr^2\cos\vartheta]} \ . \tag{5.63}$$

For a uniform disk ($I_1 = mr^2/4$, $I_3 = mr^2/2$) we have

$$\dot{\varphi}^2 = \frac{4g\cos\vartheta}{\sin\vartheta \ (6R - 5r\cos\vartheta)} \ . \tag{5.64}$$

5.7 Chapter 5 problems

5.1 A uniform sphere of mass m and radius r is uniformly charged with total charge q. It rolls on the horizontal (x_1, x_2) plane under the influence of the uniform electrostatic field $\mathbf{E} = (E, 0, 0)$.

(i) Show that its center of mass moves like a point particle of mass $7m/5$.

(ii) Show that the components ω_1 and ω_3 of the spin are constants of the motion, while $\dot{\omega}_2 = 5qE/7mr$. The ω_i's are the components of ω with respect to the fixed axes.

5.2 A sphere of radius a and mass m rolls inside a cylinder of radius $a + b$, whose axis is horizontal and coincides with the z-axis of a system of cylindrical coordinates.

Denoting by $\rho = b$, ϕ, and z the coordinates of the center of the sphere, show that

$$mb\ddot{\phi} = -(5/7)mg \sin\phi \quad , \quad m\ddot{z} = -(2/7)ma\omega_\rho\dot{\phi}$$

and

$$a\,\dot{\omega}_\rho - \dot{\phi}\,\dot{z} = 0 \quad ,$$

where $\omega_\rho = \omega \cdot \hat{\mathbf{e}}_\rho$.

5.3 Show that a body whose principal moments of inertia I_1, I_2, and I_3 are all different, can rotate uniformly around one of them, say $\hat{\mathbf{e}}'_3$. Discuss the stability of such a motion.

5.4 Discuss the regular precession of the spinning top by requiring that $E = U_\mathrm{e}(\vartheta_0)$ and $U'_\mathrm{e}(\vartheta_0) = 0$.

5.5 Study small oscillations about the regular precession of a spinning top. To simplify the calculations, assume $I_1 = I_2 = I_3 \to I$. (How would you realize this situation? Not a sphere!)

5.6 Derive the results for the sleeping top (section 5.4.2) by using the equation $\dot{u}^2 = P(u)$ with $P(u)$ given by (5.42).

5.7 Study the motion of a "skating top", a spinning symmetrical top constrained to remain in contact with a smooth horizontal surface.

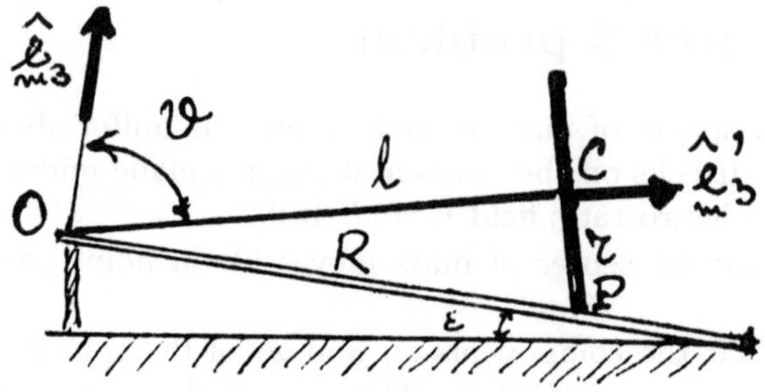

Figure 5.9: Thumbtack

5.8 (i) Find the components $\Omega_i = \omega \cdot \hat{\mathbf{b}}_i$ of the angular velocity with respect to the system $\hat{\mathbf{b}}_i$ $(i = 1, 2, 3)$. Only the third, $\hat{\mathbf{b}}_3 = \hat{\mathbf{e}}'_3$, is fixed with respect to the body.

(ii) Find the components $\Lambda_i = \mathbf{L} \cdot \hat{\mathbf{b}}_i$ of the angular momentum with respect to the center of mass for the case $I_2 = I_1$.

(iii) Express $d\mathbf{L}/dt = \tau$ in terms of the Ω_i's and of the components $\mathbf{T} = \tau \cdot \hat{\mathbf{b}}_i$ of the torque about the center of mass.

5.9 Consider the rigid body shaped like a thumbtack with a long pin shown in figure 5.9. The end of the pin is fixed at O, and the disk can roll on the inclined plane $x_3 = 0$. (i) Find an equation for φ. (ii) Find the frequency of small oscillations about the equilibrium position.

5.10 Establish the equations of motion for a uniform disk rolling on a horizontal plane. This is an extension of the special case treated in section 5.6, where the point of contact with the plane was assumed to be on a circle.

Solutions to ch. 5 problems

S5.1 Denoting by F_1 and F_2 the horizontal components of the force exerted by the plane on the sphere at the point of contact, we have

$$m\ddot{x}_1 = F_1 + qE \ , \ m\ddot{x}_2 = F_2 \ , \ I\dot{\omega}_1 = rF_2 \ , \ I\dot{\omega}_2 = -rF_1$$

with $I = 2mr^2/5$, and the rolling conditions

$$r\omega_1 = -\dot{x}_2 \text{ and } r\omega_2 = \dot{x}_1 \quad .$$

Then $F_1 = -(I/r)\dot{\omega}_2 = -(I/r^2)\ddot{x}_1$, $F_2 = (I/r)\dot{\omega}_1 = -(I/r^2)\ddot{x}_2$, and so

$$[m + (I/r^2)]\ddot{x}_1 = qE \text{ or } (7/5)m\ddot{x}_1 = qE, \text{ and } \ddot{x}_2 = 0 \quad .$$

Since there is no component of the torque in the vertical direction, ω_3=constant. We have also $\dot{\omega}_1 = -\ddot{x}_2/r = 0$ and $\dot{\omega}_2 = \ddot{x}_1/r = 5qE/7mr$.

S5.2 Denoting by F_ρ, F_ϕ, and F_z the components of the force exerted by the cylinder on the sphere at the point of contact, $m\mathbf{a} = m\mathbf{g} + \mathbf{F}$ gives

$$-bm\dot{\phi}^2 = F_\rho \ , \ bm\ddot{\phi} = -mg \sin\phi + F_\phi \ , \ m\ddot{z} = F_z \quad .$$

$I d\omega/dt = \tau$ with $\tau = a\mathbf{e}_\rho \times \mathbf{F} = a(F_\phi \hat{\mathbf{e}}_z - F_z \hat{\mathbf{e}}_\phi)$ gives

$$I(\dot{\omega}_\rho - \dot{\phi}\omega_\phi) = 0 \ , \ I(\dot{\omega}_\phi + \dot{\phi}\omega_\rho) = -aF_z \ , \ I\dot{\omega}_z = aF_\phi \quad .$$

The rolling condition $a\omega \times (-\hat{\mathbf{e}}_\rho) = \dot{\mathbf{r}}$ yields

$$-a\omega_z = b\dot{\phi} \ , \ a\omega_\phi = \dot{z} \ .$$

The final results are easily obtained by manipulating these equations.

S5.3 Euler's equations, $I_1\dot{\omega}'_1 = (I_2 - I_3)\omega'_2\omega'_3$ and cyclic permutations, are satisfied by
$\omega'_1 = \omega'_2 = 0$ and $\omega'_3 =$ constant.

For ω'_1 and ω'_2 infinitesimal, ω'_3 is constant with an error of the second order, and

$$I_1\ddot{\omega}'_1 \simeq [(I_2 - I_3)(I_3 - I_1)/I_2]\omega'^2_3\omega'_1 \quad .$$

The motion is stable if I_3 is either greater or smaller than both I_1 and I_2, unstable if I_3 is comprised between I_1 and I_2.

S5.4 It is easy to show that these conditions are equivalent to $P(u_0) = 0$ and $P'(u_0) = 0$, where $u = \cos\vartheta$ and $P(u)$ is the polynomial in equation (5.42), which in turn imply that u_0 is a double root of $P(u)$ (see figure 5.10). They give

$$\left(\frac{2E}{I_1} - \frac{I_1 a^2}{I_3} - \frac{2\ell mgu_0}{I_1}\right)(1 - u_0^2) - (b - au_0)^2 = 0 \quad ,$$

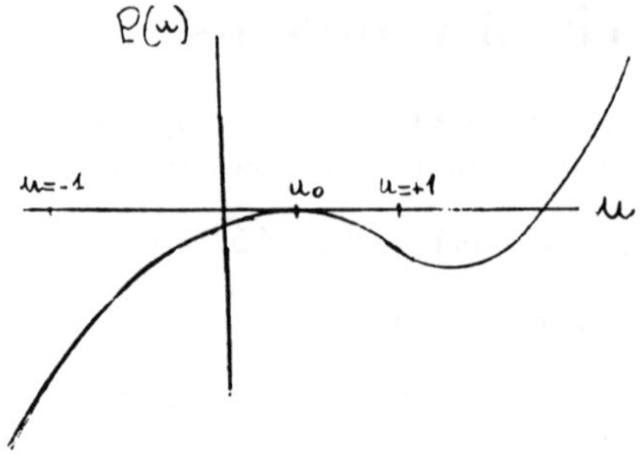

Figure 5.10: Double root of $P(u)$

$$-\frac{2\ell mg}{I_1}(1-u_0^2)-2u_0\left(\frac{2E}{I_1}-\frac{I_1a^2}{I_3}-\frac{2\ell mgu_0}{I_1}\right)+2a(b-au_0)=0 \quad .$$

Using the second of these equations in the first, one has

$$[I_1a(b-au_0)-\ell mg(1-u_0^2)](1-u_0^2)-I_1u_0(b-au_0)^2=0 \quad .$$

But now $b-au_0=\dot\varphi_0(1-u_0^2)$, $a=(I_3/I_1)(\dot\psi_0+\dot\varphi_0\cos\vartheta_0)$, and so one finds again (5.45). This is a quadratic equation for $\dot\varphi_0$,

$I_1\cos\vartheta_0\dot\varphi_0^2-I_3\omega_3'\dot\varphi_0+\ell mg=0$ or $I_1\cos\vartheta_0\dot\varphi_0^2-I_1a\dot\varphi_0+\ell mg=0$,

with the two roots

$$\dot\varphi_0=[a\pm\sqrt{a^2-(4\ell mg/I_1)\cos\vartheta_0}]/2\cos\vartheta_0 \quad .$$

These are real if $a^2>4\ell mg\cos\vartheta_0/I_1$. Assuming a very large (ω_3' very large), the roots are

$$\dot\varphi_0\simeq a/\cos\vartheta_0(\text{large}) \text{ and } \dot\varphi_0\simeq \ell mg/I_1a(\text{small}).$$

Only the second is obtained by the elementary treatment.

S5.5 Equation (5.45) for the regular precession reduces to $I\dot\varphi_0\dot\psi_0=\ell mg$, while equations (5.38) give

$$\begin{cases} I(\ddot\vartheta+\sin\vartheta\;\dot\varphi\dot\psi)=\ell mg\;\sin\vartheta \quad , \\ \mathrm{d}(\sin\vartheta\;\dot\varphi)/\mathrm{d}t=\dot\vartheta\dot\psi \quad , \\ \mathrm{d}(\dot\varphi\;\cos\vartheta+\dot\psi)/\mathrm{d}t=0 \quad . \end{cases}$$

Put $\vartheta=\vartheta_0+\delta\vartheta$, $\varphi=\varphi_0+\delta\varphi$, and $\psi=\psi_0+\delta\psi$, where ϑ_0, φ_0, and ψ_0 and their time derivatives have the steady-motion values. Then linearly in the small corrections, we have

$$\begin{cases} \delta\ddot\vartheta+\sin\vartheta_0(\dot\varphi_0\delta\dot\psi+\dot\psi_0\delta\dot\varphi)=0 \quad , \\ \sin\vartheta_0\;\delta\ddot\varphi+(\cos\vartheta_0\;\dot\varphi_0-\dot\psi_0)\delta\dot\vartheta=0 \quad , \\ \cos\vartheta_0\;\delta\ddot\varphi-\sin\vartheta_0\;\dot\varphi_0\delta\dot\vartheta+\delta\ddot\psi=0 \quad . \end{cases}$$

From the second and third equations we find that

$$\sin\vartheta_0\,\delta\dot{\varphi} = (\dot{\psi}_0 - \cos\vartheta_0\,\dot{\varphi}_0)\delta\dot{\vartheta} + \text{constant}$$

and

$$\sin\vartheta_0\,\delta\dot{\psi} = (\dot{\varphi}_0 - \cos\vartheta_0\,\dot{\psi}_0)\delta\dot{\vartheta} + \text{constant} \quad .$$

Substituting in the first equation, we have

$$\delta\ddot{\vartheta} + (\dot{\varphi}_0^2 + \dot{\psi}_0^2 - 2\dot{\varphi}_0\dot{\psi}_0\cos\vartheta_0)\delta\vartheta = \text{constant} \quad ,$$

$$I^2\dot{\varphi}_0^2\delta\ddot{\vartheta} + (I^2\dot{\varphi}_0^4 - 2I\ell mg\dot{\varphi}_0^2\cos\vartheta_0 + \ell^2m^2g^2)\delta\vartheta = \text{constant} \quad .$$

S5.6 Differentiation of this equation with respect to the time gives $\ddot{u} = P'(u)/2$. Putting $u = 1 - \delta u$ we find

$$\mathrm{d}^2\delta u/\mathrm{d}t^2 = [-P'(1) + P''(1)\delta u]/2 \quad .$$

For $b = a$ and $E = (I_1^2a^2/2I_3) + \ell mg$

$$P(u) = (1-u)^2[(2mg\ell/I_1)(1+u) - a^2] \quad .$$

The roots of this equation are $u = 1$ (double) and $u = (I_1a^2/2\ell mg) - 1$. The former tells us that the curve describing $P(u)$ is tangent to the u-axis at $u = 1$. Hence we expect $P'(1) = 0$. Infact

$$P'(u) = 2(I_1a^2 - \ell mg - 3\ell mgu)(1-u)/I_1$$

does vanish for $u = 1$. We find also that $P''(1) = 2(-I_1a^2 + 4\ell mg)/I_1$ and so

$$\mathrm{d}^2\delta u/\mathrm{d}t^2 = (P''(1)/2)\delta u = [(4\ell mg/I_1) - a^2]\delta u \quad .$$

Stability requires that $P''(1) < 0$, $a^2 > 4\ell mg/I_1$, $\omega_0 > (2/I_3)\sqrt{\ell mgI_1}$.

Note that $P'(1) = 0$ and $P''(1) < 0$ tell us that $P(u)$ has a maximum for $u = 1$. On the other hand the inequality $4\ell mg - a^2I_1 < 0$ tells us that the other root of $P(u) = 0$ is greater than unity. Figure 5.9a shows this situation, while figure 5.9b shows the reason for instability when $4\ell mg - a^2I_1 > 0$.

S5.7 The only forces acting on the top are gravity and the normal reaction from the plane, $\mathbf{N} = N\hat{\mathbf{e}}_3$. Denoting by x_i $(i = 1,2,3)$ the coordinates of the center of mass, we have $\ddot{x}_1 = \ddot{x}_2 = 0$ and $m\ddot{x}_3 = N - mg$.

The torque is $\tau = \ell N\sin\vartheta(\hat{\mathbf{e}}_1'\cos\psi - \hat{\mathbf{e}}_2'\sin\psi)$, and Euler's equations are

$$\begin{cases} I_1\dot{\omega}_1' = (I_1 - I_3)\omega_2'\omega_3' + \ell N\sin\vartheta\,\cos\psi \quad , \\ I_1\dot{\omega}_2' = (I_3 - I_1)\omega_3'\omega_1' - \ell N\sin\vartheta\,\sin\psi \quad , \\ I_3\dot{\omega}_3' = 0 \quad . \end{cases}$$

These equations are like equations (5.38) with mg replaced by N. However, while mg is constant, N is not.

The horizontal component of the velocity of the center of mass and ω_3' are constants of motion.

From the first two Euler's equations it is easy to obtain that

$$I_1(\omega_1'^2 + \omega_2'^2) + 2\ell mg\cos\vartheta + m\ell^2\sin^2\vartheta\,\dot\vartheta^2 = \text{constant} \quad .$$

If the top is released from $\vartheta = \vartheta_0$ with $\dot\vartheta = 0$ and $\dot\varphi = 0$, the "constant" has the value $2\ell mg\cos\vartheta_0$. Hence

$$(I_1 + m\ell^2\sin^2\vartheta)\dot\vartheta^2 + I_1\dot\varphi^2\sin^2\vartheta = 2\ell mg(\cos\vartheta_0 - \cos\vartheta) \quad .$$

Both ω_3' and

$$\begin{aligned} L_3 &= (I_1\sin^2\vartheta + I_3\cos^2\vartheta)\dot\varphi + I_3\cos\vartheta\,\dot\psi \\ &= I_1\sin^2\vartheta\,\dot\varphi + I_3\cos\vartheta\,\omega_3' \end{aligned}$$

are constants of motion. Hence $L_3 = I_3\cos\vartheta_0\omega_3'$,

$$\sin^2\vartheta\,\dot\varphi = a(\cos\vartheta_0 - \cos\vartheta) \quad .$$

If $x_1 = x_2 = 0$ (not skating after all, just twirling on the spot with the center of mass going up and down in the vertical direction), the energy is

$$E = [m\dot x_3^2 + I_1\dot\vartheta^2 + I_1\sin^2\vartheta\,\dot\varphi^2 + I_3(\dot\psi + \dot\varphi\,\cos\vartheta)^2]/2 + \ell mg\cos\vartheta.$$

Using

$\sin^2\vartheta\,\dot\varphi = a(\cos\vartheta_0 - \cos\vartheta)$, $x_3 = \ell\cos\vartheta$, and $\dot\psi + \dot\varphi\cos\vartheta = \omega_3' = (I_1/I_3)a$, and putting as usual $u = \cos\vartheta$, $E = (I_1^2a^2/2I_3) + \ell mg\cos\vartheta_0$ yields $\dot u^2 = Q(u)$ with

$$Q(u) = \frac{(u_0 - u)[2\ell mg(1-u^2) - I_1a^2(u_0 - u)]}{I_1 + m\ell^2(1-u^2)} \quad .$$

$Q(u)$ has two roots of interest to us, $u = u_0$ and

$$u = u_1 = [I_1a^2 - \sqrt{I_1^2a^4 - 8\ell mgI_1a^2u_0 + 16\ell^2m^2g^2}\,]/4\ell mg \quad .$$

This latter is the same value found in section 5.4.3. In both cases the center of mass G falls through $\ell(\cos\vartheta_0 - \cos\vartheta_1)$.

$[\dot\varphi]_{u=u_1} = a(\cos\vartheta_0 - \cos\vartheta_1)/(1-u_1^2)$ has the same value in both cases. It is determined solely by energy conservation.

S5.8 (i)

$$\omega = \dot\vartheta\hat{\mathbf b}_1 + \dot\varphi\sin\vartheta\,\hat{\mathbf b}_2 + (\dot\psi + \dot\varphi\cos\vartheta)\hat{\mathbf b}_3,$$

$$\Omega_1 = \dot\vartheta,\ \Omega_2 = \dot\varphi\sin\vartheta,\ \Omega_3 = \dot\psi + \dot\varphi\cos\vartheta.$$

(ii)

$$\Lambda_1 = I_1\Omega_1,\ \Lambda_2 = I_1\Omega_2,\ \Lambda_3 = I_3\Omega_3$$

(iii) Deriving $d\hat{\mathbf b}_i/dt$ $(i = 1,2,3)$ either from equations (5.31,32) or from the formula $d\hat{\mathbf b}_i/dt = (\omega - \dot\psi\hat{\mathbf b}_3)\times\hat{\mathbf b}_i$, one finds

$$\begin{cases} I_1\dot\Omega_1 - (I_1 - I_3)\Omega_2\Omega_3 + I_1\Omega_2\dot\psi = T_1 \quad , \\ I_1\dot\Omega_2 + (I_1 - I_3)\Omega_3\Omega_1 - I_1\Omega_1\dot\psi = T_2 \quad , \\ I_3\dot\Omega_3 = T_3 \quad . \end{cases}$$

S5.9 (i) Let $\hat{\mathbf{b}}_2$ point from the point of contact P to the center of the disk C, $\vec{CO} = -\ell\hat{\mathbf{b}}_3$. With the conditions $\dot{\vartheta} = 0$ and $R\dot{\varphi} = -r\dot{\psi}$, Euler's equations for the components Ω_i of the angular velocity give

$$\begin{cases} \sin\vartheta\,\dot{\varphi}^2[I_3(\cos\vartheta - R/r) - I_1\cos\vartheta] = T_1 \quad , \\ I_1\sin\vartheta\,\ddot{\varphi} = T_2 \quad , \\ I_3(\cos\vartheta - R/r)\ddot{\varphi} = T_3 \quad . \end{cases}$$

Denoting by $\mathbf{f} = f_i\hat{\mathbf{b}}_i$ the force acting on the thumbtack at O and by $\mathbf{F} = F_i\hat{\mathbf{b}}_i$ that acting at P, the components of the torque are given by

$$T_1 = \ell f_2 - rF_3,\ T_2 = -\ell f_1, \text{ and } T_3 = rF_1.$$

For the motion of the center of mass $\mathbf{r} = \ell\hat{\mathbf{b}}_3$, the equation $m\ddot{\mathbf{r}} = \mathbf{f} + \mathbf{F} + m\mathbf{g}$ yields

$$\begin{cases} m\ell\,\sin\vartheta\,\ddot{\varphi} = mg\sin\epsilon\,\cos\varphi + f_1 + F_1 \quad , \\ m\ell\,\sin\vartheta\,\cos\vartheta\,\dot{\varphi}^2 = -mg(\cos\epsilon\,\sin\vartheta + \sin\epsilon\,\cos\vartheta\,\sin\varphi) + f_2 + F_2 \quad , \\ m\ell\,\sin^2\vartheta\,\dot{\varphi}^2 = mg(\cos\epsilon\,\cos\vartheta - \sin\epsilon\,\sin\vartheta\,\sin\varphi) - f_3 - F_3 \quad . \end{cases}$$

The condition $(\mathbf{F}+\mathbf{f}+m\mathbf{g})\cdot\hat{\mathbf{e}}_3 = 0$ gives $(F_2+f_2)\sin\vartheta+(F_3+f_3)\cos\vartheta = mg\cos\epsilon$.

Eliminating the forces $\mathbf{f}$ and $\mathbf{F}$, we finally find

$$[I_1r^2 + (I_3 + mr^2)\ell^2]\ddot{\varphi} = mgRr^2\sin\epsilon\,\cos\varphi \quad .$$

(ii) In order to find the frequency of small oscillations about $\varphi = \pi/2$, we put $\delta = \pi/2 - \varphi$, obtaining

$$[I_1r^2 + (I_3 + mr^2)\ell^2]\ddot{\delta} \simeq -mgRr^2\sin\epsilon\,\delta \quad .$$

Hence

$$\omega_{\text{osc}}^2 = \frac{mgRr^2\sin\epsilon}{I_1r^2 + (I_3 + mr^2)\ell^2} \quad .$$

S5.10 Let $\mathbf{r}_c$ and P denote the position vector of the center of the disk C and the point of contact with the plane. With $\hat{\mathbf{e}}_3$ vertically up, we have

$$m\ddot{\mathbf{r}}_c = -mg\hat{\mathbf{e}}_3 + \mathbf{F} \quad ,$$

where $\mathbf{F}$ is the reaction of the plane at P.

If r is the radius of the disk, and $\hat{\mathbf{b}}_2$ is always along $\vec{PC}$, the rolling condition is

$$\dot{\mathbf{r}}_c = -r\hat{\mathbf{b}}_2 \times \omega \quad .$$

Expressing this in terms of the $\hat{\mathbf{b}}_i$ components of ω, one has

$$\dot{\mathbf{r}}_c = r(\Omega_1\hat{\mathbf{b}}_3 - \Omega_3\hat{\mathbf{b}}_1) \quad .$$

Differentiating this, after a moderate calculation $\mathbf{F} = m\ddot{\mathbf{r}}_c + mg\hat{\mathbf{e}}_3$ gives

$$\mathbf{F} = mr(-\dot{\Omega}_3+\Omega_1\Omega_2)\hat{\mathbf{b}}_1+m(g\sin\vartheta-r\Omega_1^2-r\Omega_3^2+r\Omega_3\dot{\psi})\hat{\mathbf{b}}_2+m(g\cos\vartheta+r\dot{\Omega}_1+r\Omega_2\Omega_3)\hat{\mathbf{b}}_3 \quad .$$

The components of the torque $\tau = -r\hat{\mathbf{b}}_2 \times \mathbf{F}$ are
$T_1 = -mr(g\cos\vartheta + r\dot{\Omega}_1 + r\Omega_2\Omega_3)$, $T_2 = 0$, $T_3 = mr^2(-\dot{\Omega}_3 + \Omega_1\Omega_2)$.
Euler's equation give

$$\begin{cases} (I_1 + mr^2)\dot{\Omega}_1 + (I_3 + mr^2 - I_1)\Omega_2\Omega_3 + I_1\Omega_2\dot{\psi} = -rmg\cos\vartheta \ , \\ I_1\dot{\Omega}_2 + (I_1 - I_3)\Omega_3\Omega_1 - I_1\Omega_1\dot{\psi} = 0 \ , \\ (I_3 + mr^2)\dot{\Omega}_3 = mr^2\Omega_1\Omega_2 \ . \end{cases}$$

Since $\Omega_3 = \omega_3'$, the last of these equations gives

$$(I_3 + mr^2)\dot{\omega}_3' = mr^2 \sin\vartheta \, \dot{\vartheta}\dot{\varphi} \quad .$$

The second equation gives

$$I_3\omega_3'\dot{\vartheta} - 2I_1 \cos\vartheta \, \dot{\vartheta}\dot{\varphi} - I_1 \sin\vartheta \, \ddot{\varphi} = 0 \quad .$$

The first gives

$$(I_1 + mr^2)\ddot{\vartheta} + (I_3 + mr^2)\omega_3'\dot{\varphi}\sin\vartheta - I_1 \sin\vartheta\cos\vartheta\dot{\varphi}^2 + rmg\cos\vartheta = 0 \quad .$$

Chapter 6

LAGRANGIANS

Praise be to the Lagrangian, labor-saving device and crystal ball.
A mechanical problem is encapsuled in the Lagrangian and the constraints (if any). From these the equations of motion are derived by differentiations. In many cases the existence of integrals of motion can be predicted by simple inspection of the Lagrangian.

6.1 Heuristic introduction

Consider a system of two particles, a and b, interacting by a conservative force with potential energy $U(|\mathbf{r}_a - \mathbf{r}_b|)$. Denoting by x_{ai} and x_{bi} the coordinates ($i = 1, 2, 3$), we have the equations of motion

$$m_a \ddot{x}_{ai} = -\partial U/\partial x_{ai} \quad \text{and} \quad m_b \ddot{x}_{bi} = -\partial U/\partial x_{bi} \quad . \tag{6.1}$$

From these we derive the equations of motion in terms of the coordinates of the center of mass, $X_i = (m_a x_{ai} + m_b x_{bi})/M$, and of the relative position, $x_i = x_{ai} - x_{bi}$,

$$M\ddot{X}_i = 0 \quad \text{and} \quad m\ddot{x}_i = -\partial U/\partial x_i \quad , \tag{6.2}$$

where $M = m_a + m_b$ is the total mass and $m = m_a m_b/M$ the reduced mass, and the potential energy is now written as $U(|\mathbf{r}|)$.

We may further introduce relative spherical polar coordinates, $r = |\mathbf{r}|$, θ, and ϕ, and we find

$$M\ddot{X}_i = 0 \quad , \quad ma_r = -\mathrm{d}U/\mathrm{d}r \quad , \quad ma_\theta = ma_\phi = 0 \quad , \tag{6.3}$$

where

$$\begin{cases} a_r = \ddot{r} - r\dot{\theta}^2 - r\,\sin^2\theta\;\dot{\phi}^2 \quad , \\ a_\theta = r\ddot{\theta} + 2\dot{r}\dot{\theta} - r\sin\theta\;\cos\theta\;\dot{\phi}^2 \quad , \\ a_\phi = r\sin\theta\;\ddot{\phi} + 2\sin\theta\;\dot{r}\dot{\phi} + 2r\cos\theta\;\dot{\theta}\dot{\phi} \quad . \end{cases} \tag{6.4}$$

From equations (6.2) we see that the total momentum $\mathbf{P} = M\mathbf{V}$ is a constant of motion, while equations (6.3) tell us that the 3-component of the angular momentum about the center of mass is also a constant of motion if we remember that $\mathrm{d}l_3/\mathrm{d}t = mr\sin\theta\; a_\phi$. Here $\mathbf{V} = \dot{\mathbf{R}}$ is the velocity of the center of mass. The angular momentum about the center of mass is $\mathbf{l} = m_a(\mathbf{r}_a - \mathbf{R}) \times (\mathbf{v}_a - \mathbf{V}) + m_b(\mathbf{r}_b - \mathbf{R}) \times (\mathbf{v}_b - \mathbf{V}) = m\,\mathbf{r} \times \mathbf{v}$.

Let us now introduce the Lagrangian

$$L = K - U \quad . \tag{6.5}$$

This can be expressed in terms of the various sets of coordinates used above, reading respectively

$$\begin{cases} L = (m_a\dot{\mathbf{r}}_a^2 + m_b\dot{\mathbf{r}}_b^2)/2 - U(|\mathbf{r}_a - \mathbf{r}_b|) \quad , \\ L = (M\dot{\mathbf{R}}^2 + m\dot{\mathbf{r}}^2)/2 - U(|\mathbf{r}|) \quad , \\ L = [M\dot{\mathbf{R}}^2 + m(\dot{r}^2 + r^2\dot{\theta}^2 + r^2\sin^2\theta\;\dot{\phi}^2)]/2 - U(r) \quad . \end{cases} \tag{6.6}$$

Each of equations (6.1,2,3) can be expressed in the compact form of the Lagrange equations

$$\frac{\mathrm{d}}{\mathrm{d}t}\left(\frac{\partial L}{\partial \dot{q}_i}\right) = \frac{\partial L}{\partial q_i} \qquad (i = 1, \ldots 6) \tag{6.7}$$

with L as given by the first of (6.6) and

$q_i = x_{ai}$ $(i = 1, 2, 3)$, $q_{i+3} = x_{bi}$ $(i = 1, 2, 3)$,

by the second of (6.6) and

$q_i = X_i$ $(i = 1, 2, 3)$, $q_{i+3} = x_i$ $(i = 1, 2, 3)$,

by the third and

$q_i = X_i$ $(i = 1, 2, 3)$, $q_4 = r$, $q_5 = \theta$, $q_6 = \phi$.

Furthermore the Lagrange equations tell us that from the absence of a coordinate q_i from L, so that $\partial L/\partial q_i = 0$, follows at once the conservation of the "generalized momentum"

$$p_i = \frac{\partial L}{\partial \dot{q}_i} \tag{6.8}$$

conjugate to that coordinate.

Since L in the second and third of (6.6) does not depend explicitly on X_i, the momentum p_{X_i} conjugate to X_i is seen to be conserved, $\dot{p}_{X_i} = \dot{P}_i = 0$.

Similarly, $p_\phi = l_3$ is conserved because L as given by the third of (6.6) does not depend explicitly on ϕ.

If a Lagrangian depends explicitly on the time derivative $\dot{q}_i$ of a certain coordinate, but not on the coordinate itself, it is invariant under a transformation $q_i \to q_i + \delta q_i$ (δq_i constant) of that coordinate, a translation in the case of X_i, a rotation in the case of ϕ. The invariance-connection of constants of motion will be presented in section 6.7, culminating in Noether's theorem.

Let us see whether the Lagrange technique works for a rigid body, for instance the spinning top with a fixed point of section 5.4.

Start with the Lagrangian

$$L = K - U = [I_1\dot{\vartheta}^2 + I_1 \sin^2\vartheta \, \dot{\varphi}^2 + I_3(\dot{\psi} + \dot{\varphi}\cos\vartheta)^2]/2 - \ell mg\cos\vartheta \quad , \quad (6.9)$$

and put $p_\vartheta = \partial L/\partial\dot{\vartheta}$, $p_\varphi = \partial L/\partial\dot{\varphi}$, $p_\psi = \partial L/\partial\dot{\psi}$. Then $\mathrm{d}p_\vartheta/\mathrm{d}t = \partial L/\partial\vartheta$ gives

$$I_1\ddot{\vartheta} - \sin\vartheta[I_1 \cos\vartheta \, \dot{\varphi}^2 - I_3\dot{\varphi}(\dot{\psi} + \dot{\varphi}\cos\vartheta)^2 + \ell mg] = 0 \quad , \quad (6.10)$$

while $\mathrm{d}p_\varphi/\mathrm{d}t = 0$ and $\mathrm{d}p_\psi/\mathrm{d}t = 0$ yield respectively

$$p_\varphi = (I_1 \sin^2\vartheta + I_3\cos^2\vartheta)\dot{\varphi} + I_3 \cos\vartheta \, \dot{\psi} = \text{constant} \quad , \quad (6.11)$$

and

$$p_\psi = I_3(\dot{\psi} + \dot{\varphi}\cos\vartheta) = \text{constant} \quad . \quad (6.12)$$

(We had $p_\varphi = I_1 b$, $p_\psi = I_1 a = I_3\omega_3'$.)

Multiplying equation (6.10) by $\dot{\vartheta}$, and using the other two equations, one finds easily that $\mathrm{d}E/\mathrm{d}t = 0$, where $E = I_1\dot{\vartheta}^2/2 + U_\mathrm{e}(\vartheta)$ (see equation (5.40)).

6.2 Velocity-dependent forces

Consider a particle subject to both a conservative force $\mathbf{F} = -\nabla U$ and a force of a different type $\mathbf{F}'$. Newton's second law can be written in the form

$$\frac{\mathrm{d}}{\mathrm{d}t}\left(\frac{\partial L}{\partial \dot{x}_i}\right) - \frac{\partial L}{\partial x_i} = F_i' \quad (i = 1, 2, 3) \quad . \quad (6.13)$$

In some cases F_i' can be expressed in the form

$$F_i' = -\frac{\mathrm{d}}{\mathrm{d}t}\left(\frac{\partial \delta L}{\partial \dot{x}_i}\right) + \frac{\partial \delta L}{\partial x_i} \quad , \quad (6.14)$$

and (6.13) can be replaced by

$$\frac{\mathrm{d}}{\mathrm{d}t}\left(\frac{\partial(L+\delta L)}{\partial \dot{x}_i}\right) - \frac{\partial(L+\delta L)}{\partial x_i} = 0 \quad . \tag{6.15}$$

One such case is that of an electron (charge $-e$) in a static electromagnetic field ($\mathbf{E} = -\nabla V$, $\mathbf{B} = \nabla \times \mathbf{A}$), for which

$$F_i' = -(e/c)\epsilon_{ijk}\dot{x}_j B_k \quad . \tag{6.16}$$

It is easy to verify that in this case $\delta L = -(e/c)A_j\dot{x}_j$. In fact

$$\frac{\mathrm{d}}{\mathrm{d}t}\left(\frac{\partial \delta L}{\partial \dot{x}_i}\right) - \frac{\partial \delta L}{\partial x_i} = -\frac{e}{c}\left(\frac{\mathrm{d}A_i}{\mathrm{d}t} - \frac{\partial A_j}{\partial x_i}\,\dot{x}_j\right)$$

$$= -\frac{e}{c}\left(\frac{\partial A_i}{\partial x_j} - \frac{\partial A_j}{\partial x_i}\right)\dot{x}_j = \frac{e}{c}\,\epsilon_{ijk}\dot{x}_j B_k = -F_i' \quad .$$

Therefore the Lagrangian for an electron in an electromagnetic field is

$$L = m\dot{\mathbf{r}}^2/2 + eV - (e/c)\mathbf{A}\cdot\dot{\mathbf{r}} \tag{6.17}$$

The Lagrangian

$$L = eV - (e/c)A_j\dot{x}_j - m_0c^2\sqrt{1-(\dot{\mathbf{r}}/c)^2}$$

yields the relativistic equations

$$\frac{\mathrm{d}}{\mathrm{d}t}(m_0\gamma\dot{x}_i) = e\,\frac{\partial V}{\partial x_i} - \frac{e}{c}\,\epsilon_{ijk}\dot{x}_j B_k$$

($\gamma = 1/\sqrt{1-(\dot{\mathbf{r}}/c)^2}$) for an electron (rest mass m_0, charge $-e$) in an electromagnetic field.

The quantity $\delta U = -\delta L = (e/c)A_j\dot{x}_j$ is sometimes called a velocity dependent potential. This terminology is imprecise because δU is an energy, not a potential. It is also misleading. We are used to think of a potential energy as a function whose space derivatives, taken with a minus sign, are the components of the force. Now

$$-\frac{\partial \delta U}{\partial x_i} = -\frac{e}{c}\,\frac{\partial A_j}{\partial x_i}\,\dot{x}_j$$

is not the i-th component of the Lorentz force.

Conversely, the damping force $-\gamma m\dot{x}$ can be trivially expressed as

$$-\frac{\partial \delta U}{\partial x} = -\gamma m\dot{x}$$

with $\delta U = \gamma m x \dot{x}$. However, the Lagrangian

$$L = m\dot{x}^2/2 \; + m\omega^2 x^2/2 \; - \gamma m x \dot{x}$$

yields the equation

$$m\mathrm{d}(\dot{x} - \gamma x)/\mathrm{d}t \;=\; -m\omega^2 x - \gamma m \dot{x} \quad , \quad \ddot{x} + \omega^2 x = 0$$

instead of the correct

$$\ddot{x} + \gamma \dot{x} + \omega^2 x = 0 \quad .$$

This might have been expected, since $-\gamma m x \dot{x} = \mathrm{d}\Lambda/\mathrm{d}t$ with $\Lambda = -\gamma m x^2/2$. We shall see in the following section that the addition of a total time derivative to the Lagrangian does not change the equations of motion.

A correct Lagrangian for the damped oscillator is

$$L = \mathrm{e}^{\gamma t} m(\dot{x}^2 - \omega^2 x^2)/2 \quad .$$

6.3 Equivalent Lagrangians

This may be a convenient point to mention a certain degree of arbitrariness in the choice of the Lagrangian. The reader will already have noticed that the Lagrange equations for a particle subject to conservative forces are homogeneous in L. Therefore the equations of motion yielded by the Lagrangians L and kL (k=constant) are identical.

Also identical equations of motion are yielded by the Lagrangians L and $L + \mathrm{d}\Lambda/\mathrm{d}t$, where $\Lambda(q, t)$ is an arbitrary function of the coordinates and the time. In fact, $\mathrm{d}\Lambda/\mathrm{d}t$ satisfies Lagrange's equations identically:

$$\frac{\mathrm{d}}{\mathrm{d}t}\left[\frac{\partial}{\partial \dot{q}_i}\left(\frac{\mathrm{d}\Lambda}{\mathrm{d}t}\right)\right] - \frac{\partial}{\partial q_i}\left(\frac{\mathrm{d}\Lambda}{\mathrm{d}t}\right)$$

$$= \frac{\mathrm{d}}{\mathrm{d}t}\left[\frac{\partial}{\partial \dot{q}_i}\left(\frac{\partial \Lambda}{\partial q_i}\dot{q}_i + \frac{\partial \Lambda}{\partial t}\right)\right] - \frac{\partial}{\partial q_i}\left(\frac{\partial \Lambda}{\partial q_j}\dot{q}_j + \frac{\partial \Lambda}{\partial t}\right)$$

$$= \frac{\mathrm{d}}{\mathrm{d}t}\left(\frac{\partial \Lambda}{\partial q_i}\right) - \frac{\partial^2 \Lambda}{\partial q_i \partial q_j}\dot{q}_j - \frac{\partial^2 \Lambda}{\partial q_i \partial t} = 0 \quad .$$

In section 1.1 we mentioned that the equations ($\dot{p} = 0$, $\dot{q} = p$) and ($\dot{p} = 1$, $\dot{q} = p - t$) are equivalent. The former are yielded by the Lagrangian $L = \dot{q}^2/2$, the latter by the Lagrangian

$$L' = \dot{q}^2/2 + \dot{q}t + q = L + \mathrm{d}\Lambda/\mathrm{d}t \text{ with } \Lambda = qt.$$

This property is particularly interesting when applied to equations (6.15) for an electron in a static electromagnetic field. In fact, with Λ depending only on $\mathbf{r}$, we have

$$\delta L + \mathrm{d}\Lambda/\mathrm{d}t = -(e/c)\dot{x}_i(A_i - (c/e)\partial\Lambda/\partial x_i) = -(e/c)\dot{x}_i A'_i,$$

where $\mathbf{A}' = \mathbf{A} - (c/e)\nabla\Lambda$ is the gauge-transformed vector potential corresponding to the same magnetic field as $\mathbf{A}$.

6.4 Invariance of Lagrange equations

The Lagrange equations (6.7) with the Lagrangians (6.6) were seen to yield correct results. This is an example of invariance of the Lagrange equations under time-independent coordinate transformations

$$q_i = q_i(q'_1, \dots q'_n) \ (i = 1, \dots n).$$

From

$$\frac{\mathrm{d}}{\mathrm{d}t}\left(\frac{\partial L(q,\dot{q})}{\partial \dot{q}_i}\right) - \frac{\partial L(q,\dot{q})}{\partial q_i} = 0 \tag{6.18}$$

follows that

$$\frac{\mathrm{d}}{\mathrm{d}t}\left(\frac{\partial L'(q',\dot{q}')}{\partial \dot{q}'_i}\right) - \frac{\partial L'(q',\dot{q}')}{\partial q'_i} = 0 \quad . \tag{6.19}$$

The Lagrangian is transformed as a scalar, $L'(q',\dot{q}') = L(q(q'), \dot{q}(q',\dot{q}'))$. Before presenting a general proof of invariance of the Lagrange equations under time-dependent coordinate transformations $q = q(q',t)$, we wish to verify this property in the simple case of a transformation from an inertial to a uniformly rotating frame of reference, for an electron in a uniform magnetic field in the 3-direction. We start from

$$L = m\dot{\mathbf{r}}^2/2 - (eB/2c)(x_1\dot{x}_2 - x_2\dot{x}_1) \quad , \tag{6.20}$$

the Lagrangian (6.17) with $A_1 = -x_2B/2$, $A_2 = x_1B/2$ and $A_3 = 0$, yielding the Lagrange equations

$$m\ddot{x}_1 = -(eB/c)\dot{x}_2 \quad , \quad m\ddot{x}_2 = (eB/c)\dot{x}_1 \quad , \quad m\ddot{x}_3 = 0 \ . \tag{6.21}$$

Transforming to the rotating frame,

$$\begin{cases} x_1 = x'_1\cos(\omega t) - x'_2\sin(\omega t) \quad , \\ x_2 = x'_1\sin(\omega t) + x'_2\cos(\omega t) \quad , \\ x_3 = x'_3 \quad , \end{cases}$$

with some algebra we find

$$\begin{cases} m\ddot{x}'_1 = \omega(m\omega - eB/c)x'_1 + 2m(\omega - eB/2mc)\dot{x}'_2 \quad , \\ m\ddot{x}'_2 = \omega(m\omega - eB/c)x'_2 - 2m(\omega - eB/2mc)\dot{x}'_1 \quad , \\ m\ddot{x}'_3 = 0 \ . \end{cases} \tag{6.22}$$

But these are indeed the Lagrange equations derived from the transformed Lagrangian

$$L' = m\dot{\mathbf{r}}'^2/2 + (\omega/2)(m\omega - eB/c)(x'^2_1 + x'^2_2) + (m\omega - eB/2c)(x'_1\dot{x}'_2 - x'_2\dot{x}'_1) \ . \tag{6.23}$$

Which is the more economical way of obtaining equations (6.22)?

Incidentally, Larmor's theorem states that the action of a uniform magnetic field on a charged particle can be counteracted to first order by a rotation. In fact, for $\omega = eB/2mc$ the above equations reduce to

$$m\ddot{x}'_1 = -(e^2B^2/4mc^2)x'_1 \quad , \quad m\ddot{x}'_2 = -(e^2B^2/4mc^2)x'_2 \quad , \quad m\ddot{x}'_3 = 0 \quad ,$$

namely equations for a particle acted upon by an elastic force normal to the 3-axis.

We start from a more general form of the equations,

$$\frac{\mathrm{d}}{\mathrm{d}t}\left(\frac{\partial L}{\partial \dot{q}_i}\right) - \frac{\partial L}{\partial q_i} = Q_i \quad ,$$

which will be justified later (see (6.27)). The Q_i's are generalized non-conservative forces, while $L = T(\dot{q}, q) - U(q)$ accounts for the conservative forces. Of course, nothing would prevent us from taking $L = T$ and adding to the right side $Q_i^{(\mathrm{cons})} = -\partial U/\partial q_i$.

We consider transformations of the type $q_i = q_i(q', t)$. Then $\dot{q}_i = (\partial q_i/\partial q'_j)\dot{q}'_j + \partial q_i/\partial t$, and so $\dot{q}_i$ is a function of the $\dot{q}'$, the q', and t. Conversely $q'_i = q'_i(q, t)$, $\dot{q}'_i = (\partial q'_i/\partial q_j)\dot{q}_j + \partial q'_i/\partial t$ etc. Note that even if the Lagrangian L is a function only of the q's and the $\dot{q}$'s, regarded as independent variables, the transformed Lagrangian $L'(q', \dot{q}', t) = L(q(q', t), \dot{q}(\dot{q}', q', t))$ may depend explicitly on t.

The formula $\partial \dot{q}'_i/\partial \dot{q}_j = \partial q'_i/\partial q_j$ will shortly be used together with

$$\frac{\mathrm{d}}{\mathrm{d}t}\left(\frac{\partial q'_j}{\partial q_i}\right) = \frac{\partial^2 q'_j}{\partial q_k \partial q_i}\dot{q}_k + \frac{\partial^2 q'_j}{\partial t \partial q_i} = \frac{\partial \dot{q}'_j}{\partial q_i} \quad .$$

Now

$$Q_i = \frac{\mathrm{d}}{\mathrm{d}t}\left(\frac{\partial L}{\partial \dot{q}_i}\right) - \frac{\partial L}{\partial q_i} = \frac{\mathrm{d}}{\mathrm{d}t}\left(\frac{\partial L'}{\partial \dot{q}'_j}\frac{\partial \dot{q}'_j}{\partial \dot{q}_i}\right) - \left(\frac{\partial L'}{\partial \dot{q}'_j}\frac{\partial \dot{q}'_j}{\partial q_i} + \frac{\partial L'}{\partial q'_j}\frac{\partial q'_j}{\partial q_i}\right) \quad .$$

With a little work this gives

$$Q_i = \left[\frac{\mathrm{d}}{\mathrm{d}t}\left(\frac{\partial L'}{\partial \dot{q}'_j}\right) - \frac{\partial L'}{\partial q'_j}\right]\frac{\partial q'_j}{\partial q_i} \quad .$$

Multiplying by $\partial q_i/\partial q'_k$, we find

$$\frac{\mathrm{d}}{\mathrm{d}t}\left(\frac{\partial L'}{\partial \dot{q}'_k}\right) - \frac{\partial L'}{\partial q'_k} = Q'_k$$

with $Q'_k = Q_i \, \partial q_i/\partial q'_k$.

We end this section with a few relations, which will be used in chapter 7. Note first that from $\partial \dot{q}'_i/\partial \dot{q}_j = \partial q'_i/\partial q_j$ one has

$$\frac{\partial L}{\partial \dot{q}_i} = \frac{\partial \dot{q}'_j}{\partial \dot{q}_i}\frac{\partial L'}{\partial \dot{q}'_j} = \frac{\partial q'_j}{\partial q_i}\frac{\partial L'}{\partial \dot{q}'_j} \quad .$$

Hence

$$p_i = \frac{\partial q'_j}{\partial q_i} p'_j$$

and

$$p_i \, \mathrm{d}q_i = p'_j \, \mathrm{d}q'_j \quad . \tag{6.24}$$

6.5 "Proofs" of the Lagrange equations

We have deliberately presented the Lagrange equations as just another way of expressing equations of motion which could be derived directly from Newton's second law.

In light of the previous section, and on the assumption that any physical system, however complex, can be described as a set of point particles, we may "prove" the Lagrange equations in the following simple way.

Each particle obeys Newton's second law

$$m_\alpha \ddot{\mathbf{r}}_\alpha = \mathbf{F}_\alpha \quad (\alpha = 1, \dots N) \quad ,$$

which can be expressed in the form

$$\frac{\mathrm{d}}{\mathrm{d}t}\left(\frac{\partial K}{\partial \dot{x}_{\alpha i}}\right) = F_{\alpha i} \quad ,$$

where

$$K = \sum_\alpha m_\alpha \dot{\mathbf{r}}_\alpha^2/2 \quad .$$

But this is already a Lagrange equation. If instead of the coordinates $x_{\alpha i}$ $(\alpha = 1, \dots N; i = 1, 2, 3)$ we introduce generalized coordinates q_i $(i = 1, \dots 3N))$ functions of the $\mathbf{r}_\alpha$'s, then the Lagrange equations are transformed into

$$\frac{\mathrm{d}}{\mathrm{d}t}\left(\frac{\partial K}{\partial \dot{q}_i}\right) = Q_i \quad ,$$

where

$$Q_i = \sum_\alpha F_{\alpha k} \frac{\partial x_{\alpha k}}{\partial q_i} \quad .$$

Another popular "proof" starts from Hamilton's principle derived from the d'Alembert principle, which in turn stems from the "virtual work" principle of statics.

Let $q_i(t)$ describe the real "trajectory" of a system during the time interval t_1 to t_2, and $q'_i(t) = q_i(t) + \delta q_i(t)$ a varied trajectory with $\delta q_i(t)$ infinitesimal, subject to the restriction $\delta q_i(t_1) = \delta q_i(t_2) = 0$. Thus the real and the varied motion start from, and arrive at the same points at the same times. (It is not so in the case of the Maupertuis principle, equation (1.37).)

Let

$$\delta K = [K]_{\text{varied}} - [K]_{\text{real}} = \frac{\partial K}{\partial \dot{q}_i}\delta\dot{q}_i + \frac{\partial K}{\partial q_i}\delta q_i \quad ,$$

where

$$\delta\dot{q}_i = \lim_{\Delta t \to 0}\left\{\left[\frac{q_i'(t+\Delta t) - q_i'(t)}{\Delta t}\right] - \left[\frac{q_i(t+\Delta t) - q_i(t)}{\Delta t}\right]\right\}$$

$$= \lim_{\Delta t \to 0}\frac{\delta q_i(t+\Delta t) - \delta q_i(t)}{\Delta t} = \frac{\mathrm{d}}{\mathrm{d}t}\delta q_i(t) \quad .$$

Hamilton's principle states that

$$\int_{t_1}^{t_2}(\delta K + \delta W)\mathrm{d}t = \delta\int_{t_1}^{t_2}K\mathrm{d}t + \int_{t_1}^{t_2}\delta W\ \mathrm{d}t = 0 \quad , \tag{6.25}$$

where δW is the work done by the forces for the displacements δq_i.

Notice that in general we cannot write

$$\int_{t_1}^{t_2}\delta W\ \mathrm{d}t = \delta\int_{t_1}^{t_2}W\ \mathrm{d}t \quad .$$

For conservative forces we have

$$\int_{t_1}^{t_2}\delta W^{(\text{cons})}\mathrm{d}t = -\int_{t_1}^{t_2}(\partial U/\partial q_i)\delta q_i\ \mathrm{d}t = -\delta\int_{t_1}^{t_2}U\ \mathrm{d}t \quad ,$$

while generalized non-conservative forces can be defined by writing

$$\delta W = Q_i\ \delta q_i \quad .$$

Therefore (6.25) can be expressed in the form

$$\delta\int_{t_1}^{t_2}L\mathrm{d}t + \int_{t_1}^{t_2}Q_i\delta q_i\ \mathrm{d}t = 0 \tag{6.26}$$

with $L = K - U$, or

$$\int_{t_1}^{t_2}\left(\frac{\partial L}{\partial \dot{q}_i}\delta\dot{q}_i + \frac{\partial L}{\partial q_i}\delta q_i + Q_i\delta q_i\right)\mathrm{d}t = 0 \quad .$$

A simple partial integration, using $\delta\dot{q}_i = \mathrm{d}\delta q_i/\mathrm{d}t$ and $\delta q_i(t_1) = \delta q_i(t_2) = 0$, yields

$$\int_{t_1}^{t_2} \left(-\frac{\mathrm{d}}{\mathrm{d}t}\left(\frac{\partial L}{\partial \dot{q}_i}\right) + \frac{\partial L}{\partial q_i} + Q_i \delta q_i \right) \delta q_i \mathrm{d}t = 0 \quad .$$

Since the δq_i's are arbitrary for $t_1 < t < t_2$, we infer that

$$\frac{\mathrm{d}}{\mathrm{d}t}\left(\frac{\partial L}{\partial \dot{q}_i}\right) - \frac{\partial L}{\partial q_i} = Q_i \quad . \tag{6.27}$$

A final remark. The first term in equation (6.26) is unchanged if we add $\mathrm{d}\Lambda(q,t)/\mathrm{d}t$ to the Lagrangian, since

$$\delta \int_{t_1}^{t_2} (\mathrm{d}\Lambda/\mathrm{d}t)\mathrm{d}t = (\partial\Lambda/\partial q_i)_{t_2} \delta q_i(t_2) - (\partial\Lambda/\partial q_i)_{t_1} \delta q_i(t_1) = 0.$$

This confirms what we wrote in section 6.3.

For $Q_i = 0$, Hamilton's principle (6.26) states that the real motion extremizes the action $A = \int_{t_1}^{t_2} L\, \mathrm{d}t$. Extremize, not minimize! For

$$L = (\dot{q}^2 + \alpha q^2)/2 \ (\alpha < 0 \text{ attractor}, \ \alpha > 0 \text{ repellor})$$

and $\delta q = \epsilon(t)$ ($\epsilon(t_1) = \epsilon(t_2) = 0$), using the equation of motion $\ddot{q} = \alpha q$ we have (exactly)

$$\delta A = \int_{t_1}^{t_2} [(\dot{\epsilon}^2 + \alpha\epsilon^2)/2]\mathrm{d}t \quad .$$

This is positive for $\alpha > 0$ (repellor), in which case the action is a minimum. For $\alpha < 0$ it may be positive or negative.

6.6 Constraints

In the preceding chapter we have tacitly assumed that the number of the generalized coordinates equaled the number of degrees of freedom. In deriving Lagrange's equations from Hamilton's principle, the final step required the independence of the δq_i's.

We first consider "holonomic constraints". A holonomic constraint is a geometrical restriction on the coordinates expressed by an equation

$$F(q_1, \ldots q_n; t) = 0 \quad . \tag{6.28}$$

Examples:

Spherical pendulum: A point mass moves frictionlessly on the inner surface of a hemispherical bowl (figure 6.1).

This time-independent constraint can be expressed by the equation

$$F(r) = R - r = 0 \text{ in spherical coordinates, or}$$

$$F(x_1, x_2, x_3) = R - \sqrt{x_1^2 + x_2^2 + x_3^2} = 0 \text{ in Cartesian coordinates.}$$

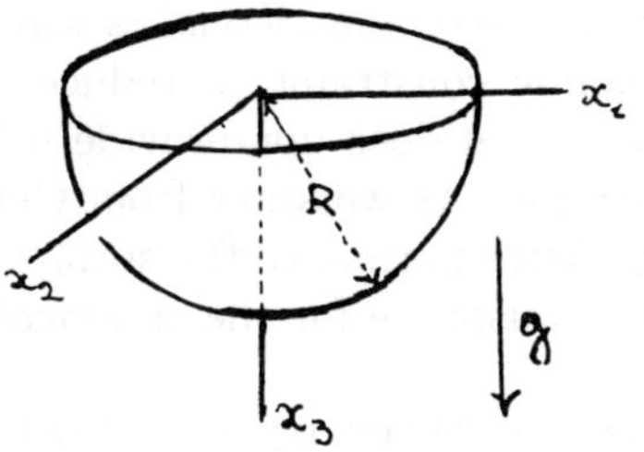

Figure 6.1: Spherical pendulum

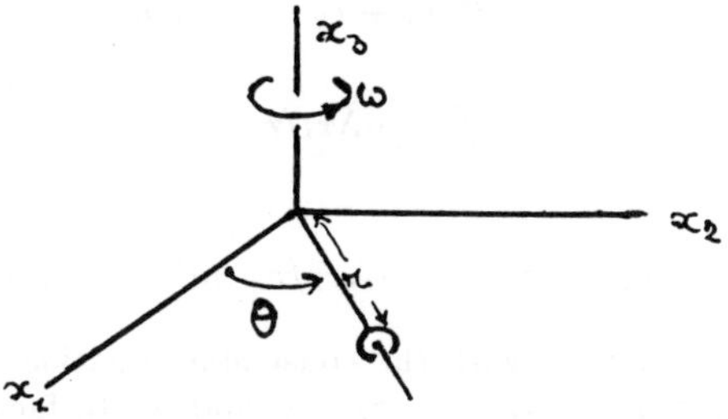

Figure 6.2: Bead on rotating wire

A time-dependent constraint: A bead can slide frictionlessly on a straight wire which is forced to rotate in the $(x_1 x_2)$-plane around the origin with constant angular velocity ω (figure 6.2).
This constraint can be expressed by

$F(\theta\ ;\ t) = \theta - \omega t = 0$ in plane polar coordinates, or by

$F(x_1, x_2; t) = -x_1 \sin(\omega t) + x_2 \cos(\omega t) = 0$ in Cartesian coordinates.

A holonomic constraint can be used to eliminate one of the coordinates from the Lagrangian or to express $q_1, \ldots q_n$ in terms of new coordinates $q'_1, \ldots q'_{n-1}$.

For instance, the Lagrangian for the spherical pendulum

$$L = (m/2)(\dot{r}^2 + r^2\dot{\theta}^2 + r^2 \sin^2\theta\ \dot{\phi}^2) + mgr\cos\theta$$

reduces to

$$L = (mR^2/2)(\dot{\theta}^2 + \sin^2\theta\ \dot{\phi}^2) + mgR\cos\theta \quad .$$

With $x_1 = r\cos(\omega t)$ and $x_2 = r\sin(\omega t)$, the Lagrangian $L = (m/2)(\dot{x}_1^2 + \dot{x}_2^2)$ for a free particle in the (x_1, x_2)-plane reduces to

$$L = (m/2)(\dot{r}^2 + r^2\omega^2)$$

for the bead constrained to slide on the frictionless wire.

The use of a holonomic constraint to reduce by one the number of coordinates in the Lagrangian does not provide information on the force implementing the constraint. As we know from General Physics, this is a reaction of varying magnitude normal to the surface $F(q_1, \ldots q_n; t) = 0$, for instance normal to the hemisphere for the spherical pendulum and to the wire for the bead.

For the spherical pendulum we may try to use the Lagrange equations in the form (6.27), writing

$$\frac{\mathrm{d}}{\mathrm{d}t}\left(\frac{\partial L}{\partial \dot{x}_i}\right) - \frac{\partial L}{\partial x_i} = Q_i' \quad ,$$

where

$$L = (m/2)(\dot{x}_1^2 + \dot{x}_2^2 + \dot{x}_3^2) + mgx_3$$

and

$$Q_i' = -\lambda x_i/r \quad .$$

We find

$$m\ddot{x}_1 = -\lambda x_1/r \ , \ m\ddot{x}_2 = -\lambda x_2/r \ , \ m\ddot{x}_3 = mg - \lambda x_3/r \ . \tag{6.29}$$

These equations, together with the constraint equation $r = R$, are sufficient to determine the four unknowns x_1, x_2, x_3, and λ. In fact, differentiating the constraint equation $r^2 = R^2$ twice with respect to t we obtain

$$x_1\ddot{x}_1 + x_2\ddot{x}_2 + x_3\ddot{x}_3 = -(\dot{x}_1^2 + \dot{x}_2^2 + \dot{x}_3^2) = -R^2(\dot{\theta}^2 + \sin^2\theta \ \dot{\phi}^2) = Ra_r \quad .$$

Multiplying the first of equations (6.29) by x_1, the second by x_2, the third by x_3, and summing we find

$$m(x_1\ddot{x}_1 + x_2\ddot{x}_2 + x_3\ddot{x}_3) = -\lambda R + mgx_3 \quad ,$$

$$ma_r = -\lambda + mg \ \cos\theta \quad ,$$

showing that λ is the magnitude of the reaction of the constraint.

From the constraint $-x_1 \sin(\omega t) + x_2 \cos(\omega t) = 0$ for the bead, we see that in this case we must take

$$Q_1' = -\lambda \sin(\omega t) \quad , \quad Q_2' = \lambda \cos(\omega t) \quad .$$

Hence the Lagrange equations are

$$m\ddot{x}_1 = -\lambda \sin(\omega t) \quad , \quad m\ddot{x}_2 = \lambda \cos(\omega t) \quad ,$$

to be used in conjunction with the constraint.

Double differentiation of the constraint with respect to t, and use of the constraint itself, yields

$$-\ddot{x}_1 \sin(\omega t) + \ddot{x}_2 \cos(\omega t) = 2\omega[\dot{x}_1 \cos(\omega t) + \dot{x}_2 \sin(\omega t)] \quad ,$$

and so

$$\lambda = 2m\omega[\dot{x}_1 \cos(\omega t) + \dot{x}_2 \sin(\omega t)] = 2m\omega(x_1\dot{x}_1 + x_2\dot{x}_2)/r = 2m\omega\dot{r} \quad .$$

Supposing $\dot{r} > 0$, this is the magnitude of the Coriolis force, which the constraint must counteract (see bug with $\theta = \pi/2$ problem 4.5).

In general, taking the variation of a holonomic constraint $F(q_1, \ldots q_n; t) = 0$ at a certain time t, we have

$$a_1\delta q_1 + \ldots + a_n\delta q_n = 0$$

with

$$a_i = \partial F/\partial q_i \quad .$$

The components of the reaction force are proportional to $\partial F/\partial q_i$ and so we can write [1]

$$Q_i' = \lambda a_i \quad .$$

Note that the corresponding work δW in Hamilton's principle is $\delta W^{(\mathrm{react})} = Q_i'\delta q_i = \lambda a_i\delta q_i = 0$, as we expect since the reaction and the "virtual displacement" are at right angles. On the other hand, when we differentiate the constraint with respect to t to supplement Lagrange's equations, we have $(\partial F/\partial q_i)\dot{q}_i + (\partial F/\partial t) = 0$, $a_i\dot{q}_i + a_t = 0$,

$$a_i \mathrm{d}q_i + a_t \mathrm{d}t = 0 \quad ,$$

with

$$a_i = \partial F/\partial q_i \quad , \quad a_t = \partial F/\partial t \quad .$$

Figure 1.3 emphasizes the distinction between the δq_i and the $\mathrm{d}q_i = \dot{q}_i\mathrm{d}t$.

In general we may have more than one, say r holonomic constraints

$$F_\alpha(q_1, \ldots q_n; t) = 0 \quad (\alpha = 1, \ldots r)$$

or, in differential form,

$$a_{\alpha i}\mathrm{d}q_i + a_{\alpha t}\mathrm{d}t = 0$$

with

$$a_{\alpha i} = \partial F_\alpha/\partial q_i \quad \text{and} \quad a_{\alpha t} = \partial F_\alpha/\partial t \quad .$$

[1] In the examples, the constraint equation was written in such a way that the a_i's turned out to be the components of unit vectors normal to the sphere and to the wire, respectively. Thus λ was the magnitude of the reaction. In general we can expect λ to be only proportional to the magnitude of the reaction.

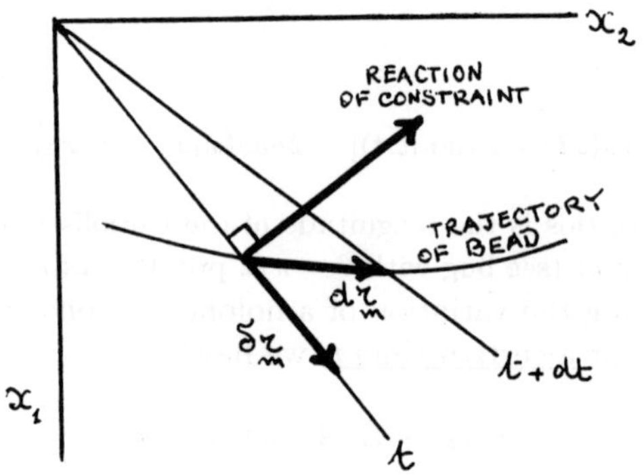

Figure 6.3: $\mathrm{d}q_i$ and δq_i

Then the Lagrange equations have the general form

$$\frac{\mathrm{d}}{\mathrm{d}t}\left(\frac{\partial K}{\partial \dot{q}_i}\right) - \frac{\partial K}{\partial q_i} = Q_i + Q_i' \quad ,$$

where

$$Q_i' = \lambda_\alpha a_{\alpha i} \quad (\text{sum over } \alpha) \quad .$$

These equations, together with the r constraints, are sufficient to determine the q_i and the r "Lagrange multipliers" λ_α.

A constraint in differential form

$$a_i \mathrm{d}q_i + a_t \mathrm{d}t = 0$$

is "non-holonomic" if it is not integrable, i.e. cannot be expressed in the form $\mathrm{d}F(q_1, \ldots q_n; t) = 0$.

In order to determine the integrability of a differential relation $A_i \mathrm{d}x_i = 0$ it is not sufficient to calculate the components of the generalized "curl", $(\partial A_i/\partial x_j) - (\partial A_j/\partial x_i)$. Even if one or more of these are different from zero, the differential relation may be integrable after multiplication by an "integrating factor".

For instance, the adiabatic condition $C_\mathrm{v}\mathrm{d}T + (RT/V)\mathrm{d}V = 0$ for an ideal gas, after multiplication by $1/T$ becomes $\mathrm{d}S = 0$, where S is the entropy.

A rolling disk provides a good example of non-holonomic constraint. The rolling condition $\dot{\mathbf{r}}_\mathrm{c} = -r\hat{\mathbf{b}}_2 \times \omega$ can be expressed in terms of components with respect to the fixed axes

$$\begin{cases} \dot{x}_1 = -r\dot{\varphi}\cos\vartheta\cos\varphi + r\dot{\vartheta}\sin\vartheta\sin\varphi - r\dot{\psi}\cos\varphi \quad , \\ \dot{x}_2 = -r\dot{\varphi}\cos\vartheta\sin\varphi - r\dot{\vartheta}\sin\vartheta\cos\varphi - r\dot{\psi}\sin\varphi \quad , \\ \dot{x}_3 = r\dot{\vartheta}\cos\vartheta = \mathrm{d}(r\sin\vartheta)/\mathrm{d}t \quad , \end{cases}$$

where x_1, x_2, and x_3 are the coordinates of the center of the disk. The third constraint is, of course, holonomic, $\mathrm{d}(x_3 - r\sin\vartheta) = 0$.

We have

$$\begin{cases} a_{1c}\mathrm{d}x_1 + a_{1\varphi}\mathrm{d}\varphi + a_{1\vartheta}\mathrm{d}\vartheta + a_{1\psi}\mathrm{d}\psi = 0 \ , \\ a_{2c}\mathrm{d}x_2 + a_{2\varphi}\mathrm{d}\varphi + a_{2\vartheta}\mathrm{d}\vartheta + a_{2\psi}\mathrm{d}\psi = 0 \ , \\ a_{3c}\mathrm{d}x_3 + a_{3\vartheta}\mathrm{d}\vartheta = 0 \ , \end{cases}$$

with $a_{ic} = 1$ $(i = 1, 2, 3)$, $a_{1\varphi} = r\cos\vartheta\cos\varphi$, $a_{2\varphi} = r\cos\vartheta\sin\varphi$, $a_{1\vartheta} = -r\sin\vartheta\sin\varphi$, $a_{2\vartheta} = r\sin\vartheta\cos\varphi$, $a_{3\vartheta} = -r\cos\vartheta$, $a_{1\psi} = r\cos\varphi$, $a_{2\psi} = r\sin\varphi$.

With the Lagrangian

$$L = (m/2)(\dot{x}_1^2 + \dot{x}_2^2 + \dot{x}_3^2) + (mr^2/8)(\dot{\vartheta}^2 + \sin^2\vartheta\ \dot{\varphi}^2) \\ +(mr^2/4)(\dot{\psi} + \dot{\varphi}\cos\vartheta)^2 - rmg\sin\vartheta$$

we find

$$\begin{cases} m\ddot{x}_1 = \lambda_1 \ , \ m\ddot{x}_2 = \lambda_2 \ , \ m\ddot{x}_3 = \lambda_3 \ , \\ (mr^2/4)(\ddot{\vartheta} - \sin\vartheta\cos\vartheta\ \dot{\varphi}^2) + (mr^2/2)\omega_3'\dot{\varphi}\sin\vartheta \\ +rmg\cos\vartheta = \lambda_1 a_{1\vartheta} + \lambda_2 a_{2\vartheta} + \lambda_3 a_{3\vartheta} \ , \\ (mr^2/4)\mathrm{d}[\sin^2\vartheta\ \dot{\varphi} + 2\cos\vartheta\ \omega_3']/\mathrm{d}t = \lambda_1 a_{1\varphi} + \lambda_2 a_{2\varphi} \ , \\ (mr^2/2)\mathrm{d}\omega_3'/\mathrm{d}t = \lambda_1 a_{1\psi} + \lambda_2 a_{2\psi} \ . \end{cases}$$

The last equation then gives

$$(mr^2/2)\mathrm{d}\omega_3'/\mathrm{d}t = r(\lambda_1\cos\varphi + \lambda_2\sin\varphi) = mr^2(-\ddot{\varphi}\cos\vartheta - \ddot{\psi} + 2\dot{\vartheta}\dot{\varphi}\sin\vartheta) \ .$$

In a similar way one obtains two more equations. Compare with problem 5.10.

The first and the second of the constraints are non-holonomic. For the rolling disk with horizontal axis ($\vartheta = \pi/2$) problem 6.17 gives a direct proof of the non integrability of the conditions [2]

$$\mathrm{d}x_1 + r\cos\varphi\ \mathrm{d}\psi = 0 \text{ and } \mathrm{d}x_2 + r\sin\varphi\ \mathrm{d}\psi = 0.$$

Non-holonomic constraints are restrictions on the infinitesimal changes of the coordinates, whose range of allowed values is unrestricted. A rolling disk can reach any point of the (x_1x_2)-plane, and ϑ, φ, and ψ can attain any of their allowed values, in spite of the non-holonomic constraints.

6.7 Invariance of L and constants of motion

Consider an infinitesimal transformation

$$q_i'(t') = q_i(t) + \delta q_i(t) \quad , \quad t' = t + \delta t \quad .$$

We say that the Lagrangian L is invariant under this transformation if

$$\delta L = L(q_i'(t'), \mathrm{d}q_i'(t')/\mathrm{d}t', t') - L(q_i(t), \mathrm{d}q_i(t)/\mathrm{d}t, t) = 0$$

[2] See also A. Sommerfeld, *Mechanics* (Academic Press, 1964), Problem II.1, pp. 244, 261.

up to the second order in δq_i and δt. The first term is obtained from the second by simply replacing $q_i(t)$ and $\mathrm{d}q_i(t)/\mathrm{d}t$ by $q_i'(t')$ and $\mathrm{d}q_i'(t')/\mathrm{d}t'$, and t by t' if the Lagrangian depends explicitly on the time.

Coordinate transformations

Consider transformations $\delta t = 0$, $q_i'(t) = q_i(t) + \epsilon f_i(q, \dot{q}, t)$ with ϵ an infinitesimal parameter.

If $\delta L = O(\epsilon^2)$, then

$$I = \frac{\partial L}{\partial \dot{q}_i}$$

is a constant of motion.

Proof:

$$0 = \delta L = \frac{\partial L}{\partial q_i}\delta q_i + \frac{\partial L}{\partial \dot{q}_i}\delta \dot{q}_i = \epsilon\left(\frac{\partial L}{\partial q_i} f_i + \frac{\partial L}{\partial \dot{q}_i}\frac{\mathrm{d}f_i}{\mathrm{d}t}\right)$$

$$= \epsilon\left[\left(\frac{\mathrm{d}}{\mathrm{d}t}\left(\frac{\partial L}{\partial \dot{q}_i}\right)\right) f_i + \frac{\partial L}{\partial \dot{q}_i}\frac{\mathrm{d}f_i}{\mathrm{d}t}\right] = \epsilon\frac{\mathrm{d}}{\mathrm{d}t}\left(\frac{\partial L}{\partial \dot{q}_i} f_i\right) \quad .$$

Example: The Lagrangian

$$L = (m_a \dot{\mathbf{r}}_a^2 + m_b \dot{\mathbf{r}}_b^2)/2 - U(|\mathbf{r}_a - \mathbf{r}_b|)$$

is (exactly) invariant under the transformation

$$\mathbf{r}_a \to \mathbf{r}_a + \delta\mathbf{r} \quad , \quad \mathbf{r}_b \to \mathbf{r}_b + \delta\mathbf{r} \quad ,$$

where $\delta\mathbf{r} = \epsilon\hat{\mathbf{n}}$ ($\hat{\mathbf{n}}$ constant unit vector). In this case

$$I = n_i(m_a \dot{x}_{ai} + m_b \dot{x}_{bi}) = M\hat{\mathbf{n}} \cdot \mathbf{V} = \hat{\mathbf{n}} \cdot \mathbf{P} \quad .$$

The component of the total momentum in an arbitrary direction is conserved.

Example: The Lagrangian

$$L = (m\dot{\mathbf{r}}^2 + k\mathbf{r}^2)/2$$

is invariant under the transformation

$$\mathbf{r} \to \mathbf{r} + \epsilon\hat{\mathbf{n}} \times \mathbf{r} \quad , \quad x_i \to x_i + \epsilon\, \epsilon_{ijk} n_j x_k$$

$((\mathbf{r} + \epsilon\hat{\mathbf{n}} \times \mathbf{r})^2 = \mathbf{r}^2 + O(\epsilon^2))$. The conserved quantity is

$$I = m\dot{x}_i(\epsilon_{ijk} n_j x_k) = n_j[\epsilon_{ijk} x_k (m\dot{x}_i)] = \hat{\mathbf{n}} \cdot \mathbf{l} \quad ,$$

the component of the angular momentum in an arbitrary direction.

Example: The Lagrangian

$$L = m\dot{\mathbf{r}}^2/2 - c\tan^{-1}(x_2/x_1) + x_3$$

(c = constant) is (exactly) invariant under the helical transformation

$$x_1' = x_1\cos s - x_2\sin s \ , \ x_2' = x_1\sin s + x_2\cos s \ , \ x_3' = x_3 + cs \ ,$$

where s is a constant parameter.

In fact, $\dot{x}_1'^2 + \dot{x}_2'^2 = \dot{x}_1^2 + \dot{x}_2^2$ and $\dot{x}_3' = \dot{x}_3$, while in cylindrical coordinates $(x_1 = \rho\cos\phi, x_2 = \rho\sin\phi,\ x_3)$ one has $x_1' = \rho\cos(\phi+s)$, $x_2' = \rho\sin(\phi+s)$,
$c\tan^{-1}(x_2'/x_1') - x_3' = c\tan^{-1}(\tan(\phi+s)) - x_3 - cs = c\phi - x_3$
$= c\tan^{-1}(x_2/x_1) - x_3$.

For $s \to \epsilon$ infinitesimal, we have $f_1 = -x_2$, $f_2 = x_1$, $f_3 = c$, and the conserved quantity is

$$I = (\partial L/\partial \dot{x}_i) f_i = m(-x_2\dot{x}_1 + x_1\dot{x}_2 + c\dot{x}_3) = l_3 + cp_3 \quad .$$

Time translations

For $t' = t - \epsilon$ we have

$$\delta L = L(q(t'), \mathrm{d}q(t')/\mathrm{d}t', t') - L(q(t), \mathrm{d}q(t)/\mathrm{d}t, t) = -\epsilon \mathrm{d}L/\mathrm{d}t \quad ,$$

$$\delta L = \frac{\partial L}{\partial q_i}(q_i(t') - q_i(t)) + \frac{\partial L}{\partial \dot{q}_i}(\dot{q}_i(t') - \dot{q}_i(t)) + \frac{\partial L}{\partial t}(-\epsilon)$$

$$= -\frac{\partial L}{\partial q_i}\epsilon\dot{q}_i - \frac{\partial L}{\partial \dot{q}_i}\epsilon\frac{\mathrm{d}}{\mathrm{d}t}\dot{q}_i - \epsilon\frac{\partial L}{\partial t} = -\epsilon\frac{\mathrm{d}}{\mathrm{d}t}\left(\frac{\partial L}{\partial \dot{q}_i}\dot{q}_i\right) - \epsilon\frac{\partial L}{\partial t} \quad .$$

$\delta L = -\epsilon \mathrm{d}L/\mathrm{d}t$ gives

$$\frac{\mathrm{d}}{\mathrm{d}t}\left(\frac{\partial L}{\partial \dot{q}_i}\dot{q}_i - L\right) = -\frac{\partial L}{\partial t} \quad .$$

If a Lagrangian does not depend explicitly on the time, then

$$I = \frac{\partial L}{\partial \dot{q}_i}\dot{q}_i - L$$

is a constant of motion, which will be seen to be the Hamiltonian.

This is a special case of a general property of the Euler-Lagrange equation of a variational principle, $\mathrm{d}(\partial f/\partial y')/\mathrm{d}x = \partial f/\partial y$. One has

$$\mathrm{d}(y'\partial f/\partial y' - f)/\mathrm{d}x = -\partial f/\partial x.$$

If $\partial f/\partial x = 0$, one has the Jacobi integral of motion

$$y'\partial f/\partial y' - f = \text{constant} \quad .$$

As a preparation for Noether's theorem, we want to derive this result in a slightly more complicated way.

Assume that the Lagrangian does not depend explicitly on the time [3] and that

$$L(q'(t'), \mathrm{d}q'(t')/\mathrm{d}t') = L(q(t), \mathrm{d}q(t)/\mathrm{d}t) + O(\epsilon^2) \quad ,$$

[3]For instance $L = (1/2)a_{ij}\dot{q}_i\dot{q}_j - U(q)$, with a_{ij} and U functions of the q_i's, but not explicitly dependent on t.

where $t' = t - \epsilon$ (ϵ infinitesimal) and $q'_i(t') = q_i(t)$.

We write

$$\begin{aligned} 0 &= L(q'(t'), \mathrm{d}q'(t')/\mathrm{d}t') - L(q(t), \mathrm{d}q(t)/\mathrm{d}t) \\ &= [L(q'(t'), \mathrm{d}q'(t')/\mathrm{d}t') - L(q(t'), \mathrm{d}q(t')/\mathrm{d}t')] \\ &+ [L(q(t'), \mathrm{d}q(t')/\mathrm{d}t') - L(q(t), \mathrm{d}q(t)/\mathrm{d}t)] \quad . \end{aligned}$$

The difference in the first set of square brackets represents the variation of the Lagrangian due to the change in form of the q_i's [4],

$$\bar{\delta}q_i(t) = q'_i(t) - q_i(t) = q'_i(t') - q_i(t') + O(\epsilon^2) \quad .$$

Thus we have

$$\begin{aligned} 0 &= \frac{\partial L}{\partial q_i}\bar{\delta}q_i + \frac{\partial L}{\partial \dot{q}_i}\frac{\mathrm{d}}{\mathrm{d}t}\bar{\delta}q_i - \epsilon\frac{\mathrm{d}L}{\mathrm{d}t} \\ &= \left(\frac{\mathrm{d}}{\mathrm{d}t}\frac{\partial L}{\partial \dot{q}_i}\right)\bar{\delta}q_i + \frac{\partial L}{\partial \dot{q}_i}\frac{\mathrm{d}}{\mathrm{d}t}\bar{\delta}q_i - \epsilon\frac{\mathrm{d}L}{\mathrm{d}t} \quad . \end{aligned}$$

Since $q'_i(t') = q_i(t) = q_i(t' + \epsilon)$, we have

$$\bar{\delta}q_i = q'_i(t) - q_i(t) = q_i(t+\epsilon) - q_i(t) = \epsilon\dot{q}_i(t),$$

$$0 = \epsilon\frac{\mathrm{d}}{\mathrm{d}t}\left(\frac{\partial L}{\partial \dot{q}_i}\dot{q}_i - L\right) \quad ,$$

and we find again the integral of motion $I = (\partial L/\partial \dot{q}_i)\dot{q}_i - L$ [5].

6.7.1 Invariance of L up to total time derivative

If under an infinitesimal transformation $q_i \to q_i + \delta q_i$ with $\delta q_i = \epsilon f_i(q, \dot{q}, t)$ the Lagrangian changes by

$$\delta L = \epsilon\frac{\mathrm{d}\Lambda(q,t)}{\mathrm{d}t} + O(\epsilon^2) \quad ,$$

then

$$I = \frac{\partial L}{\partial \dot{q}_i}f_i - \Lambda \tag{6.30}$$

is a constant of motion.

Proof:

$$\epsilon\,\frac{\mathrm{d}\Lambda}{\mathrm{d}t} = \left(\frac{\partial L}{\partial q_i}\right)\delta q_i + \frac{\partial L}{\partial \dot{q}_i}\delta\dot{q}_i = \epsilon\,\frac{\mathrm{d}}{\mathrm{d}t}\left(\frac{\partial L}{\partial \dot{q}_i}f_i\right)$$

---

[4] For instance, if $q(t) = t^2$ then $q'(t') = q(t) = (t'+\epsilon)^2$, $\bar{\delta}q(t) = q'(t) - q(t) = (t+\epsilon)^2 - t^2 = 2\epsilon t + \epsilon^2$, while $q'(t') - q(t') = t^2 - (t-\epsilon)^2 = 2\epsilon t - \epsilon^2$.

[5] For the example in footnote [3], $I = (1/2)a_{ij}\dot{q}_i\dot{q}_j + U$ is clearly the energy.

Example: The variation of the Lagrangian

$$L = (m/2)(\dot{x}^2 - \omega^2 x^2)$$

under the transformation

$$x' = x + \epsilon \sin(\omega t)$$

is

$$\begin{aligned}\delta L &= (m/2)\{[\dot{x}'^2 - \dot{x}^2] - \omega^2[x'^2 - x^2]\} \\ &= \epsilon m\{\omega\dot{x}\cos(\omega t) - \omega^2 x \sin(\omega t)\} + O(\epsilon^2) = \epsilon \, \mathrm{d}\Lambda/\mathrm{dt} + O(\epsilon^2)\end{aligned}$$

with

$$\Lambda = m\omega x \cos(\omega t) \quad .$$

The integral of motion $I = m\dot{x}\sin(\omega t) - m\omega x\cos(\omega t)$,

$$I = m\begin{vmatrix} \sin(\omega t) & x \\ \mathrm{d}\sin(\omega t)/\mathrm{dt} & \mathrm{d}x/\mathrm{dt} \end{vmatrix} ,$$

is proportional to the Wronskian determinant of the solutions $\sin(\omega t)$ and $x(t)$ of the equations of motion.

Example:

Consider an infinitesimal Galilean transformation

$$\mathbf{r}_a \to \mathbf{r}_a + t\epsilon\hat{\mathbf{n}} \quad , \quad \mathbf{r}_b \to \mathbf{r}_b + t\epsilon\hat{\mathbf{n}}$$

where $\hat{\mathbf{n}}$ is a unit vector and ϵ is an infinitesimal parameter with the dimensions of a velocity.

The variation of the Lagrangian given by the first of (6.6) is

$$\delta L = \epsilon\hat{\mathbf{n}} \cdot (m_a\dot{\mathbf{r}}_a + m_b\dot{\mathbf{r}}_b) + O(\epsilon^2) = \epsilon \, \mathrm{d}\Lambda/\mathrm{dt} + O(\epsilon^2) ,$$

$$\Lambda = \hat{\mathbf{n}} \cdot (m_a\mathbf{r}_a + m_b\mathbf{r}_b) = M\hat{\mathbf{n}} \cdot \mathbf{R} \quad ,$$

where $\mathbf{R}$ is the position of the center of mass. The corresponding integral of motion is

$$\begin{aligned}I &= [t(m_a\dot{\mathbf{r}}_a + m_b\dot{\mathbf{r}}_b) - (m_a\mathbf{r}_a + m_b\mathbf{r}_b)] \cdot \hat{\mathbf{n}} \\ &= M(t\mathbf{V} - \mathbf{R}) \cdot \hat{\mathbf{n}} = -M\mathbf{R}(0) \cdot \hat{\mathbf{n}} \quad ,\end{aligned}$$

where $\mathbf{R}(0)$ is the position of the center of mass at $t = 0$.

6.7.2 Noether's theorem

The following is a version of the theorem for Classical Mechanics.

Consider an infinitesimal transformation

$$q_i'(t') = q_i(t) + \delta q_i(t) \quad , \quad t' = t + \delta t \quad ,$$

where

$$\delta q_i(t) = \epsilon f_i(q, \dot{q}, t) \quad , \quad \delta t = \epsilon g(q, t) \quad .$$

Assume that the action integral is invariant under this transformation with an error of second order in ϵ [6]

$$\int_{t_1'}^{t_2'} L\left(q'(t'), \frac{\mathrm{d}q'(t')}{\mathrm{d}t'}\right) \mathrm{d}t' = \int_{t_1}^{t_2} L\left(q(t), \frac{\mathrm{d}q(t)}{\mathrm{d}t}\right) \mathrm{d}t \quad .$$

Then with a second order error in ϵ we have

$$0 = \int_{t_1}^{t_2} \left[L\left(q'(t'), \frac{\mathrm{d}q'(t')}{\mathrm{d}t'}\right) \frac{\mathrm{d}t'}{\mathrm{d}t} - L\left(q(t), \frac{\mathrm{d}q(t)}{\mathrm{d}t}\right)\right] \mathrm{d}t$$

giving

$$0 = \frac{\partial L}{\partial q_i}(q_i'(t') - q_i(t)) + \frac{\partial L}{\partial \dot{q}_i}\left(\frac{\mathrm{d}q_i'(t')}{\mathrm{d}t'} - \frac{\mathrm{d}q_i(t)}{\mathrm{d}t}\right) + L\frac{\mathrm{d}\delta t}{\mathrm{d}t} \quad .$$

We already know that

$$q_i'(t') - q_i(t) = (q_i'(t') - q_i(t')) + (q_i(t') - q_i(t)) = \bar{\delta} q_i + \dot{q}_i \delta t \quad .$$

On the other hand, using

$$\frac{\mathrm{d}}{\mathrm{d}t} = \frac{\mathrm{d}t'}{\mathrm{d}t}\frac{\mathrm{d}}{\mathrm{d}t'} = \left(1 + \frac{\mathrm{d}\delta t}{\mathrm{d}t}\right)\frac{\mathrm{d}}{\mathrm{d}t'}$$

and

$$\frac{\mathrm{d}}{\mathrm{d}t'} \simeq \left(1 - \frac{\mathrm{d}\delta t}{\mathrm{d}t}\right)\frac{\mathrm{d}}{\mathrm{d}t} \quad ,$$

we have

$$\begin{aligned}\frac{\mathrm{d}}{\mathrm{d}t'}q_i'(t') - \frac{\mathrm{d}}{\mathrm{d}t}q_i(t) &\simeq \frac{\mathrm{d}}{\mathrm{d}t}[q_i'(t') - q_i(t)] - \frac{\mathrm{d}\delta t}{\mathrm{d}t}\frac{\mathrm{d}}{\mathrm{d}t}q_i(t) \\ &= \frac{\mathrm{d}}{\mathrm{d}t}(\bar{\delta} q_i + \dot{q}_i\delta t) - \frac{\mathrm{d}\delta t}{\mathrm{d}t}\dot{q}_i = \frac{\mathrm{d}}{\mathrm{d}t}\bar{\delta} q_i + \delta t\frac{\mathrm{d}}{\mathrm{d}t}\dot{q}_i \quad .\end{aligned}$$

Hence

$$0 = \frac{\partial L}{\partial q_i}(\bar{\delta} q_i + \dot{q}_i\delta t) + \frac{\partial L}{\partial \dot{q}_i}\left(\frac{\mathrm{d}}{\mathrm{d}t}\bar{\delta} q_i + \delta t\frac{\mathrm{d}}{\mathrm{d}t}\dot{q}_i\right) + L\frac{\mathrm{d}\delta t}{\mathrm{d}t}$$

[6] Do not confuse this with the exact, but trivial identity

$$\int_{t_1'}^{t_2'} L(q(t(t')), (\mathrm{d}q(t(t'))/\mathrm{d}t')(\mathrm{d}t/\mathrm{d}t')\mathrm{d}t' = \int_{t_1}^{t_2} L(q(t), \mathrm{d}q(t)/\mathrm{d}t)\mathrm{d}t \quad .$$

$$= \frac{\mathrm{d}}{\mathrm{d}t}\left(\frac{\partial L}{\partial \dot{q}_i}\bar{\delta} q_i + L\delta t\right) = \epsilon \frac{\mathrm{d}}{\mathrm{d}t}\left(\frac{\partial L}{\partial \dot{q}_i}(f_i - \dot{q}_i g) + Lg\right) \quad .$$

The quantity

$$I = \frac{\partial L}{\partial \dot{q}_i}(f_i - g\dot{q}_i) + Lg \tag{6.31}$$

is a constant of motion. For time translations one has $f_i = 0$ and $g = -1$.

Example: The action

$$\int_{t_1}^{t_2} (\dot{q}^2/2 + kq^{-2})\mathrm{d}t$$

is invariant under

$$q'(t') = \sqrt{\lambda}\, q(t) \quad , \quad t' = \lambda t \quad .$$

For $\lambda = 1 + \epsilon$ we have $f = q/2$ and $g = t$.

The integral of motion is

$$I = \frac{\partial L}{\partial \dot{q}}(f - \dot{q}g) + Lg = -Et + \dot{q}q/2 \quad ,$$

where $E = \dot{q}^2/2 - k/q^2$ is the energy integral of motion.

6.8 Chapter 6 problems

6.1 Derive the equation for the thumbtack (problem 5.9) by using a Lagrangian involving only φ, $\dot{\varphi}$, and the fixed angle ϑ.

6.2 A uniform plank initially leans against a smooth wall and rests on a frictionless floor.

Discuss its motion (i) by an elementary method, and (ii) by using a Lagrangian and constraints.

$x = (\ell/2)\cos\theta$, $y = (\ell/2)\sin\theta$, $I_{\text{cm}} = m\ell^2/12$, $I_A = I_B = m\ell^2/3 \to I$

6.3 A particle of charge q moves in a static electromagnetic field with potentials $V = -a \ln\rho$, $\mathbf{A} = (-By/2, Bx/2, 0)$, where $\rho = \sqrt{x^2+y^2}$, and a and B are constants.

(i) Express the fields and the force acting on the particle in terms of the unit vectors $\hat{\mathbf{e}}_\rho$, $\hat{\mathbf{e}}_\phi$, and $\hat{\mathbf{e}}_z$.

(ii) Write $\mathbf{F} = m\mathbf{a}$ in cylindrical coordinates.

(iii) Write the Lagrangian and derive the equations of motion. Compare with (ii).

(iv) Do you expect constants of motion other than the energy? Why?

(v) What are they?

6.4 Consider a system of N particles, interacting with one another with velocity-independent forces, Lagrangian

$$L = \sum_{\alpha=1}^{N} m_\alpha \dot{\mathbf{r}}_\alpha^2/2 \ + U(\mathbf{r}_1, \ldots \mathbf{r}_N) \quad .$$

Let ϕ be an angle about the unit vector $\hat{\mathbf{n}}$. We know that

$$p_\phi = \hat{\mathbf{n}} \cdot \mathbf{l} = \epsilon_{ijk} n_i x_{\alpha j} \frac{\partial L}{\partial \dot{x}_{\alpha k}}$$

(sum convention also for α).

(i) What is $p'_\phi = p_\phi + \delta p_\phi$ for the Lagrangian $L' = L + \delta L$, where δL is a function of the $\mathbf{r}_\alpha$'s and their time derivatives $\dot{\mathbf{r}}_\alpha$'s ?

(ii) What is δp_ϕ if δL is the Lagrangian $\delta L = c^{-1} q_\alpha A_i \dot{x}_{\alpha i}$?

6.5 Obtain the equation of problem 3.10 by a Lagrangian method.

6.6 Consider the block of mass m on the frictionless floor of a cart which moves according to $X(t)$ as shown in figure 6.4. Denote by k the elastic constant of the spring, and by ℓ its natural length.

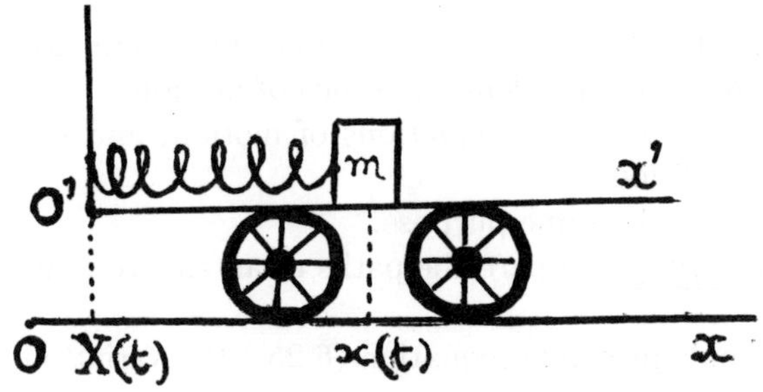

Figure 6.4: Block on accelerated cart

(i) Write the Lagrangian for the coordinate x of m with respect to a fixed x-axis.

(ii) Change coordinate from x (with respect to the fixed axis) to x' (with respect to the cart), find the new Lagrangian and the equation of motion with respect to the cart.

(iii) Incorporate the inertial force found in (ii) in the Lagrangian (i) as a potential energy, and find again the equation of motion for x'.

6.7 Let $L(q, \dot{q}; t)$ be a Lagrangian. A "complementary" Lagrangian $L_c(p, \dot{p}, t)$, can be defined by the Legendre transformation

$$L_c(Q, \dot{Q}; t) + \alpha L(q, \dot{q}; t) = P\dot{Q} + \alpha p\dot{q} \quad ,$$

where $Q = p$, $P = \alpha q$, and $\alpha = \pm 1$. Only the case $\alpha = 1$ seems to have been considered in the literature.

(i) Show that $P = \partial L_c/\partial \dot{Q}$, $\dot{P} = \partial L_c/\partial Q$, $\partial L_c/\partial t = -\alpha \partial L/\partial t$.

(ii) (a) Find L_c for the harmonic oscillator, $L = (m\dot{q}^2 - kq^2)/2$.
(b) Find the corresponding Lagrange equations.

(iii) (a) Find L_c for the Kepler problem, $L = m\dot{\mathbf{q}}^2/2 + k/|\mathbf{q}|$ $(k > 0)$.
(b) Find the corresponding Lagrange equations.

6.8 At $t = 0$ a particle of charge q and mass m starts from rest at the origin of a Cartesian coordinate system, and moves under the influence of the electromagnetic field $\mathbf{E} = E\hat{\mathbf{e}}_y$ and $\mathbf{B} = B\hat{\mathbf{e}}_z$ (E and B constant). While the electrostatic potential is unique up to a constant, $\phi = -Ey$, the vector potential is determined up to a gauge transformation.

(i) Find the gauge transformation connecting the two vector potentials $\mathbf{A}' = -By\hat{\mathbf{e}}_x$ and $\mathbf{A}'' = Bx\hat{\mathbf{e}}_y$, both giving the same $\mathbf{B}$.

(ii) Write the Lagrangians L' and L'' for the two cases.

(iii) Check that these Lagrangians differ by a term $\mathrm{d}F/\mathrm{d}t$, and so can be expected to yield equivalent equations of motion.

(iv) Use L' to write the equations of motion, and solve them for the given initial conditions.

(v) Use L'' for the same purpose.

(vi) Qualitatively compare the paths of an electron and a proton.

6.9 In Hamilton's principle (equation (6.25)) the variation $q(t) \to q'(t) = q(t) + \delta q(t)$ is synchronous, i.e. the system is at P in the real motion and at P' in the varied motion at the same time t.

On the other hand, if the time is also varied, it is at P' at a time $\bar{t}(t) = t + \delta t$, and the varied motion is described by a function of $\bar{t}$, $q''(\bar{t}) = q'(t(\bar{t}))$. In this case, one speaks of asynchronous variations.

(i) Show that, while $\delta q = q'(t) - q(t)$ and $\delta^\star q = q''(\bar{t}) - q(t)$ are equal, one has $\delta^\star \dot{q} = \delta \dot{q} - \dot{q}\mathrm{d}\delta t/\mathrm{d}t$. Here

$$\delta^\star \dot{q} \equiv \lim_{\Delta t \to 0} \frac{q''(\bar{t}(t+\Delta t)) - q''(\bar{t}(t))}{\bar{t}(t+\Delta t) - \bar{t}(t)} - \lim_{\Delta t \to 0} \frac{q(t+\Delta t) - q(t)}{\Delta t} \quad .$$

(ii) Show that Hamilton's principle for motion under conservative forces takes the form

$$\int_{t_1}^{t_2} (\delta^\star K + 2K(\mathrm{d}\delta t/\mathrm{d}t) - \delta^\star U)\mathrm{d}t = 0$$

in terms of asynchronous variations.

(iii) The varied motion is necessarily asynchronous if it is required to take place with the same total energy as the original motion. From (ii), derive Maupertuis' principle for isoenergetic variations.

6.10 We have a Lagrangian $L(q, \dot{q}, t)$, but want to restrict the coordinates and their derivatives by a holonomic constraint $F(q, \dot{q}, t) = 0$, more general than (6.28), since F depends also on $\dot{q}$. Show that the equations of motion can be derived from the modified Lagrangian

$$L' = L + \lambda(t)F \quad ,$$

where $\lambda(t)$ is regarded as a new coordinate.

6.11 Assume that the holonomic constraint F in the preceding problem does not depend explicitly on t and is of the form $F = (a_{ij}\dot{q}_i\dot{q}_j - b)/2 = 0$, where the symmetric coefficients a_{ij} and b are constants.

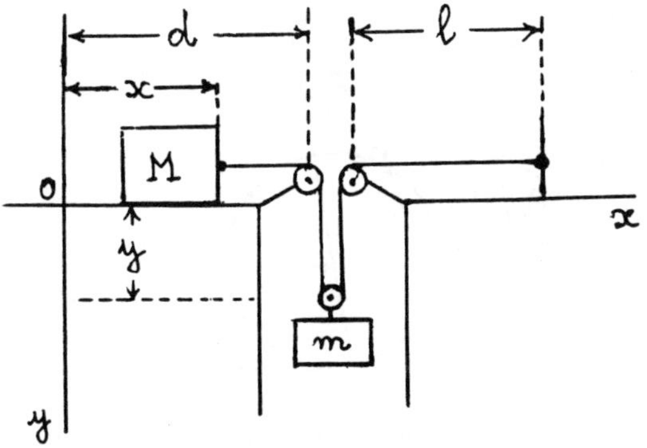

Figure 6.5: Problem 6.13

Show that one finds

$$\frac{\mathrm{d}}{\mathrm{d}t}\left(\frac{\partial L}{\partial \dot{q}_i}\right) - \frac{\partial L}{\partial q_i} + \frac{\mathrm{d}}{\mathrm{d}t}\left[\frac{1}{b}\left(L - \dot{q}_j\,\frac{\partial L}{\partial \dot{q}_j}\right) a_{ik}\dot{q}_k\right] = 0 \quad .$$

6.12 A bead of mass m slides without friction on the curve $y = f(x)$. The y-axis is vertically up. Find the equation of motion for x

(i) by using the holonomic constraint,

(ii) by using a Lagrange multiplier λ,

and

(iii) find the reaction of the constraint and compare it with λ.

6.13 Consider the system shown in figure 6.5. The pulleys are massless and frictionless. Assuming that the acceleration A of M is known, find the friction force F acting on M and the tension T of the cord

(i) by an elementary method,

(ii) by a Lagrangian method with a holonomic constraint relating the position coordinate x of M with the position coordinate y of m,

(iii) by expressing the constraint in non-holonomic form and using a Lagrange multiplier.

6.14 A solid cylinder of radius R_2 and mass m (moment of inertia $I = mR_2^2/2$) rolls without slipping inside a cylindrical surface of radius R_1 as shown in figure 6.6.

(i) Write the Lagrangian in terms of the angles θ_1 and θ_2 shown in figure 6.8.

(ii) Derive the equation for θ_1 using the holonomic constraint

$$R_1\theta_1 = R_2(\theta_1 + \theta_2).$$

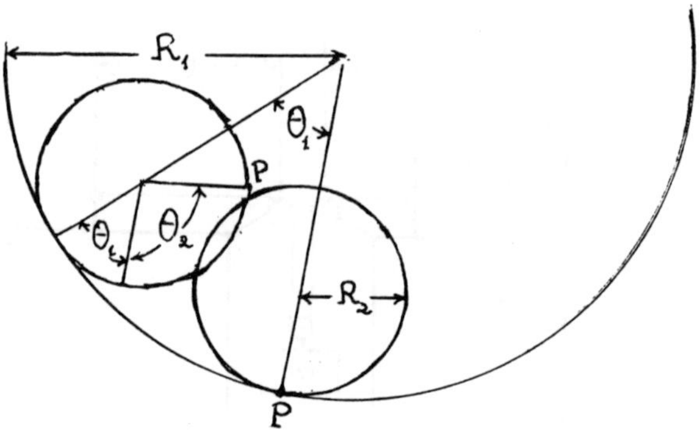

Figure 6.6: Cylinder rolling inside cylinder

(iii) Use Lagrange multipliers to find the forces exerted by the cylindrical surface on the rolling cylinder.

6.15 A particle of mass m moves under the action of gravity and of the constraint

$$\cos\alpha \, \mathrm{d}\rho - (\sin^2\alpha/\cos\alpha)(z \, \mathrm{d}z/\rho) = 0 \quad ,$$

where α is a fixed angle, and ρ and z are cylindrical coordinates. The z-axis is vertically upward.

(i) Write the Lagrange equation in Cartesian coordinates with a Lagrange multiplier.

(ii) Show that the constraint, though couched in non-holonomic form, is, in fact, holonomic, and represents the frictionless motion on any of a family of hyperboloids.

(iii) For the $z > 0$ branch of the degenerate hyperboloid that passes through the origin, show the relation of the Lagrange multiplier with the reaction of the frictionless surface on the particle.

6.16 Consider the system of figure 6.7, two wheels of equal radius rigidly attached to an axle, suggested by Goldstein. Find the constraints.

6.17 Show that the constraints for a rolling disk with horizontal axis,

$$\mathrm{d}x_1 + r\cos\varphi \, \mathrm{d}\psi = 0 \text{ and } \mathrm{d}x_2 + r\sin\varphi \, \mathrm{d}\psi = 0.$$

(section 6.6) are non-integrable.

6.18 Derive the results of problem 5.1 by the method of Lagrange multipliers, using as initial variables the Euler angles and (x_1, x_2).

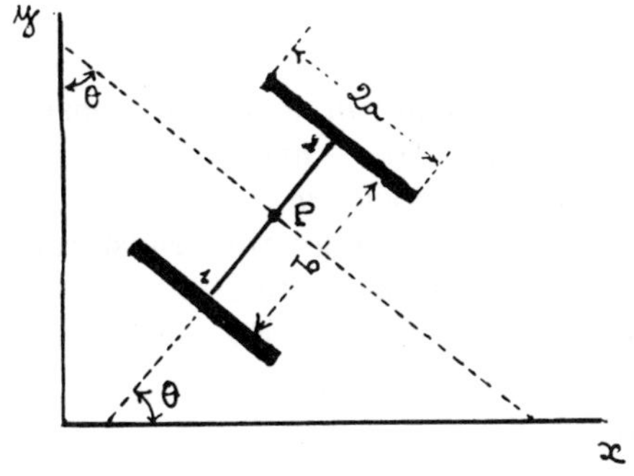

Figure 6.7: Wheels and axle

6.19 The use of Euler angles is not convenient in treating the motion of a uniform sphere. Derive the results of problem 6.18 by the Lagrange multiplier technique without using Euler angles.

6.20 Derive the results of problem 5.2 by Lagrange equations with multipliers.

6.21 Give a Lagrangian formulation of the problem of a rolling disk (problem 5.10).

6.22 Find the variation of the harmonic oscillator Lagrangian $L = m(\dot{q}^2 - \omega^2 q^2)/2$ under the infinitesimal transformation $q \to q + \delta q$ with $\delta q = \mathrm{i}\epsilon/2m(\dot{q} - \mathrm{i}\omega q)$, and determine the corresponding integral of motion according to section 6.7.1.

6.23 Find the variation of the Toda Lagrangian

$$L = (\dot{x}^2 + \dot{y}^2)/2 - [\exp(2(x - y\sqrt{3})) + \exp(2(x + y\sqrt{3})) + \exp(-4x)]/24 + 1/8$$

under the infinitesimal transformation

$$\begin{cases} \delta x = \epsilon[-48\dot{x}\dot{y} + \sqrt{3}(\exp(2(x - y\sqrt{3})) - \exp(2(x + y\sqrt{3}))] \ , \\ \delta y = \epsilon[16\dot{y}^2 - 24\dot{x}^2 + \exp(2(x - y\sqrt{3})) + \exp(2(x + y\sqrt{3})) - 2\exp(-4x)] \ , \end{cases}$$

and derive an integral of motion according to section 6.7.1.

6.24 Apply the transformation $x_k \to x_k + \delta x_k$ with $\delta x_k = \eta[-2x_i\dot{x}_k + (x_j\dot{x}_j)\delta_{ik} + x_k\dot{x}_i]/k$ (η infinitesimal parameter) to the Lagrangian $L = m\dot{\mathbf{r}}^2/2 + k/r$. Show that $\delta L = \eta \mathrm{d}\Lambda/\mathrm{d}t$ with $\Lambda = -2x_i/r$.

(ii) Show that the ensuing constant of motion is the i-th component of the L-R-L vector.

6.25 Consider a particle of mass m in one dimension with Lagrangian $L = m\dot{x}^2/2 - U(x - ut)$, where $U(x - ut)$ represents the interaction with a field travelling with constant velocity u.

(i) Show that L is invariant under the transformation $x' = x + \epsilon u$, $t' = t + \epsilon$, where ϵ is a constant parameter.

(ii) From this invariance property, derive a constant of motion.

6.26 The Lagrangian

$$L = (m_a\dot{\mathbf{r}}_a^2 + m_b\dot{\mathbf{r}}_b^2)/2 \; - U(|\mathbf{r}_a - \mathbf{r}_b|)$$

is invariant under the seven transformations of the Galilean group (space and time translations: 4 parameters; space rotations: 3 parameters). Use (6.31),

$$I = \frac{\partial L}{\partial \dot{q}_i}\,(f_i - \dot{q}_i g) + Lg \quad ,$$

to find the corresponding seven integrals of motion.

6.27 Find the integral of motion at the end of section 6.7.1 by transforming both coordinates and time so that the action is invariant.

6.28 In problem 6.8 the Lagrangian

$$L' = m(\dot{x}^2 + \dot{y}^2 + \dot{z}^2)/2 \; + qEy - (qB/c)y\dot{x}$$

yields the constants of motion

$$m\dot{x} - qBy/c \; = \text{constant}\;, \; m\dot{y} + qBx/c \; - qEt \; = \text{constant}\;, \; m\dot{z} = \text{constant}\;.$$

While the first and the third originate from the absence of x and z, respectively, from L', the second has a different origin. Derive it by considering the result of the transformation $y \to y + \epsilon$ (ϵ infinitesimal constant).

Solutions to ch. 6 problems

S6.1

$$K = [m(\dot{x}_1^2 + \dot{x}_2^2) + I_1\dot{\varphi}^2\sin^2\vartheta + I_3(\cos\vartheta\ \dot{\varphi} + \dot{\psi})^2]/2 \quad ,$$

$$U = -mg(x_1 \sin\epsilon - x_3\cos\epsilon) \quad ,$$

where (x_1, x_2, x_3) are the coordinates of the center of mass with respect to the fixed axes.
Since $x_1 = (R - r\ \cos\vartheta)\sin\varphi$, $x_2 = -(R - r\ \cos\vartheta)\cos\varphi$, $x_3 = r\ \sin\vartheta$, $R - r\ \cos\vartheta = \ell\ \sin\vartheta$, $\sin\vartheta = \ell/R$, we have

$$L = (1/2)(m\ell^2 + I_1 + \ell^2 I_3/r^2)\dot{\varphi}^2\sin^2\vartheta + mg(\ell\ \sin\vartheta\ \sin\epsilon\ \sin\varphi - r\ \sin\vartheta\ \cos\epsilon) \quad .$$

The Lagrange equation for φ,

$$\frac{\mathrm{d}}{\mathrm{d}t}\left(\frac{\partial L}{\partial\dot{\varphi}}\right) = \frac{\partial L}{\partial\varphi} \quad ,$$

gives

$$[I_1 r^2 + (I_3 + mr^2)\ell^2]\ddot{\varphi} = mgRr^2\sin\epsilon\ \cos\varphi \quad .$$

S6.2 (i) Elementary:

$m\ddot{x} = F_A$, $m\ddot{y} = F_B - mg$, $I_{\mathrm{cm}}\ddot{\theta} = (\ell/2)(F_A\sin\theta - F_B\cos\theta)$

Eliminating F_A and F_B, and expressing x and y in terms of θ, we obtain

$$I_{\mathrm{cm}}\ddot{\theta} = -(\ell mg/2)\cos\theta - (m\ell^2/4)\ddot{\theta} \quad ,$$

$$I\ddot{\theta} = (-\ell mg/2)\cos\theta \quad .$$

Puzzling? No, A is not a fixed point. Energy conservation gives

$$\frac{I\dot{\theta}^2}{2} + \frac{\ell mg\ \sin\theta}{2} = \frac{\ell mg\ \sin\theta_0}{2} \quad ,$$

$$I\dot{\theta}^2 + \ell mg(\sin\theta - \sin\theta_0) = 0 \quad .$$

$F_A = 0$ when $\ddot{x} = 0$, $\sin\theta\ \ddot{\theta} + \cos\theta\ \dot{\theta}^2 = 0$,

$$\cos\theta\ (-3\sin\theta + 2\sin\theta_0) = 0 \quad .$$

Either $\theta = \pi/2$ (trivial unstable equilibrium) or $\theta = \sin^{-1}((2/3)\sin\theta_0)$.

(ii)

$$L = [m(\dot{x}^2 + \dot{y}^2) + I_{\mathrm{cm}}\dot{\theta}^2]/2 \ - mgy$$

Constraints:

$$f_A = -x + \frac{\ell}{2}\cos\theta = 0 \quad , \quad f_B = -y + \frac{\ell}{2}\sin\theta = 0$$

The Lagrange equation

$$\frac{\mathrm{d}}{\mathrm{d}t}\left(\frac{\partial L}{\partial\dot{x}}\right) - \frac{\partial L}{\partial x} + \lambda_A\frac{\partial f_A}{\partial x} + \lambda_B\frac{\partial f_B}{\partial x} = 0 \quad ,$$

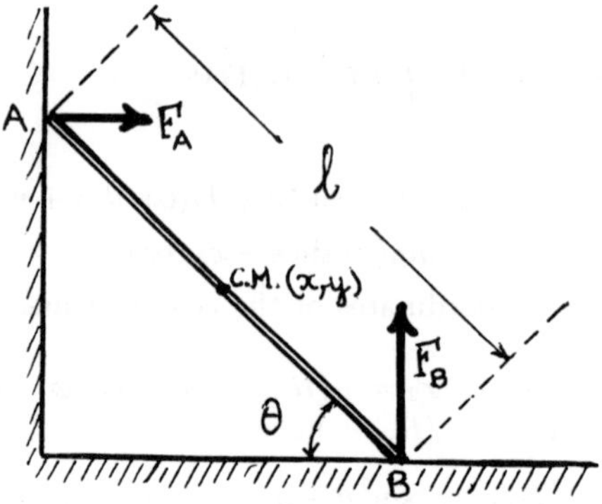

Figure 6.8: Plank, problem 6.2

another with $x \to y$, and another with $x \to \theta$, give

$$m\ddot{x} - \lambda_A = 0 \quad , \quad m\ddot{y} + mg - \lambda_B = 0 \quad ,$$

$$I_{\rm cm}\ddot{\theta} = \ell(\lambda_A \sin\theta - \lambda_B \cos\theta)/2 \quad .$$

In the last equation we substitute for λ_A and λ_B their expressions in terms of θ,

$$\lambda_A = m\ddot{x} = -m\ell(\sin\theta\,\ddot{\theta} + \cos\theta\,\dot{\theta}^2)/2 \quad ,$$

$$\lambda_B = mg + m\ddot{y} = mg + m\ell(\cos\theta\,\ddot{\theta} - \sin\theta\,\dot{\theta}^2)/2 \quad ,$$

obtaining the same equation as in the elementary treatment. λ_A and λ_B coincide with F_A and F_B of the elementary treatment.

S6.3 (i)

$$\mathbf{E} = a\hat{\mathbf{e}}_\rho/\rho,\ \mathbf{B} = B\hat{\mathbf{e}}_z,\ \mathbf{F} = q[a\hat{\mathbf{e}}_\rho/\rho + B(-\dot{\rho}\hat{\mathbf{e}}_\phi + \rho\dot{\phi}\hat{\mathbf{e}}_\rho)/c]$$

(ii)

$$m(\ddot{\rho} - \rho\dot{\phi}^2) = q\left(\frac{a}{\rho} + \frac{B\rho\dot{\phi}}{c}\right) \quad ,$$

$$m(\rho\ddot{\phi} + 2\dot{\rho}\dot{\phi}) = \frac{m}{\rho}\frac{\mathrm{d}}{\mathrm{d}t}(\rho^2\dot{\phi}) = -\frac{qB}{c}\dot{\rho} \quad ,$$

$$m\ddot{z} = 0 \quad .$$

(iii)

$$L = m(\dot{\rho}^2 + \rho^2\dot{\phi}^2 + \dot{z}^2)/2 + q[a\ln\rho + B\rho^2\dot{\phi}/2c]$$

(iv) Yes. The Lagrangian does not depend on ϕ and z

(v)

$$m\rho^2\dot{\phi} + qB\rho^2/2c = l_3 + qB\rho^2/2c = \text{constant and } \dot{z} = \text{constant}$$

S6.4

(i) $\delta p_\phi = \epsilon_{ijk} n_i x_{\alpha j}\ \partial\delta L/\partial \dot{x}_{\alpha k}$, (ii) $\delta p_\phi = c^{-1} q_\alpha \epsilon_{ijk} n_i x_{\alpha j} A_k$

S6.5

With $x = \ell\ \sin\phi$, $y = a\ \cos(\Omega t) + \ell\ \cos\phi$ the Lagrangian

$$L = (\dot{x}^2 + \dot{y}^2)/2 + mg\ell\ \cos\phi$$

becomes

$$L = m\ell^2\dot{\phi}^2/2\ + ma\ell\Omega\dot{\phi}\ \sin(\Omega t)\sin\phi + mg\ell\ \cos\phi + a^2\Omega^2\sin^2(\Omega t)/2 \quad .$$

This can be replaced by

$$L' = m\ell^2\dot{\phi}^2/2\ + ma\ell\Omega^2\cos(\Omega t)\cos\phi + mg\ell\ \cos\phi$$

since $L - L' = \mathrm{d}\Lambda/\mathrm{d}t$ with $\Lambda = -ma\ell\Omega\ \sin(\Omega t)\cos\phi + a^2\Omega[2\Omega t - \sin(2\Omega t)]/8$.

The Lagrange equation for L' is $\ddot{\phi} + \ell^{-1}[g + a\Omega^2\cos(\Omega t)]\sin\phi = 0$.

S6.6

(i) $L = m\dot{x}^2/2\ - k(x - X - \ell)^2/2$

(ii)

$$x = x' + X,\ L = m(\dot{x}'^2 + \dot{X}^2 + 2\dot{X}\dot{x}')/2\ - k(x' - \ell)^2/2,$$
$$p' = \partial L/\partial \dot{x}' = m(\dot{x}' + \dot{X}),\ \dot{p}' = \partial L/\partial x',\ m\ddot{x}' = -k(x' - \ell) - m\ddot{X}.$$

(iii) Add to the Lagrangian the term $-\mathrm{d}(m\dot{X}x')/\mathrm{d}t = -m(\ddot{X}x' + \dot{X}\dot{x}')$. Then

$$L \to \bar{L} = m(\dot{x}'^2 + \dot{X}^2)/2\ - m\ddot{X}x' - k(x' - \ell)^2/2,$$

$\bar{p}' = \partial\bar{L}/\partial\dot{x}' = m\dot{x}'$, $\dot{\bar{p}}' = \partial\bar{L}/\partial x'$, giving the same equation as in (ii).

S6.7

$$(\partial L_c/\partial Q)\mathrm{d}Q + (\partial L_c/\partial\dot{Q})\mathrm{d}\dot{Q} + (\partial L_c/\partial t)\mathrm{d}t + \alpha[(\partial L/\partial q)\mathrm{d}q + (\partial L/\partial\dot{q})\mathrm{d}\dot{q} + (\partial L/\partial t)\mathrm{d}t]$$
$$= P\ \mathrm{d}\dot{Q} + \dot{Q}\ \mathrm{d}P + \alpha(p\ \mathrm{d}\dot{q} + \dot{q}\ \mathrm{d}p)$$

We find at once $\partial L_c/\partial t = -\alpha\partial L/\partial t$, $P = \partial L_c/\partial\dot{Q}$, and, using $p = \partial L/\partial\dot{q}$, we are left with

$$(\partial L_c/\partial Q)\mathrm{d}Q + \alpha(\partial L/\partial q)\mathrm{d}q = \dot{Q}\ \mathrm{d}P + \alpha\dot{q}\ \mathrm{d}p.$$

At this point, using $Q = p$, $P = \alpha q$, and $\dot{p} = \partial L/\partial q$, we have $\dot{P} = \partial L_c/\partial Q$.

(ii) (a) $$L_c = -\alpha(m\dot{q}^2 - kq^2)/2\ + \alpha q\dot{p} + \alpha p\dot{q}$$

Since $p = m\dot{q}$, $\dot{p} = m\ddot{q} = -kq$, $q = -\dot{p}/k$, this gives

$$L_c = \alpha\left[\frac{m\dot{q}^2}{2} - \frac{kq^2}{2}\right] = \alpha\left[\frac{p^2}{2m} - \frac{\dot{p}^2}{2k}\right] = \frac{\alpha}{2}\left[\frac{Q^2}{m} - \frac{\dot{Q}^2}{k}\right] \quad .$$

(b)

$$\frac{\mathrm{d}}{\mathrm{d}t}\left(\frac{\partial L_c}{\partial\dot{Q}}\right) = \frac{\partial L_c}{\partial Q} \quad , \quad -\frac{\mathrm{d}}{\mathrm{d}t}\left(\frac{\dot{Q}}{k}\right) = \frac{Q}{m} \quad ,$$

$$m\ddot{Q} = -kQ,\ m\ddot{p} = -kp,\ \mathrm{d}(m\ddot{q} + kq)/\mathrm{d}t = 0.$$

(iii) (a)

$$L_c == -\alpha(m\dot{\mathbf{q}}^2/2 \ + k/|\mathbf{q}|) + \alpha\mathbf{q}\cdot\dot{\mathbf{p}} + \alpha\mathbf{p}\cdot\dot{\mathbf{q}}$$

Using $\dot{\mathbf{p}} = -k\mathbf{q}/|\mathbf{q}|^3$ and $\mathbf{p} = m\dot{\mathbf{q}}$, we find

$$L_c = +\alpha\left(\frac{\mathbf{p}^2}{2m} - 2\sqrt{k|\dot{\mathbf{p}}|}\right) = -\alpha\left(\frac{\mathbf{Q}^2}{2m} - 2\sqrt{k|\dot{\mathbf{Q}}|}\right) \quad .$$

(b)

$$P_i = \partial L_c/\partial\dot{Q}_i = \alpha\sqrt{k}\dot{Q}_i/|\dot{\mathbf{Q}}|^{\frac{3}{2}} \text{ and } \dot{P}_i = \partial L_c/\partial Q_i,$$

give

$$\alpha\sqrt{k}\mathrm{d}(\dot{Q}_i/|\dot{\mathbf{Q}}|^{\frac{3}{2}})/\mathrm{d}t = -\alpha Q_i/m,\ \mathrm{d}(\dot{p}_i/|\dot{\mathbf{p}}|^{\frac{3}{2}}) = -p_i/m\sqrt{k}.$$

S6.8

(i)

$$\mathbf{A}'' = \mathbf{A}' + \nabla\Lambda \text{ with } \Lambda = Bxy.$$

In fact,

$$\mathbf{A}' + \nabla\Lambda = -By\hat{\mathbf{e}}_x + B(y\hat{\mathbf{e}}_x + x\hat{\mathbf{e}}_y) = Bx\hat{\mathbf{e}}_y = \mathbf{A}''$$

(ii)

$$L' = m(\dot{x}^2 + \dot{y}^2 + \dot{z}^2)/2 \ + qEy - (q/c)By\dot{x}$$

$$L'' = m(\dot{x}^2 + \dot{y}^2 + \dot{z}^2)/2 \ + qEy + (q/c)Bx\dot{y}$$

(iii)

$$L'' - L' = \mathrm{d}F/\mathrm{d}t \text{ with } F = (q/c)Bxy = (q/c)\Lambda$$

(iv)

$$\mathrm{d}(m\dot{x} - qBy/c)/\mathrm{d}t = 0 \ ,$$
$$\mathrm{d}(m\dot{y})/\mathrm{d}t = qE - qB\dot{x}/c \ ,$$
$$\mathrm{d}(m\dot{z})/\mathrm{d}t = 0 \ .$$

From the last, and the initial conditions, we have $z = 0$. From the first, and the initial conditions, $\dot{x} = (qB/mc)y$. Substituting in the second equation, we have

$$\ddot{y} + \omega^2 y = qE/m \quad \text{with} \quad \omega = |qB|/mc$$

$$y = (qE/m\omega^2)[1 - \cos(\omega t)],\ \dot{x} = (Ec/B)[1 - \cos(\omega t)],\ x = (cE/B)[t - \sin(\omega t)]/\omega].$$

(v)

$$m\ddot{x} = (q/c)B\dot{y} \ ,$$
$$\mathrm{d}(m\dot{y} + qBx/c)/\mathrm{d}t = qE \ ,$$
$$m\ddot{z} = 0 \ .$$

From the first and initial conditions, $m\dot{x} = qBy/c$. Using the in second, we have again $\ddot{y} + \omega^2 y = qE/m$ etc. as in (iv).

(vi) Assuming $E > 0$ and $B > 0$, a proton (charge $+e$) is at time t on a circle of center $(Ect/B,\ Em_pc^2/eB^2)$ and radius Em_pc^2/eB^2, while an electron will be on a circle of center $(Ect/B,\ -Em_ec^2/eB^2)$ and radius Em_ec^2/eB^2. No wonder! The proton's trajectory is the cycloid (see figure 6.9)

$$x = \ (Em_pc^2/eB^2)(\tau - \sin\tau) \ , \ y = \ (Em_pc^2/eB^2)(1 - \cos\tau) \ ,$$

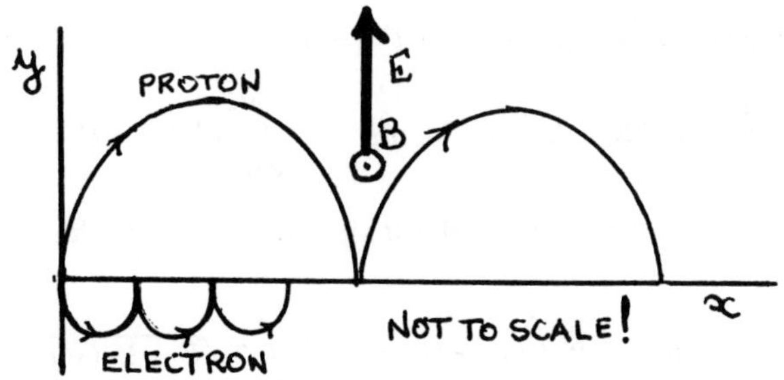

Figure 6.9: Electron and proton trajectories

while the electron's is the cycloid

$$x = (Em_ec^2/eB^2)(\tau - \sin\tau) \quad , \quad y = -(Em_ec^2/eB^2)(1 - \cos\tau) \quad .$$

From the equations and the initial conditions, we have

$$m(\dot{x}^2 + \dot{y}^2)/2 \ - qEy = \text{constant} \to 0 \quad ,$$

$$v_{\text{max}}^2 = (2E/m)|qy|_{\text{max}} \quad .$$

S6.9

(i) While $\delta\dot{q} = \mathrm{d}\delta q/\mathrm{d}t$, we have

$$\delta^\star\dot{q} = \lim_{\Delta t=0} \frac{q'(t+\Delta t) - q'(t)}{\Delta t \ (\mathrm{d}\bar{t}/\mathrm{d}t)} - \lim_{\Delta t=0} \frac{q(t+\Delta t) - q(t)}{\Delta t} \quad .$$

Since $\mathrm{d}\bar{t}/\mathrm{d}t \ = 1 + \mathrm{d}\delta t/\mathrm{d}t$, for δt infinitesimal this gives

$$\delta^\star\dot{q} = \lim_{\Delta t=0} \left[\frac{q'(t+\Delta t) - q'(t)}{\Delta t} \left(1 - \frac{\mathrm{d}\delta t}{\mathrm{d}t}\right) - \frac{q(t+\Delta t) - q(t)}{\Delta t} \right]$$

$$= \delta\dot{q} - \frac{\mathrm{d}\delta t}{\mathrm{d}t}\,\frac{\mathrm{d}q'(t)}{\mathrm{d}t} \simeq \delta\dot{q} - \dot{q}\frac{\mathrm{d}\delta t}{\mathrm{d}t} \quad .$$

(ii) For $K = c(q)\dot{q}^2$ one has

$$\delta^\star K = \frac{\mathrm{d}c}{\mathrm{d}q}\,\delta^\star q\,\dot{q}^2 + 2c\dot{q}\,\delta^\star\dot{q}$$

$$= \frac{\mathrm{d}c}{\mathrm{d}q}\,\delta q\,\dot{q}^2 + 2c\dot{q}\left(\delta\dot{q} - \dot{q}\,\frac{\mathrm{d}\delta t}{\mathrm{d}t}\right) = \delta K - 2K\,\frac{\mathrm{d}\delta t}{\mathrm{d}t} \quad .$$

On the other hand, $\delta U = \delta^\star U$. Therefore

$$0 = \int_{t_1}^{t_2} (\delta K - \delta U)\mathrm{d}t = \int_{t_1}^{t_2} (\delta^\star K + 2K\ \mathrm{d}\delta t/\mathrm{d}t \ - \delta^\star U)\ \mathrm{d}t \quad .$$

(iii) For isoenergetic variations, $\delta^*(K+U)=0$, $\delta^* U = -\delta^* K$,

$$0=\int_{t_1}^{t_2}(2\,\delta^* K + 2K\;\mathrm{d}\delta t/\mathrm{d}t)\mathrm{d}t \quad , \quad \int_{t_1}^{t_2}\delta^*(2K\;\mathrm{d}t)=0 \quad ,$$

since $\delta^*\mathrm{d}t = \bar{t}(t+\mathrm{d}t)-\bar{t}(t)-\mathrm{d}t = (\mathrm{d}\bar{t}/\mathrm{d}t)\mathrm{d}t - \mathrm{d}t = (\mathrm{d}\delta t/\mathrm{d}t)\mathrm{d}t$.
Then

$$0=\delta^*\int 2K\;\mathrm{d}t = \delta^* 2\int (E-U)\mathrm{d}t = \delta^*\int 2\sqrt{E-U}\,(\sqrt{E-U}\;\mathrm{d}t)$$

$$=\delta^*\int 2\sqrt{E-U}\;\sqrt{m/2}\;v\;\mathrm{d}t = \delta^*\int\sqrt{2m(E-U)}\;\mathrm{d}s \quad .$$

In chapter 1, equation (1.37), we wrote δ instead of δ^* in order to keep the notation simple at that early stage.

S6.10

$$\int_{t_1}^{t_2}\left(\frac{\partial L'}{\partial q_i}\,\delta q_i + \frac{\partial L'}{\partial \dot{q}_i}\,\delta\dot{q}_i + \frac{\partial L'}{\partial\lambda}\,\delta\lambda\right)\mathrm{d}t \;=0 \quad ,$$

$$\frac{\mathrm{d}}{\mathrm{d}t}\left(\frac{\partial L'}{\partial\dot{q}_i}\right)-\frac{\partial L'}{\partial q_i}\;=0 \quad , \quad \frac{\partial L'}{\partial\lambda}\;=0 \quad .$$

The second equation is nothing but $F(q,\dot{q},t)=0$. The first gives

$$\frac{\mathrm{d}}{\mathrm{d}t}\left(\frac{\partial L}{\partial\dot{q}_i}\right)-\frac{\partial L}{\partial q_i}+\frac{\mathrm{d}}{\mathrm{d}t}\left(\lambda\,\frac{\partial F}{\partial\dot{q}_i}\right)-\lambda\,\frac{\partial F}{\partial q_i}\;=0 \quad .$$

For the simple case $F(q,t)=0$, we have

$$\frac{\mathrm{d}}{\mathrm{d}t}\left(\frac{\partial L}{\partial\dot{q}_i}\right)-\frac{\partial L}{\partial q_i}\;=\lambda\,\frac{\partial F}{\partial q_i}$$

as expected. The generalization to the case of r constraints $F_\alpha(q,\dot{q},t)=0$ $(\alpha=1,\dots r)$ is trivial.

S6.11 One has

$$\frac{\mathrm{d}}{\mathrm{d}t}\left(\frac{\partial L}{\partial\dot{q}_i}\right)-\frac{\partial L}{\partial q_i}\;+\dot{\lambda}a_{ij}\dot{q}_j+\lambda a_{ij}\ddot{q}_j=0 \quad ,$$

$$a_{ij}\dot{q}_i\dot{q}_j=b \quad , \quad a_{ij}\dot{q}_i\ddot{q}_j=0 \quad .$$

Multiply the first equation by $\dot{q}_i$ and integrate with respect to t finding

$$b\lambda\;=\int\frac{\partial L}{\partial q_i}\;\dot{q}_i\mathrm{d}t\;-\int\dot{q}_i\frac{\mathrm{d}}{\mathrm{d}t}\left(\frac{\partial L}{\partial\dot{q}_i}\right)\mathrm{d}t \quad .$$

Since in the present case $\mathrm{d}L/\mathrm{d}t = (\partial L/\partial q_i)\dot{q}_i + (\partial L/\partial \dot{q}_i)\ddot{q}_i$, we have

$$b\lambda = \int \frac{\mathrm{d}L}{\mathrm{d}t}\,\mathrm{d}t - \int \frac{\partial L}{\partial \dot{q}_i}\,\ddot{q}_i\mathrm{d}t - \int \dot{q}_i \frac{\mathrm{d}}{\mathrm{d}t}\left(\frac{\partial L}{\partial q_i}\right)\mathrm{d}t$$

$$= L - \int \frac{\mathrm{d}}{\mathrm{d}t}\left(\dot{q}_i \frac{\partial L}{\partial \dot{q}_i}\right)\mathrm{d}t = L - \dot{q}_i \frac{\partial L}{\partial \dot{q}_i} \quad .$$

S6.12

(i) $$L = m(\dot{x}^2 + \dot{y}^2)/2 - mgy = m[1 + f'(x)^2]\dot{x}^2 - mgf(x)$$

and the Lagrange equation gives

$$\ddot{x} = -\frac{f'(x)(f''(x)\dot{x}^2 + g)}{1 + f'(x)^2} \quad .$$

(ii) $$\dot{y} = f'(x)\dot{x},\ f'(x)\,\mathrm{d}x - \mathrm{d}y = 0$$

$$m\ddot{x} - \lambda f'(x) = 0,\ m\ddot{y} + \lambda = -mg \text{ and } \ddot{y} - f'(x)\ddot{x} - f''(x)\dot{x}^2 = 0$$

$$\lambda = -m(\ddot{y} + g) = -m[f'(x)\ddot{x} + f''(x)\dot{x}^2 + g]$$

Use this expression for λ in $m\ddot{x} = \lambda f'(x)$ etc.

(iii) $m\ddot{x} = R_x$, $m\ddot{y} = R_y - mg$, and so $\lambda = -R_y$. One has also

$$|\mathbf{R}| = m|f''\dot{x}^2 + g|/\sqrt{1 + f'^2} \quad .$$

Special case:

$$f(x) = -\sqrt{\ell^2 - x^2}\ ,\ f'(x) = x/\sqrt{\ell^2 - x^2} = \tan\theta\ ,$$

$$f''(x) = \ell^2/(\ell^2 - x^2)^{\frac{3}{2}} = 1/\ell\cos^3\theta,\ \dot{x} = \ell\dot{\theta}\ \cos\theta\ ,$$

$$|\mathbf{R}| = m\left(\frac{1}{\ell\,\cos^3\theta}\,\ell^2\dot{\theta}^2\,\cos^2\theta + g\right)/\sqrt{1 + \frac{\sin^2\theta}{\cos^2\theta}} = m(g\,\cos\theta + \ell\dot{\theta}^2)\ ,$$

$$|\mathbf{R}| - mg\,\cos\theta = m\ell\dot{\theta}^2 = ma_c \quad .$$

S6.13

(i) A: acceleration of M, a: acceleration of m, $MA = T - F$, $ma = mg - 2T$, where F is the friction force and T is the tension. Since $a = A/2$, we write the second equation in the form $mA/4 = mg/2 - T$. Summing to the first equation we find

$$F = mg/2 - (M + m/4)A,\ T = m(g - A/2)/2.$$

(ii) If ℓ' is the length of the cord, $\ell' = d - x + 2y + \ell$, $y = (x + \ell' - \ell - d)/2$, $\dot{y} = \dot{x}/2$, $U = -mgy$,

$$L = M\dot{x}^2/2 + m\dot{y}^2/2 + mgy = (M + m/4)\dot{x}^2/2 + mg(x + \text{constant})/2$$

$$\delta W = -F\,\delta x,\ \mathrm{d}(\partial L/\partial \dot{x})/\mathrm{d}t - \partial L/\partial x = Q = -F$$

$$(M + m/4)\ddot{x} - mg/2 = -F,\ \ddot{x} = A, \text{ see (i).}$$

(iii) From the constraint we have $\dot{x} - 2\dot{y} = 0$ and $\ddot{x} - 2\ddot{y} = 0$, this last to be used together with the Lagrange equations

$$M\ddot{x} + \lambda + F = 0 \quad , \quad m\ddot{y} - mg - 2\lambda = 0 \quad .$$

Eliminating λ we find $(M + m/4)\ddot{x} - mg/2 + F = 0$. Clearly $\lambda = -T$.

S6.14

(i) $L = m(R_1 - R_2)^2\dot{\theta}_1^2/2 + (mR_2^2/2)\dot{\theta}_2^2/2 - mg(R_1 - R_2)(1 - \cos\theta_1)$

(ii)
$$L = (3m/4)(R_1 - R_2)^2\dot{\theta}_1^2 - mg(R_1 - R_2)(1 - \cos\theta_1),$$
$$(3m/2)(R_1 - R_2)^2\ddot{\theta}_1 = -mg(R_1 - R_2)\sin\theta_1$$

Frequency of small oscillations about lowest position:
$$\omega^2 = 2g/3(R_1 - R_2),\ T = 2\pi/\omega == \pi\sqrt{6(R_1 - R_2)/g}$$

(iii) $L = m(\dot{r}^2 + r^2\dot{\theta}_1^2)/2 + (mR_2^2/2)\dot{\theta}_2^2/2 - mg(R_1 - r\cos\theta_1)$+constant

Constraints: $r = R_1 - R_2$, $(R_1 - R_2)\theta_1 = R_2\theta_2$

$$\begin{cases} m\ddot{r} = mr\dot{\theta}_1^2 + mg\cos\theta_1 + \lambda_1 \ , \\ \mathrm{d}(mr^2\dot{\theta}_1)/\mathrm{d}t = -mgr\sin\theta_1 + \lambda_2(R_1 - R_2) \ , \\ mR_2^2\ddot{\theta}_2/2 \ = -R_2\lambda_2 \ . \end{cases}$$

$-\lambda_1 = +N$ is the normal reaction. Since the first constraint implies $\ddot{r} = 0$, we find $N = mg\cos\theta_1 + m(R_1 - R_2)\dot{\theta}_1^2$.

λ_2, from the third equation, is the frictional force.

S6.15

(i) Since $\mathrm{d}\rho = (x\ \mathrm{d}x + y\ \mathrm{d}y)/\rho$, the constraint is $a_x\ \mathrm{d}x + a_y\ \mathrm{d}y + a_z\ \mathrm{d}z = 0$ with $a_x = \cos\alpha\ x/\rho$, $a_y = \cos\alpha\ y/\rho$, $a_z = -(\sin^2\alpha/\cos\alpha)z/\rho$.

If the equations of motions are written as $m\ddot{x} = -\lambda a_x$, $m\ddot{y} = -\lambda a_y$, and $m\ddot{z} = -\lambda a_z - mg$, then λ turns out to be positive (see (iii)).

(ii) By integrating $\rho\ \mathrm{d}\rho - \tan^2\alpha\ z\ \mathrm{d}z = 0$, one finds $\rho^2 - z^2\tan^2\alpha$ =constant.

(iii) For constant = 0 we have a cone $x^2 + y^2 - \tan^2\alpha\ z^2 = 0$. The unit vector cormal to the cone has components $n_x = \cos\alpha\ x/\rho$, $n_y = \cos\alpha\ y/\rho$, $n_z = -(\sin^2\alpha/\cos\alpha)z/\rho$. Then

$$m\ddot{\mathbf{r}}\cdot\hat{\mathbf{n}} = m[\cos\alpha(x\ddot{x} + y\ddot{y})/\rho - (\sin^2\alpha/\cos\alpha)z\ddot{z}/\rho]$$

$$= mg(\sin^2\alpha/\cos\alpha)z/\rho - \lambda\cos^2\alpha - \lambda(\sin^4\alpha/\cos^2\alpha)z^2/\rho^2 = mg\sin\alpha - \lambda \ ,$$

having used $\rho = \tan\alpha\ z$.

S6.16

x, y: coordinates of midpoint P, x_i, y_i: coordinates of point i ($i = 1, 2$)

ϕ_i: angle of rotation of i-th wheel, positive direction clockwise

$$\dot{x}_i = a\dot{\phi}_i\ \sin\theta,\ \dot{y}_i = -a\dot{\phi}_i\ \cos\theta$$

Check: If ϕ_i increases, the point of contact moves left to right, x_i increases, y_i decreases.

$$\dot{x} = \frac{\dot{x}_1 + \dot{x}_2}{2} = \frac{a}{2}\ \sin\theta\ (\dot{\phi}_1 + \dot{\phi}_2) \ ,$$

$$\dot{y} = \frac{\dot{y}_1 + \dot{y}_2}{2} = -\frac{a}{2}\ \cos\theta\ (\dot{\phi}_1 + \dot{\phi}_2) \ .$$

Hence $\mathrm{d}x = (a/2)\sin\theta\ (\mathrm{d}\phi_1 + \mathrm{d}\phi_2)$, $\mathrm{d}y = -(a/2)\cos\theta\ (\mathrm{d}\phi_1 + \mathrm{d}\phi_2)$, from which

$$\cos\theta\ \mathrm{d}x + \sin\theta\ \mathrm{d}y = 0 \ ,$$

$$\sin\theta\ \mathrm{d}x - \cos\theta\ \mathrm{d}y = a(\mathrm{d}\phi_1 + \mathrm{d}\phi_2)/2 \quad .$$

On the other hand, from $\theta = \tan^{-1}[(y_2 - y_1)/(x_2 - x_1)]$ one finds

$$\mathrm{d}\theta = \frac{(x_2 - x_1)(\mathrm{d}y_2 - \mathrm{d}y_1) - (y_2 - y_1)(\mathrm{d}x_2 - \mathrm{d}x_1)}{(x_2 - x_1)^2 + (y_2 - y_1)^2}$$

$$= -\frac{a}{b}\,(\mathrm{d}\phi_2 - \mathrm{d}\phi_1) = \frac{a}{b}\mathrm{d}(\phi_1 - \phi_2) \text{ integrable, holonomic} \quad .$$

All the above can be condensed as follows:

$z = x + \mathrm{i}y$, $z_i = x_i + \mathrm{i}y_i$, $u = \exp(\mathrm{i}\theta)$, $z_1 = z - bu/2$, $z_2 = z + bu/2$

Rolling without slipping of each wheel: $\mathrm{Im}(u\ \mathrm{d}\bar{z}_i) = a\ \mathrm{d}\phi_i$

Displacement of midpoint is normal to axle: $\mathrm{Re}(u\ \mathrm{d}\bar{z}) = 0$

Rotation of axle: $\mathrm{d}\theta = -\mathrm{Im}(u\ \mathrm{d}\bar{u}) = -\mathrm{Im}(u[\mathrm{d}\bar{z}_2 - \mathrm{d}\bar{z}_1]/b) = (a/b)\mathrm{d}(\phi_1 - \phi_2)$

S6.17

(i) Assume the existence of two integrating factors $f(x_1, x_2, \varphi, \psi)$ and $g(x_1, x_2, \varphi, \psi)$ for the two conditions. Write

$$f\ \mathrm{d}x_1 + 0 \times \mathrm{d}x_2 + 0 \times \mathrm{d}\varphi + fr\cos\varphi\ \mathrm{d}\psi = 0 \quad ,$$

$$0 \times \mathrm{d}x_1 + g\ \mathrm{d}x_2 + 0 \times \mathrm{d}\varphi + gr\sin\varphi\ \mathrm{d}\psi = 0 \quad .$$

Write "curl= 0" conditions: $\partial f/\partial x_2 = 0$, $\partial f/\partial\varphi = 0$, $\partial f/\partial\psi = r\cos\varphi\ \partial f/\partial x_1$, $\partial f/\partial\varphi\ r\cos\varphi - fr\sin\varphi = 0$.
The second and the fourth imply $f = 0$, except for the trivial case $\varphi = 0$, rolling along the x_1-axis.
Similarly
$\partial g/\partial x_1 = 0$, $\partial g/\partial\varphi = 0$, $\partial g/\partial\psi = r\sin\varphi\ \partial g/\partial x_2$, $\partial g/\partial\varphi\ r\sin\varphi + gr\cos\varphi = 0$.
The second and the fourth imply $g = 0$, except for the trivial case $\varphi = \pi/2$, rolling along the x_2-axis.

$$\text{Rolling along the } x_1\text{-axis: } \overbrace{\mathrm{d}x_1 + r\ \mathrm{d}\psi = 0}^{\text{integrable}},\ \overbrace{\mathrm{d}x_2}^{\text{zero}} + r\ \overbrace{\sin\varphi}^{\text{zero}}\ \mathrm{d}\psi = 0$$

$$\text{Rolling along the } x_2\text{-axis: } \underbrace{\mathrm{d}x_1}_{\text{zero}} + r\ \underbrace{\cos\varphi}_{\text{zero}}\ \mathrm{d}\psi = 0,\ \underbrace{\mathrm{d}x_2 + r\ \mathrm{d}\psi = 0}_{\text{integrable}}$$

S6.18 Using the expressions for the components of ω with respect to <u>fixed</u> Cartesian axes (see (5.35)), we have

$$L = I[\dot{\vartheta}^2 + \dot{\varphi}^2\sin^2\vartheta + (\dot{\psi} + \dot{\varphi}\ \cos\vartheta)^2]/2\ + m(\dot{x}_1^2 + \dot{x}_2^2)/2\ + qEx_1$$

and the constraints

$$\dot{x}_1 - r(\dot{\vartheta}\ \sin\varphi - \dot{\psi}\ \sin\vartheta\ \cos\varphi) = 0 \quad (C1) \quad ,$$

$$\dot{x}_2 + r(\dot{\vartheta}\ \cos\varphi + \dot{\psi}\ \sin\vartheta\ \sin\varphi) = 0 \quad (C2) \quad .$$

Hence the equations

$$m\ddot{x}_1 + \lambda_1 = qE \ \ (1) \ \ , m\ddot{x}_2 + \lambda_2 = 0 \ \ (2) \quad ,$$

$$I(\ddot{\vartheta} + \dot{\varphi}\dot{\psi}\ \sin\vartheta) - \lambda_1 r\ \sin\varphi + \lambda_2 r\ \cos\varphi = 0 \quad (3) \quad ,$$
$$I(\ddot{\psi} + \ddot{\varphi}\ \cos\vartheta - \dot{\varphi}\dot{\vartheta}\ \sin\vartheta) + \lambda_1 r\ \sin\vartheta\ \cos\varphi + \lambda_2 r\ \sin\vartheta\ \sin\varphi = 0 \quad (4) \quad ,$$
$$\mathrm{d}(\dot{\varphi} + \dot{\psi}\ \cos\vartheta)/\mathrm{d}t = 0 \quad (5) \quad ,$$

namely

$$\ddot{\varphi} + \ddot{\psi}\ \cos\vartheta - \dot{\psi}\dot{\vartheta}\ \sin\vartheta \quad (5') \quad .$$

Equation (5) tells us that $\omega_3 =$constant.
Multiplying (4) by $\cos\varphi$, and subtracting (3) multiplied by $\sin\vartheta\ \sin\varphi$, we have

$$\lambda_1 r\ \sin\vartheta + I(\cos\varphi\ \ddot{\psi} + \cos\vartheta\ \cos\varphi\ \ddot{\varphi} - \sin\vartheta\ \cos\varphi\ \dot{\varphi}\dot{\vartheta} - \sin\vartheta\ \sin\varphi\ \ddot{\vartheta} - \sin^2\vartheta\ \sin\varphi\ \dot{\varphi}\dot{\psi}) = 0$$

Subtracting from this the time derivative of (C1) multiplied by $I\ \sin\vartheta/r$, we have

$$\sin\vartheta\ (\lambda_1 r - I\ddot{x}_1/r) + I\ \cos\varphi\ \cos\vartheta\ (\ddot{\varphi} + \ddot{\psi}\ \cos\vartheta - \dot{\psi}\dot{\vartheta}\ \sin\vartheta) = 0 \quad ,$$

which by (5') gives

$$\lambda_1 = (I/r^2)\ddot{x}_1 \quad .$$

In a similar way we find $\lambda_2 = (I/r^2)\ddot{x}_2$.
Thus we have $(7m/5)\ddot{x}_1 = qE$, $\ddot{x}_2 = 0$, $\dot{\omega}_1 = 0$, and $\dot{\omega}_2 = 5qE/7mr$.

S6.19 With $\omega_1 = \dot{\alpha}$, $\omega_2 = \dot{\beta}$, $\omega_3 = \dot{\gamma}$, we have the Lagrangian

$$L = m(\dot{x}_1^2 + \dot{x}_2^2)/2 \ + I(\dot{\alpha}^2 + \dot{\beta}^2 + \dot{\gamma}^2)/2 + qEx_1$$

with the constraints

$$\dot{x}_2 + r\dot{\alpha} = 0 \quad , \quad \dot{x}_1 - r\dot{\beta} = 0 \quad .$$

The Lagrange equations with multipliers λ and μ read

$$m\ddot{x}_1 + \mu = qE \quad , \quad m\ddot{x}_2 + \lambda = 0 \quad ,$$
$$I\ddot{\alpha} + \lambda r = 0 \quad , \quad I\ddot{\beta} - \mu r = 0 \quad , \quad I\ddot{\gamma} = 0 \quad .$$

The first and the fourth give $m\ddot{x}_1 + (I/r)\ddot{\beta} = qE$, and this, using the second constraint, gives $(m + I/r^2)\ddot{x}_1 = qE$.
The second and the third give $m\ddot{x}_2 - (I/r)\ddot{\alpha} = 0$, and, using the first constraint, $\ddot{x}_2 = 0$. The last equation gives $\omega_3 = \dot{\gamma}$ =constant.

S6.20

With $\dot{\alpha} = \omega_1$, $\dot{\beta} = \omega_2$, $\dot{\gamma} = \omega_3$, we have

$$L = m(b^2\dot{\phi}^2 + \dot{z}^2)/2 \; + I(\dot{\alpha}^2 + \dot{\beta}^2 + \dot{\gamma}^2)/2 \; - bmg(1 - \cos\phi) \quad ,$$

$$\begin{cases} \omega_\rho = \dot{\alpha}\, \cos\phi + \dot{\beta}\, \sin\phi \;, \\ \omega_\phi = -\dot{\alpha}\, \sin\phi + \dot{\beta}\, \cos\phi \;, \\ \omega_z = \dot{\gamma} \;. \end{cases}$$

Constraints:

$$b\dot{\phi} + a\dot{\gamma} = 0 \quad , \quad \dot{z} + a\dot{\alpha}\, \sin\phi - a\dot{\beta}\, \cos\phi = 0 \quad .$$

Lagrange equations:

$$mb^2\ddot{\phi} + b\lambda = -bmg\, \sin\phi \quad , \quad m\ddot{z} + \mu = 0 \quad ,$$

$$I\ddot{\alpha} + \mu a\, \sin\phi = 0 \; , \; I\ddot{\beta} - \mu a\, \cos\phi = 0 \; , \; I\ddot{\gamma} + \lambda a = 0 \quad .$$

Time derivative of constraints:

$$b\ddot{\phi} + a\ddot{\gamma} = 0 \; , \; \ddot{z} + a\ddot{\alpha}\, \sin\phi - a\ddot{\beta}\, \cos\phi + a\dot{\alpha}\dot{\phi}\, \cos\phi + a\dot{\beta}\dot{\phi}\, \sin\phi = 0 \quad .$$

We have

$$\lambda a = -I\ddot{\gamma} = (bI/a)\ddot{\phi} \;,$$

$$b[m\ddot{\phi} + (I/a^2)\ddot{\phi}] = -mg\, \sin\phi \;,$$

$$\mu a = -I(\ddot{\alpha}\, \sin\phi - \ddot{\beta}\, \cos\phi) = I(\ddot{z}/a \; + \dot{\alpha}\dot{\phi}\, \cos\phi + \dot{\beta}\dot{\phi}\, \sin\phi) \;,$$

$$m\ddot{z} + (I/a)(\ddot{z}/a \; + \dot{\alpha}\dot{\phi}\, \cos\phi + \dot{\beta}\dot{\phi}\, \sin\phi) = 0 \;,$$

$$(m + I/a^2)\ddot{z} = -(I/a)\dot{\phi}\omega_\rho \;.$$

From the third and fourth Lagrange equations we have $\ddot{\alpha}\, \cos\phi + \ddot{\beta}\, \sin\phi = 0$. Hence

$$\dot{\omega}_\rho = (-\dot{\alpha}\, \sin\phi + \dot{\beta}\, \cos\phi)\dot{\phi} = \dot{z}\dot{\phi}/a \quad .$$

S6.21

$$L = m(\dot{x}_1^2 + \dot{x}_2^2 + \dot{x}_3^2)/2 \; - mgx_3 + [I_1(\dot{\varphi}^2\, \sin^2\vartheta + \dot{\vartheta}^2) + I_3(\dot{\psi} + \dot{\varphi}\, \cos\vartheta)^2]/2$$

Constraints: $x_3 = r\, \sin\vartheta$ (holonomic), and the nonholonomic

$$\dot{x}_1 = -r(\hat{\mathbf{b}}_2 \times \omega) \cdot \hat{\mathbf{e}}_1 = -r(\hat{\mathbf{e}}_1 \times \hat{\mathbf{b}}_2) \cdot \omega$$

$$= -r([\hat{\mathbf{b}}_1\, \cos\varphi + \hat{\mathbf{b}}_3\, \sin\vartheta\, \sin\varphi] \times \hat{\mathbf{b}}_2) \cdot \omega$$

$$= -r(\hat{\mathbf{b}}_3\, \cos\varphi - \hat{\mathbf{b}}_1\, \sin\vartheta\, \sin\varphi) \cdot \omega$$

$$= -r(\Omega_3\, \cos\varphi - \Omega_1\, \sin\vartheta\, \sin\varphi)$$

and similarly

$$\dot{x}_2 = -r(\Omega_3\, \sin\varphi + \Omega_1\, \sin\vartheta\, \cos\varphi) \;,$$

where $\Omega_i = \omega \cdot \hat{\mathbf{b}}_i$ (see problem 5.8).
Since $\Omega_1 = \dot{\vartheta}$ and $\Omega_3 = \dot{\psi} + \dot{\varphi}\ \cos\vartheta$, we have the constraints

$$\mathrm{d}x_1 + r\ \cos\varphi\ \mathrm{d}\psi + r\ \cos\varphi\ \cos\vartheta\ \mathrm{d}\varphi - r\ \sin\vartheta\ \sin\varphi\ \mathrm{d}\vartheta = 0 \quad ,$$
$$\mathrm{d}x_2 + r\ \sin\varphi\ \mathrm{d}\psi + r\ \sin\varphi\ \cos\vartheta\ \mathrm{d}\varphi + r\ \sin\vartheta\ \cos\varphi\ \mathrm{d}\vartheta = 0 \quad ,$$

for the Lagrangian

$$L = m(\dot{x}_1^2 + \dot{x}_2^2 + r^2\dot{\vartheta}^2\ \cos^2\vartheta)/2\ - mgr\ \sin\vartheta$$
$$+[I_1(\dot{\varphi}^2\ \sin^2\vartheta + \dot{\vartheta}^2) + I_3(\dot{\psi} + \dot{\varphi}\ \cos\vartheta)^2]/2 \quad .$$

The Lagrange equations

$$\frac{\mathrm{d}}{\mathrm{d}t}\left(\frac{\partial L}{\partial \dot{x}_1}\right) - \frac{\partial L}{\partial x_1} + \lambda_1 = 0 \quad ,$$
$$\frac{\mathrm{d}}{\mathrm{d}t}\left(\frac{\partial L}{\partial \dot{x}_2}\right) - \frac{\partial L}{\partial x_2} + \lambda_2 = 0 \quad ,$$
$$\frac{\mathrm{d}}{\mathrm{d}t}\left(\frac{\partial L}{\partial \dot{\vartheta}}\right) - \frac{\partial L}{\partial \vartheta} - \lambda_1 r\ \sin\vartheta\ \sin\varphi + \lambda_2 r\ \sin\vartheta\ \cos\varphi = 0 \quad ,$$
$$\frac{\mathrm{d}}{\mathrm{d}t}\left(\frac{\partial L}{\partial \dot{\varphi}}\right) - \frac{\partial L}{\partial \varphi} + \lambda_1 r\ \cos\vartheta\ \cos\varphi + \lambda_2 r\ \cos\vartheta\ \sin\varphi = 0 \quad ,$$
$$\frac{\mathrm{d}}{\mathrm{d}t}\left(\frac{\partial L}{\partial \dot{\psi}}\right) - \frac{\partial L}{\partial \psi} + \lambda_1 r\ \cos\varphi + \lambda_2 r\ \sin\varphi = 0 \quad ,$$

give

$$m\ddot{x}_1 + \lambda_1 = 0 \ \ (1), \quad m\ddot{x}_2 + \lambda_2 = 0 \ \ (2),$$
$$(I_1 + mr^2\ \cos^2\vartheta)\ddot{\vartheta} - 2mr^2\dot{\vartheta}^2\ \sin\vartheta\ \cos\vartheta + mgr\ \cos\vartheta$$
$$-I_1\dot{\varphi}^2\ \sin\vartheta\ \cos\vartheta + I_3\omega_3'\dot{\varphi}\ \sin\vartheta + r\ \sin\vartheta(-\lambda_1\ \sin\varphi + \lambda_2\ \cos\varphi) = 0 \ \ (3),$$
$$I_1\ \sin\vartheta(\ddot{\varphi}\ \sin\vartheta + 2\dot{\varphi}\dot{\vartheta}\ \cos\vartheta) + I_3\dot{\omega}_3'\ \cos\vartheta$$
$$-I_3\omega_3'\dot{\vartheta}\ \sin\vartheta + r\ \cos\vartheta(\lambda_1\ \cos\varphi + \lambda_2\ \sin\varphi) = 0 \ \ (4),$$
$$I_3\dot{\omega}_3' + \lambda_1 r\ \cos\varphi + \lambda_2 r\ \sin\varphi = 0 \ \ (5).$$

Differentiating

$$\dot{x}_1 = -r\ \cos\varphi(\dot{\psi} + \dot{\varphi}\ \cos\vartheta) + r\dot{\vartheta}\ \sin\vartheta\ \sin\varphi \quad ,$$
$$\dot{x}_2 = -r\ \sin\varphi(\dot{\psi} + \dot{\varphi}\cos\vartheta) - r\dot{\vartheta}\ \sin\vartheta\ \cos\varphi \quad ,$$

we have

$$\lambda_1 = -m\ddot{x}_1 = mr(\ddot{\psi}\ \cos\varphi + \ddot{\varphi}\ \cos\varphi\ \cos\vartheta - \ddot{\vartheta}\ \sin\vartheta\ \sin\varphi$$
$$-\dot{\varphi}\dot{\psi}\ \sin\varphi - \dot{\varphi}^2\ \sin\varphi\ \cos\vartheta - \dot{\vartheta}\dot{\varphi}\ \cos\varphi\ \sin\vartheta - \dot{\vartheta}^2\ \cos\vartheta\ \sin\varphi - \dot{\varphi}\dot{\vartheta}\sin\vartheta\ \cos\varphi) \quad ,$$
$$\lambda_2 = -m\ddot{x}_2 = mr(\ddot{\psi}\ \sin\varphi + \ddot{\varphi}\ \sin\varphi\ \cos\vartheta + \ddot{\vartheta}\ \sin\vartheta\ \cos\varphi$$
$$+\dot{\varphi}\dot{\psi}\ \cos\varphi + \dot{\varphi}^2\ \cos\varphi\ \cos\vartheta - \dot{\vartheta}\dot{\varphi}\ \sin\varphi\ \sin\vartheta + \dot{\vartheta}^2\ \cos\vartheta\ \cos\varphi - \dot{\varphi}\dot{\vartheta}\sin\vartheta\ \sin\varphi) \quad ,$$

$r\ \sin\vartheta(-\lambda_1\ \sin\varphi + \lambda_2\ \cos\varphi) = mr^2\ \sin\vartheta(\ddot{\vartheta}\ \sin\vartheta + \dot{\varphi}\dot{\psi} + \dot{\varphi}^2\ \cos\vartheta + \dot{\vartheta}^2\ \cos\vartheta)$.
Therefore (3) becomes

$$(I_1 + mr^2)\ddot{\vartheta} + (I_3 + mr^2)\omega_3'\dot{\varphi}\ \sin\vartheta - I_1\dot{\varphi}^2\ \sin\vartheta\ \cos\vartheta + mgr\ \cos\vartheta = 0 \quad .$$

Similarly we have

$$r(\lambda_1\ \cos\varphi + \lambda_2\ \sin\varphi) = mr^2(\ddot{\psi} + \ddot{\varphi}\ \cos\vartheta - 2\dot{\vartheta}\dot{\varphi}\ \sin\vartheta) \quad ,$$

and so (5) becomes

$$(I_3 + mr^2)\dot{\omega}_3' = mr^2\dot{\varphi}\dot{\vartheta}\ \sin\vartheta \quad .$$

Finally, using (5) equation (4) becomes

$$I_3\omega_3'\dot{\vartheta} - 2I_1\dot{\varphi}\dot{\vartheta}\ \cos\vartheta - I_1\ddot{\varphi}\ \sin\vartheta = 0 \quad .$$

S6.22
$\delta L = \epsilon\ \mathrm{d}\Lambda/\mathrm{d}t\ + O(\epsilon^2)$ with

$$\Lambda = \frac{\mathrm{i}}{2}\left[\frac{\dot{q}}{\dot{q} - \mathrm{i}\omega q} - \mathrm{i}\omega t - \ln\left(\frac{\dot{q} - \mathrm{i}\omega q)}{C}\right)\right] \quad .$$

Hence

$$I = \epsilon^{-1}\frac{\partial L}{\partial \dot{q}}\ \delta q - \Lambda = \frac{\mathrm{i}}{2}\left[\mathrm{i}\omega t + \ln\left(\frac{\dot{q} - \mathrm{i}\omega q}{C}\right)\right]$$

is an integral of motion.
For $q(t) = A\ \cos(\omega t + \alpha)$, choosing $C = -\dot{q}(0) - \mathrm{i}\omega q(0)$, one has $I = \alpha$.

S6.23

$$\delta L = 48\ \epsilon\frac{\mathrm{d}}{\mathrm{d}t}\left(\frac{\dot{y}^3}{3} - \dot{x}^2\dot{y}\right) + O(\epsilon^2)$$

Hence

$$\begin{aligned} I &= \frac{\partial L}{\partial \dot{x}}\ \delta x + \frac{\partial L}{\partial \dot{y}}\ \delta y - 48\ \epsilon\left(\frac{\dot{y}^3}{3} - \dot{x}^2\dot{y}\right) \\ &= \epsilon[8\dot{y}(\dot{y}^2 - 3\dot{x}^2) + (\dot{y} + \sqrt{3}\ \dot{x})\exp(2(x - y\sqrt{3})) \\ &\quad +(\dot{y} - \sqrt{3}\ \dot{x})\exp(2(x + y\sqrt{3})) - 2\dot{y}\exp(-4x)] \end{aligned}$$

is an integral of motion. This integral of motion was found by Hénon.

In the limit of small displacements $I \simeq 12\epsilon(x\dot{y} - y\dot{x})$(angular momentum conservation), as one might expect since in the same limit

$$L \simeq (\dot{x}^2 + \dot{y}^2)/2\ - (x^2 + y^2)/2$$

is invariant under rotations.

S6.24 (i)

$$\mathrm{d}\delta x_k/\mathrm{dt} = (\eta/k)[-\dot{x}_i\dot{x}_k + \dot{\mathbf{r}}^2\delta_{ik} - 2x_i\ddot{x}_k + \mathbf{r}\cdot\ddot{\mathbf{r}}\ \delta_{ik} + x_k\ddot{x}_i]$$

Using the equations of motion, we have

$$\mathrm{d}\delta x_k/\mathrm{dt} = (\eta/k)[-\dot{x}_i\dot{x}_k + (\dot{\mathbf{r}}^2 - k/mr)\delta_{ik} + kx_ix_k/mr^3].$$

With $\delta\dot{x}_k \equiv \mathrm{d}\delta x_k/\mathrm{dt}$, we have

$$\delta L = m\dot{x}_k\delta\dot{x}_k + k\ \delta r^{-1} = (2\eta/r^3)[-r^2\dot{x}_i + \mathbf{r}\cdot\dot{\mathbf{r}}\ x_i] = \eta\ \mathrm{d}(-2x_i/r)/\mathrm{dt}.$$

(ii) Integral of motion:

$$\partial L/\partial\dot{x}_k\ \delta x_k - \Lambda = -2\eta[-x_i/r\ + mx_i|\dot{\mathbf{r}}|^2/k - m\mathbf{r}\cdot\dot{\mathbf{r}}\dot{x}_i/k] = -2\eta A_i$$

S6.25 (i)

$$m(\mathrm{d}x'/\mathrm{d}t')^2/2\ - U(x' - ut') = m(\mathrm{d}x/\mathrm{d}t)^2/2\ - U(x - ut)$$

(ii) Since $\mathrm{d}t' = \mathrm{d}t$, the action is also invariant. Using equation (6.31) with $f = 1$ and $g = 1$, we have

$$I = m\dot{x}(u - \dot{x}) + m\dot{x}^2/2\ - U(x - ut)$$
$$= m(u\dot{x} - \dot{x}^2)/2\ - U(x - ut) = \text{constant}\quad .$$

The reader who wishes to work from first principles may proceed as follows:

$$0 = \delta L = \partial L/\partial x\ \delta x + \partial L/\partial\dot{x}\ \delta\dot{x} + \partial L/\partial t\ \delta t = \partial L/\partial x\ \epsilon u + 0 - \partial U(x - ut)/\partial t\ \epsilon$$
$$= \partial L/\partial x\ \epsilon u - \epsilon\ \mathrm{d}U(x - ut)/\mathrm{dt}\ + \epsilon\dot{x}\ \partial U(x - ut)/\partial x$$

But $\partial U(x - ut)/\partial x = \partial(-L + m\dot{x}^2/2)/\partial x = -\partial L/\partial x = -\mathrm{d}(\partial L/\partial\dot{x})/\mathrm{dt}$, and so

$$0 = \delta L = \epsilon\mathrm{d}[um\dot{x} - U(x - ut) - m\dot{x}^2/2]/\mathrm{dt}\quad .$$

S6.26

$$I = \frac{\partial L}{\partial\dot{x}_{ai}}(f_{ai} - g\dot{x}_{ai}) + \frac{\partial L}{\partial\dot{x}_{bi}}(f_{bi} - g\dot{x}_{bi}) + Lg$$
$$= m_a\dot{x}_{ai}(f_{ai} - g\dot{x}_{ai}) + m_b\dot{x}_{bi}(f_{bi} - g\dot{x}_{bi})$$
$$+g[(m_a\dot{\mathbf{r}}_a^2 + m_b\dot{\mathbf{r}}_b^2)/2\ - U(|\mathbf{r}_a - \mathbf{r}_b|)]$$
$$= m_a\dot{x}_{ai}f_{ai} + m_b\dot{x}_{bi}f_{bi} - gE\quad ,$$

where E is the energy $E = K + U$.
Space translations: $f_{ai} = f_{bi} = 1,\ g = 0$

$$m_a\dot{x}_{ai} + m_b\dot{x}_{bi} = \text{constant}$$

(i-th component of total momentum conserved)
Time translations: $f_{ai} = f_{bi} = 0,\ g = -1$

$$E = (m_a\dot{\mathbf{r}}_a^2 + m_b\dot{\mathbf{r}}_b^2)/2\ + U(|\mathbf{r}_a - \mathbf{r}_b|) = \text{constant}$$

(energy conservation)
Space rotations: $f_{ai} = \epsilon_{ijk}x_{ak},\ f_{bi} = \epsilon_{ijk}x_{bk},\ g = 0$

$$\epsilon_{jki}(m_ax_{ak}\dot{x}_{ai} + m_bx_{bk}\dot{x}_{bi}) = \text{constant}$$

(j-th component of total angular momentum conserved)

S6.27

$$L = (m_a\dot{\mathbf{r}}_a^2 + m_b\dot{\mathbf{r}}_b^2)/2 \ - U(|\mathbf{r}_a - \mathbf{r}_b|)$$

Consider the transformation $\mathbf{r}'_a = \mathbf{r}_a + \epsilon\hat{\mathbf{n}}t$, $\mathbf{r}'_b = \mathbf{r}_b + \epsilon\hat{\mathbf{n}}t$, $t' = t + (\epsilon/E)(m_a\mathbf{r}_a + m_b\mathbf{r}_b)\cdot\hat{\mathbf{n}}$, where $E = K_a + K_b + U$ is the total energy, treated as a constant. We have

$$\frac{\mathrm{d}\mathbf{r}'_a}{\mathrm{d}t'} = \frac{\mathrm{d}\mathbf{r}_a + \epsilon\hat{\mathbf{n}}\mathrm{d}t}{\mathrm{d}t\,[1 + (\epsilon/E)(m_a\dot{\mathbf{r}}_a + m_b\dot{\mathbf{r}}_b)\cdot\hat{\mathbf{n}}]}$$
$$\simeq (\dot{\mathbf{r}}_a + \epsilon\hat{\mathbf{n}})(1 - (\epsilon/E)[m_a\dot{\mathbf{r}}_a + m_b\dot{\mathbf{r}}_b]\cdot\hat{\mathbf{n}})$$

(and similarly for b),

$$K'_a = \frac{m_a}{2}\left(\frac{\mathrm{d}\mathbf{r}'_a}{\mathrm{d}t'}\right)^2 \simeq \frac{m_a}{2}\left[\dot{\mathbf{r}}_a^2 + 2\epsilon\left(\hat{\mathbf{n}}\cdot\dot{\mathbf{r}}_a - \frac{\dot{\mathbf{r}}_a^2}{E}[m_a\dot{\mathbf{r}}_a + m_b\dot{\mathbf{r}}_b]\cdot\dot{\mathbf{n}}\right)\right] \quad ,$$

(and similarly for b)

$$K'_a + K'_b \simeq K_a + K_b + \epsilon[1 - 2(K_a + K_b)/E](m_a\dot{\mathbf{r}}_a + m_b\dot{\mathbf{r}}_b)\cdot\hat{\mathbf{n}} \quad ,$$

$$\delta L = (\epsilon/E)(E - 2K_a - 2K_b)(m_a\dot{\mathbf{r}}_a + m_b\dot{\mathbf{r}}_b)\cdot\hat{\mathbf{n}} \quad .$$

One sees at once that the action is invariant,

$$\delta L\ \mathrm{d}t + L\ \delta\mathrm{d}t = 0 \quad .$$

Hence the constant of motion

$$I = \left(\frac{\partial L}{\partial \dot{x}_{ai}} + \frac{\partial L}{\partial \dot{x}_{bi}}\right)\epsilon n_i t - \left(\frac{\partial L}{\partial \dot{x}_{ai}}\dot{x}_{ai} + \frac{\partial L}{\partial \dot{x}_{bi}}\dot{x}_{bi}\right) g + Lg$$
$$= \epsilon[(m_a\dot{\mathbf{r}}_a + m_b\dot{\mathbf{r}}_b)\cdot\hat{\mathbf{n}}t] - (2K_a + 2K_b - L)g$$
$$= \epsilon(M\mathbf{V}t - M\mathbf{R})\cdot\hat{\mathbf{n}}$$

S6.28

$$\delta L' = qE\epsilon - (qB/c)\epsilon\dot{x} = \epsilon\ \mathrm{d}\Lambda/\mathrm{d}t$$

with $\Lambda = qEt - (qB/c)x$. Then, according to equation (6.30),

$$\frac{\partial L'}{\partial \dot{y}} - \Lambda = m\dot{y} + qBx/c \ - qEt$$

is a constant of motion.

Chapter 7

HAMILTONIANS

The evolution of a system with n degrees of freedom is described as the motion of a point in a $2n$-dimensional space, governed by Hamilton's equations. These are of the first order in the time.
Canonical transformations are discussed at length.
The use of Cartan's differential forms in providing a concise description of mechanics is extolled with missionary zeal.

7.1 First look. 1

Using the notation of chapter 3, the Lagrangian

$$L = a_{ij}(q)\dot{q}_i\dot{q}_j/2 - U(q) \quad (a_{ji} = a_{ij}) \tag{7.1}$$

can be written more concisely in the form

$$L = \dot{\mathsf{q}}^{\mathrm{T}}\mathsf{A}\mathsf{q}/2 - U(q) \quad , \tag{7.2}$$

where the coefficients a_{ij} may be functions of the coordinates.

According to the program outlined in section 1.1, we wish to replace the Lagrange equations

$$\frac{\mathrm{d}}{\mathrm{d}t}(a_{kj}\dot{q}_j) = \frac{1}{2}\frac{\partial a_{ij}(q)}{\partial q_k}\dot{q}_i\dot{q}_j - \frac{\partial U(q)}{\partial q_k} \tag{7.3}$$

with first-order equations.

We note first that by inverting the equations

$$p_j = \frac{\partial L}{\partial \dot{q}_j} = a_{ji}(q)\dot{q}_i \tag{7.4}$$

we have

$$\dot{q}_i = a_{ij}^{-1} p_j \quad , \tag{7.5}$$

where a_{ij}^{-1} are the elements of the matrix A^{-1} [1].

Then we rewrite the Lagrange equations in the form

$$\dot{p}_k = \frac{1}{2}(a_{im}^{-1}p_m)\frac{\partial a_{ij}}{\partial q_k}(a_{jn}^{-1}p_n) - \frac{\partial U}{\partial q_k} = \frac{1}{2}\mathsf{p}^{\mathrm{T}}\mathsf{A}^{-1}\frac{\partial \mathsf{A}}{\partial q_k}\mathsf{A}^{-1}\mathsf{p} - \frac{\partial U}{\partial q_k} \quad ,$$

$$\dot{p}_k = -\frac{1}{2}\mathsf{p}^{\mathrm{T}}\frac{\partial \mathsf{A}^{-1}}{\partial q_k}\mathsf{p} - \frac{\partial U}{\partial q_k} \quad . \tag{7.6}$$

Defining the Hamiltonian

$$H(p,q) = \mathsf{p}^{\mathrm{T}}\mathsf{A}^{-1}\mathsf{p}/2 + U \tag{7.7}$$

we have

$$\dot{p}_k = -\frac{\partial H(p,q)}{\partial q_k} \quad . \tag{7.8}$$

Going back to equation (7.5), we see that it can be written in the form

$$\dot{q}_k = \frac{\partial H(p,q)}{\partial p_k} \quad . \tag{7.9}$$

Equations (7.8) and (7.9) are Hamilton's equations.

Example: $L = m\dot{\mathbf{r}}^2/2 + k/r$, $\mathsf{A} = m\mathsf{I}$, $\mathsf{A}^{-1} = m^{-1}\mathsf{I}$,

$$H = \frac{1}{2m}\mathbf{p}^2 - \frac{k}{r} \quad \text{with} \quad \mathbf{p} = m\dot{\mathbf{r}} \quad . \tag{7.10}$$

Hamilton's equations yield

$$\dot{p}_i = -k\frac{x_i}{r^3} \quad , \quad \dot{x}_i = \frac{1}{m}p_i \quad . \tag{7.11}$$

In spherical coordinates $L = m(\dot{r}^2 + r^2\dot{\theta}^2 + r^2\sin^2\theta\,\dot{\phi}^2)/2 + k/r$,

$$\mathsf{A} = \begin{pmatrix} m & 0 & 0 \\ 0 & mr^2 & 0 \\ 0 & 0 & mr^2\sin^2\theta \end{pmatrix} \quad , \quad \mathsf{A}^{-1} = \begin{pmatrix} 1/m & 0 & 0 \\ 0 & 1/mr^2 & 0 \\ 0 & 0 & 1/mr^2\sin^2\theta \end{pmatrix} \quad ,$$

$$H = \frac{1}{2m}\left(p_r^2 + \frac{1}{r^2}p_\theta^2 + \frac{1}{r^2\sin^2\theta}p_\phi^2\right) - \frac{k}{r} \quad , \tag{7.12}$$

$$\begin{pmatrix} p_r \\ p_\theta \\ p_\phi \end{pmatrix} = \mathsf{A}\begin{pmatrix} \dot{r} \\ \dot{\theta} \\ \dot{\phi} \end{pmatrix} \quad ,$$

---

[1] Of course, in general $a_{ij}^{-1} \neq (a_{ij})^{-1}$, although the equality may be true in some cases for some matrix elements.

$$p_r = m\dot{r} \quad , \quad p_\theta = mr^2\dot{\theta} \quad , \quad p_\phi = mr^2\sin^2\theta\ \dot{\phi} \quad . \tag{7.13}$$

Hamilton's equations

$$\dot{p}_r = -\partial H/\partial r \quad , \quad \dot{p}_\theta = -\partial H/\partial \theta \quad , \quad \dot{p}_\phi = -\partial H/\partial \phi \tag{7.14}$$

give [2]

$$m\ddot{r} = \frac{\mathbf{l}^2}{mr^3} - \frac{k}{r^2} \quad , \quad \frac{\mathrm{d}}{\mathrm{d}t}(mr^2\dot{\theta}) = \frac{\cos\theta}{mr^2\sin^3\theta}\, p_\phi^2 \quad , \quad \dot{p}_\phi = 0 \quad . \tag{7.15}$$

Since ϕ does not appear in the Hamiltonian - it is "ignorable" - the momentum conjugate to ϕ, p_ϕ, is a constant of motion.

Example:

The Lagrangian for the spinning top can be written in the form

$$L = \frac{1}{2}\begin{pmatrix} \dot{\vartheta} & \dot{\varphi} & \dot{\psi} \end{pmatrix} \mathsf{A} \begin{pmatrix} \dot{\vartheta} \\ \dot{\varphi} \\ \dot{\psi} \end{pmatrix} - \ell mg\cos\vartheta \quad , \tag{7.16}$$

where

$$\mathsf{A} = \begin{pmatrix} I_1 & 0 & 0 \\ 0 & I_1\sin^2\vartheta + I_3\cos^2\vartheta & I_3\cos\vartheta \\ 0 & I_3\cos\vartheta & I_3 \end{pmatrix} \quad .$$

The Hamiltonian will be

$$H = \frac{1}{2}\begin{pmatrix} p_\vartheta & p_\varphi & p_\psi \end{pmatrix} \mathsf{A}^{-1} \begin{pmatrix} p_\vartheta \\ p_\varphi \\ p_\psi \end{pmatrix} + \ell mg\cos\vartheta \tag{7.17}$$

with

$$\mathsf{A}^{-1} = \begin{pmatrix} 1/I_1 & 0 & 0 \\ 0 & 1/I_1\sin^2\vartheta & -\cos\vartheta/I_1\sin^2\vartheta \\ 0 & -\cos\vartheta/I_1\sin^2\vartheta & \cos^2\vartheta/I_1\sin^2\vartheta + 1/I_3 \end{pmatrix} \quad .$$

Therefore

$$H = \frac{1}{2I_1}p_\vartheta^2 + \frac{1}{2I_1\sin^2\vartheta}(p_\varphi - p_\psi\cos\vartheta)^2 + \frac{1}{2I_3}p_\psi^2 + \ell mg\cos\vartheta \quad . \tag{7.18}$$

Since φ and ψ are ignorable, both p_φ and p_ψ are constants of motion.
The Hamilton equation $\dot{p}_\vartheta = -\partial H/\partial\vartheta$ for $p_\vartheta = I_1\dot{\vartheta}$ gives

$$I_1\ddot{\vartheta} = -\frac{(p_\varphi - p_\psi\cos\vartheta)(p_\psi - p_\varphi\cos\vartheta)}{I_1\sin^3\vartheta} + \ell mg\sin\vartheta \tag{7.19}$$

---

[2] From the identity $\dot{\mathbf{r}}^2 = \dot{r}^2 + (\mathbf{r}\times\dot{\mathbf{r}})^2/r^2$ one can write the kinetic energy in the form $K = m\dot{r}^2/2 + \mathbf{l}^2/2mr^2 = (p_r^2 + \mathbf{l}^2/r^2)/2m$. Comparing with the expression for K as given by equation (7.12), one finds

$$\mathbf{l}^2 = p_\theta^2 + p_\phi^2/\sin^2\theta \quad .$$

or, with the notation of section 5.4 ($p_\varphi = I_1 b$, $p_\psi = I_1 a$),

$$\ddot{\vartheta} = -\frac{(b - a\cos\vartheta)(a - b\cos\vartheta)}{\sin^3\vartheta} + \frac{\ell m g \sin\vartheta}{I_1} \quad . \tag{7.20}$$

7.2 First look. 2

Canonical transformations

Suppose we are given a system of Hamilton's equations

$$\dot{p}_i = -\partial H/\partial q_i \quad , \quad \dot{q}_i = \partial H/\partial p_i \quad . \tag{7.21}$$

We express the "canonical" variables p and q in terms of new variables P and Q, $p = p(P,Q)$ and $q = q(P,Q)$, and define a new Hamiltonian $K(P,Q) = H(p(P,Q), q(P,Q))$. It is unlikely that the reader will confuse this K with a kinetic energy.

Will Hamilton's equations

$$\dot{P}_i = -\partial K/\partial Q_i \quad , \quad \dot{Q}_i = \partial K/\partial P_i \tag{7.22}$$

hold?

In general, they will not, unless the transformation $(p,q) \to (P,Q)$ is suitably restricted. Then it is a "canonical transformation".

In one dimension, let us express P and Q in the above equations in terms of p and q, and viceversa. Using the scalar transformation property of the Hamiltonian and Hamilton's equations for H, we obtain the system

$$\left(\frac{\partial P}{\partial q} + \frac{\partial p}{\partial Q}\right)\dot{q} + \left(\frac{\partial P}{\partial p} - \frac{\partial q}{\partial Q}\right)\dot{p} = 0 \quad ,$$

$$\left(\frac{\partial Q}{\partial q} - \frac{\partial p}{\partial P}\right)\dot{q} + \left(\frac{\partial Q}{\partial p} + \frac{\partial q}{\partial P}\right)\dot{p} = 0 \quad .$$

Regard this as a system of equations for $\dot{p}$ and $\dot{q}$. Requiring that the determinant of the coefficients be zero, we find

$$[Q,P]_{p,q} + [q,p]_{P,Q} = 2 \quad , \tag{7.23}$$

where

$$[Q,P]_{p,q} = \frac{\partial Q}{\partial q}\frac{\partial P}{\partial p} - \frac{\partial Q}{\partial p}\frac{\partial P}{\partial q}$$

and

$$[q,p]_{P,Q} = \frac{\partial q}{\partial Q}\frac{\partial p}{\partial P} - \frac{\partial q}{\partial P}\frac{\partial p}{\partial Q} \quad .$$

As we shall see later, these are "Poisson brackets". In the present one-dimensional case, they are equal to Jacobian determinants,

$$[Q,P]_{p,q} = \frac{\partial(Q,P)}{\partial(q,p)}$$

and

$$[q,p]_{P,Q} = \frac{\partial(q,p)}{\partial(Q,P)} = \left(\frac{\partial(Q,P)}{\partial(q,p)}\right)^{-1} = \left([Q,P]_{p,q}\right)^{-1} \quad .$$

Using this in equation (7.23), we have

$$\left([Q,P]_{p,q}\right)^2 - 2\,[Q,P]_{p,q} + 1 = 0$$

with the double root

$$[Q,P]_{p,q} = 1 \quad .$$

With the Hamiltonian transforming as a scalar, equations (7.22) follow from equations (7.21) if the Poisson bracket $[Q,P]_{p,q}$ is unity like that for the identity transformation $[q,p]_{p,q} = 1$.
The transformation $Q = p$, $P = -q$ is canonical, and so is also $Q = -p$, $P = q$.

Generalizing to a system with n degrees of freedom, define the Poisson bracket

$$[A,B]_{p,q} = \sum_{k=0}^{n}\left(\frac{\partial A}{\partial q_k}\frac{\partial B}{\partial p_k} - \frac{\partial A}{\partial p_k}\frac{\partial B}{\partial q_k}\right) \quad ,$$

$$[q_i,q_j]_{p,q} = 0 \quad , \quad [p_i,p_j]_{p,q} = 0 \quad ,$$

$$[q_i,p_j]_{p,q} = \delta_{ij} \quad .$$

A condition for the transformation $(p,q) \to (P,Q)$ to be canonical is

$$[Q_i,Q_j]_{p,q} = 0 \quad , \quad [P_i,P_j]_{p,q} = 0 \quad , \quad [Q_i,P_j]_{p,q} = \delta_{ij} \quad .$$

Let us consider a region "S" of the (p,q) plane bounded by a closed curve "c". If the transformation $(p,q) \to (P,Q)$ is canonical, we have

$$\int_S \mathrm{d}P\mathrm{d}Q = \int_S \frac{\partial(Q,P)}{\partial(q,p)}\mathrm{d}p\,\mathrm{d}q = \int_S [Q,P]_{p,q}\mathrm{d}p\,\mathrm{d}q = \int_S \mathrm{d}p\,\mathrm{d}q \quad ,$$

$$\oint_c (p\mathrm{d}q - P\mathrm{d}Q) = 0 \quad .$$

This last tells us that the integrand must be a perfect differential,

$$p\,\mathrm{d}q - P\,\mathrm{d}Q = \mathrm{d}G_1 \quad ,$$

where the "generating function" is a function of q and Q. Thus

$$p = \frac{\partial G_1}{\partial q} \quad , \quad P = -\frac{\partial G_1}{\partial Q} \quad . \tag{7.24}$$

In section 7.8 we shall encounter three more types of generating functions.

Example:

We check all the above on the harmonic oscillator, $H = (p^2/m + m\omega^2 q^2)/2$, Hamilton's equations $\dot{p} = -\partial H/\partial q = -m\omega^2 q$, $\dot{q} = \partial H/\partial p = p/m$.

Let $p = \sqrt{2m\omega P}\,\cos Q$, $q = \sqrt{2P/m\omega}\sin Q$, $Q = \tan^{-1}(m\omega q/p)$, $P = (p^2/m + m\omega^2 q^2)/2\omega = H/\omega$. This transformation is canonical since $[Q, P]_{p,q} = 1$ as is easy to verify.

Since the transformed Hamiltonian is $K = P\omega$, Hamilton's equations read

$$\dot{P} = -\partial K/\partial Q = 0 \quad \text{and} \quad \dot{Q} = \partial K/\partial P = \omega \quad .$$

These tell us that $P =$ constant, namely $H = E =$ constant, and $Q = \omega t + \alpha$, $q = A\,\sin(\omega t + \alpha)$, where $A = \sqrt{2P/m\omega} = \sqrt{2E/m\omega^2}$.

The generating function is

$$G_1 = (m\omega/2)q^2\,\text{cotan}Q \quad .$$

In fact, $\partial G_1/\partial q = m\omega q\,\text{cotan}Q = p$ and $\partial G_1/\partial Q = -P$.

In most textbooks, canonical transformations are introduced starting from the Lagrangian. If $Q_i = Q_i(q, t)$, the Lagrangians $L(q, \dot{q}, t)$ and

$$M(Q, \dot{Q}, t) = L(q, \dot{q}, t) + \mathrm{d}\Lambda(q, Q, t)/\mathrm{d}t$$

are equivalent. Differentiating we have

$$\frac{\partial M}{\partial Q_i}\delta Q_i + P_i\,\delta\dot{Q}_i + \frac{\partial M}{\partial t}\delta t = \frac{\partial L}{\partial q_i}\delta q_i + p_i\,\delta\dot{q}_i + \frac{\partial L}{\partial t}\delta t$$

$$+\frac{\mathrm{d}}{\mathrm{d}t}\left(\frac{\partial\Lambda}{\partial q_i}\delta q_i + \frac{\partial\Lambda}{\partial Q_i}\delta Q_i + \frac{\partial\Lambda}{\partial t}\delta t\right) \quad .$$

Equating coefficients of $\delta\dot{Q}_i$ and $\delta\dot{q}_i$, we find

$$P_i = \partial\Lambda/\partial Q_i \quad \text{and} \quad p_i = -\partial\Lambda/\partial q_i \quad .$$

Clearly $\Lambda = -G_1$, but owing to the time dependence of the coordinate transformation, the Hamiltonian does not transform as a scalar, but instead

$$K(P, Q) = P_i\dot{Q}_i - M = H(p, q) - \partial\Lambda/\partial t$$

as is easy to verify.

In section 6.4, equation (6.24), we saw that if $q = q(Q)$ then $p = (\partial Q/\partial q)P$ and $p\,\mathrm{d}q - P\,\mathrm{d}Q = 0$. The transformation $(p, q) \to (P, Q)$ is canonical since $[Q, P]_{p,q} = (\mathrm{d}Q/\mathrm{d}q)(\mathrm{d}q/\mathrm{d}Q) = 1$. In this case, however, G_1 would seem to be zero, and so equations 7.24 would give $p = 0$ and $P = 0$. However, write $G_1 = G_2 - PQ$ with $G_2 = PQ(q)$, $\mathrm{d}G_1 = \mathrm{d}G_2 - P\mathrm{d}Q - Q\mathrm{d}P$. Then $p\,\mathrm{d}q - P\,\mathrm{d}Q = \mathrm{d}G_1 = 0$ gives

$$p\,\mathrm{d}q + Q\,\mathrm{d}P = \mathrm{d}G_2 \quad .$$

This tells us that $p = \partial G_2/\partial q = P\,\partial Q/\partial q$ and $Q = \partial G_2/\partial P = Q(q)$ as we want.

Constants of motion

For a function $f(p(t), q(t))$ one has

$$\frac{\mathrm{d}f}{\mathrm{d}t} = \frac{\partial f}{\partial p}\dot{p} + \frac{\partial f}{\partial q}\dot{q} = -\frac{\partial f}{\partial p}\frac{\partial H}{\partial q} + \frac{\partial f}{\partial q}\frac{\partial H}{\partial p} \quad ,$$

$$\frac{\mathrm{d}f}{\mathrm{d}t} = [f, H]_{p.q} \quad ,$$

a formula valid also in more than one dimension. Hence $f(p(t), q(t))$ is a constant of motion if its Poisson bracket with the Hamiltonian is zero.

Before proceeding, the reader should prove that

$$[A, B + C] = [A, B] + [A, C] \quad , \quad [AB, C] = [A, C]B + A[B, C]$$

and

$$[f(A), B] = \frac{\mathrm{d}f(A)}{\mathrm{d}A}\,[A, B] \quad .$$

For brevity's sake, we omit the subscripts "p, q" in the Poisson brackets.

Using these formulae one finds

$$[x_i, l_j] = \epsilon_{ijk}x_k \ , \ [p_i, l_j] = \epsilon_{ijk}p_k \ , \ [l_i, l_j] = \epsilon_{ijk}l_k \ ,$$

where $l_i = \epsilon_{ijk}x_jp_k$ is the i-th component of the angular momentum. We prove the last one:

$$[l_i, l_j] = \epsilon_{iab}\epsilon_{jcd}[x_ap_b, x_cp_d] = \epsilon_{iab}\epsilon_{jcd}(\delta_{ad}x_cp_b - \delta_{bc}x_ap_b) = -\epsilon_{iba}\epsilon_{ajc}x_cp_b + \epsilon_{iab}\epsilon_{bjd}x_ap_d$$

$$= -(\delta_{ij}\delta_{bc} - \delta_{ic}\delta_{bj})x_cp_b + (\delta_{ij}\delta_{ad} - \delta_{id}\delta_{aj})x_ap_d = x_ip_j - x_jp_i = \epsilon_{ijk}l_k \quad .$$

The important "Jacobi identity" for three functions (A, B, C)

$$[[A, B], C] + [[B, C], A] + [[C, A], B] = 0$$

can be proved by work. In section 7.14 we shall present a neat proof learned from Arnold.

The following is an important application. Let A and B be constants of motion, so that

$$[A,H]=0 \quad , \quad [B,H]=0 \quad .$$

Then the Jacobi identity tells us that $[[A,B],H]=0$, and so $[A,B]$ is also a constant of motion.

Example: If the components l_i and l_j of the angular momentum of a particle are constants of motion (Hamiltonian invariant under rotations around x_i and x_j axes), so is also their Poisson bracket $[l_i,l_j]=\epsilon_{ijk}l_k$.

7.3 H as Legendre transform of L

We pulled out of a hat the expression (7.7) for the Hamiltonian corresponding to a Lagrangian which depends bilinearly on the $\dot{q}_i$'s. Some readers may ask for a prescription by which the Hamiltonian can be constructed from the Lagrangian.

The prescription is a simple Legendre transformation from the $\dot{q}_i$'s to the p_i's.

Let $A(x,z,t)$ and $B(y,z,t)$ two functions satisfying the relation

$$A(x,z,t)+B(y,z,t)=xy \quad . \tag{7.25}$$

Partial differentiation with respect to x and y gives

$$\frac{\partial A}{\partial x}=y \quad , \quad \frac{\partial B}{\partial y}=x \quad ,$$

and with respect to z and t

$$\frac{\partial B}{\partial z}=-\frac{\partial A}{\partial z} \quad , \frac{\partial B}{\partial t}=-\frac{\partial A}{\partial t} \quad .$$

Replace [3] (x,y,z,t) by $(\dot{q},p,q,t)$, A by L and B by H. We have at once

$$p=\frac{\partial L}{\partial \dot{q}} \ , \ \dot{q}=\frac{\partial H}{\partial p} \ , \ \frac{\partial H}{\partial q}=-\frac{\partial L}{\partial q}\rightarrow -\frac{\mathrm{d}}{\mathrm{d}t}\left(\frac{\partial L}{\partial \dot{q}}\right)=-\dot{p} \ , \ \frac{\partial H}{\partial t}=-\frac{\partial L}{\partial t} \ ,$$

where the arrow denotes equality by Lagrange's equation.

If there are several coordinates, then

$$H=p_i\dot{q}_i-L \quad , \tag{7.26}$$

[3] Omitting t, replacing (x,y,z) by (v=volume of environment, p=pressure, s=entropy), A by (u=internal energy) and B by ($-h$=minus enthalpy), we have $u-h=pv$, $\partial u/\partial v=p$, $\partial h/\partial p=-v$, $\partial h/\partial s=\partial u/\partial s(=T)$, as is known from Thermodynamics.

$$\dot{p}_i = -\frac{\partial H}{\partial q_i} \quad , \quad \dot{q}_i = \frac{\partial H}{\partial p_i} \quad , \quad \frac{\partial H}{\partial t} = -\frac{\partial L}{\partial t} \quad . \tag{7.27}$$

The reader should check that these formulae yield those in section 7.1.

Example:

Electron in time-dependent electromagnetic field.

$$L = m\dot{\mathbf{r}}^2/2 - (e/c)\dot{\mathbf{r}} \cdot \mathbf{A}(\mathbf{r}, t) + eV(\mathbf{r}, t),\ p_i = \partial L/\partial \dot{x}_i = m\dot{x}_i - (e/c)A_i,$$

$$H = p_i\dot{x}_i - L = p_i(p_i + (e/c)A_i)/m - (m/2)[(\mathbf{p} + (e/c)\mathbf{A})/m]^2$$
$$+(e/mc)\mathbf{A} \cdot (\mathbf{p} + e\mathbf{A}/c) - eV \ ,$$

$$H = \frac{1}{2m}[\mathbf{p} + (e/c)\mathbf{A}]^2 - eV \quad , \tag{7.28}$$

$$\dot{p}_i = -\frac{\partial H}{\partial x_i} = -\frac{1}{m}\left(p_j + \frac{e}{c}A_j\right)\frac{e}{c}\frac{\partial A_j}{\partial x_i} + e\frac{\partial V}{\partial x_i} \quad ,$$

$$\dot{x}_i = \frac{\partial H}{\partial p_i} = \frac{1}{m}\left(p_i + \frac{e}{c}A_i\right) \quad ,$$

$$\frac{\mathrm{d}}{\mathrm{d}t}\left(m\dot{x}_i - \frac{e}{c}A_i\right) = -\frac{e}{c}\dot{x}_j\frac{\partial A_j}{\partial x_i} + e\frac{\partial V}{\partial x_i} \quad ,$$

$$m\ddot{x}_i = \frac{e}{c}\left(\frac{\partial A_i}{\partial x_j} - \frac{\partial A_j}{\partial x_i}\right)\dot{x}_j + e\frac{\partial V}{\partial x_i} + \frac{e}{c}\frac{\partial A_i}{\partial t}$$
$$= -eE_i - (e/c)\epsilon_{ijk}\dot{x}_j B_k \quad ,$$
$$m\ddot{\mathbf{r}} = -e\mathbf{E} - (e/c)\dot{\mathbf{r}} \times \mathbf{B} \quad .$$

From the last equation we obtain

$$\frac{\mathrm{d}K}{\mathrm{d}t} = \frac{\mathrm{d}}{\mathrm{d}t}\left(\frac{m}{2}\dot{\mathbf{r}}^2\right) == -e\dot{\mathbf{r}} \cdot \mathbf{E} \quad , \tag{7.29}$$

where the right side is the rate of work done by the electric field.

We now have

$$\frac{\mathrm{d}H}{\mathrm{d}t} = \frac{\partial H}{\partial p_i}\dot{p}_i + \frac{\partial H}{\partial q_i}\dot{q}_i + \frac{\partial H}{\partial t} = \frac{\partial H}{\partial t} = -\frac{\partial L}{\partial t} \quad , \tag{7.30}$$

where we have used Hamilton's equations. Thus

$$\frac{\mathrm{d}H}{\mathrm{d}t} = \frac{e}{c}\dot{\mathbf{r}} \cdot \frac{\partial \mathbf{A}}{\partial t} - e\frac{\partial V}{\partial t} = e\dot{\mathbf{r}} \cdot (-\nabla V - \mathbf{E}) - e\frac{\partial V}{\partial t} = -e\dot{\mathbf{r}} \cdot \mathbf{E} - e\left(\frac{\partial V}{\partial t} + \frac{\partial V}{\partial x_i}\dot{x}_i\right) \quad ,$$

$$\frac{\mathrm{d}H}{\mathrm{d}t} = \frac{\mathrm{d}(K+U)}{\mathrm{d}t} \quad , \tag{7.31}$$

where $U = -eV$ is the potential energy.

The Legendre transform of the relativistic Lagrangian

$$L = -m_0c^2\sqrt{1 - (\dot{\mathbf{r}}/c)^2} - (e/c)A_j\dot{x}_j + eV$$

is the Hamiltonian

$$H = c\sqrt{m_0^2c^2 + (\mathbf{p} + e\mathbf{A}/c)^2} - eV \quad .$$

Example:
According to equation (6.23) (but omitting the primes), the Lagrangian for the motion of a particle in a rotating frame is

$$L = [m\dot{\mathbf{r}}^2 + m\omega^2(x_1^2 + x_2^2)]/2 + m\omega(x_1\dot{x}_2 - x_2\dot{x}_1) \quad .$$

Since $p_1 = m(\dot{x}_1 - \omega x_2)$, $p_2 = m(\dot{x}_2 + \omega x_1)$, $p_3 = m\dot{x}_3$, a straightforward calculation gives

$$H = \mathbf{p}^2/2m + \omega(p_1 x_2 - p_2 x_1)$$

or, expressing H in terms of the coordinates and their time derivatives,

$$H = m[\dot{\mathbf{r}}^2 - \omega^2(x_1^2 + x_2^2)]/2 \quad .$$

It is easy to verify that Hamilton's equations give $m\ddot{x}_1 = 2m\omega\dot{x}_2 + m\omega^2 x_1$ etc.

7.4 Liouville's theorem

The conservation of phase space volumes under Hamiltonian flow will be expressed in the next section in the language of forms as the invariance of one of Poincaré's integrals. However, we now wish to present the standard proof, which is a generalization of that for $n = 1$ given in the chapter 1.

In an N-dimensional space, let $\mathbf{x}(t) = (x_1(t), \ldots x_N(t))$ be the position vector of a point P at time t, $\mathbf{x}(t+\mathrm{d}t)$ the position vector at the time $t+\mathrm{d}t$ of P which has moved under a flow

$$\mathbf{x}(t+\mathrm{d}t) = \mathbf{x}(t) + \mathbf{A}(\mathbf{x}(t))\mathrm{d}t + \ldots$$

Let $D(t)$ be a region in the N-dimensional space at time t, and $D(t+\mathrm{d}t)$ the region occupied at the time $t+\mathrm{d}t$ by the points of $D(t)$ if they have moved under the flow.

The volumes $v(t)$ and $v(t+\mathrm{d}t)$ of $D(t)$ and $D(t+\mathrm{d}t)$ are related by the formula

$$v(t+\mathrm{d}t) = v(t) + \left(\int_{D(t)} \mathrm{div}\mathbf{A}\, \mathrm{d}x_1(t)\ldots\mathrm{d}x_N(t)\right)\mathrm{d}t + \ldots$$

Proof:

$$v(t+\mathrm{d}t) = \int_{D(t+\mathrm{d}t)} \mathrm{d}x_1(t+\mathrm{d}t)\ldots\mathrm{d}x_N(t+\mathrm{d}t)$$

$$= \int_{D(t)} \left|\frac{\partial x_i(t+\mathrm{d}t)}{\partial x_j(t)}\right| \mathrm{d}x_1(t)\ldots\mathrm{d}x_N(t) \quad .$$

But now to the first order in dt the Jacobian determinant is

$$\frac{\partial x_i(t+\mathrm{d}t)}{\partial x_j(t)} = \begin{vmatrix} 1+(\partial A_1/\partial x_1)\mathrm{d}t & (\partial A_1/\partial x_2)\mathrm{d}t & \cdots \\ (\partial A_2/\partial x_1)\mathrm{d}t & 1+(\partial A_2/\partial x_2)\mathrm{d}t & \cdots \\ \cdots & \cdots & \cdots \end{vmatrix}$$

$$= 1 + \left(\sum_i \frac{\partial A_i}{\partial x_i}\right) \mathrm{d}t + \ldots$$

If div$\mathbf{A} = 0$, then $v(t+\mathrm{d}t) = v(t) + O(\mathrm{d}t^2)$.

In phase space ($N = 2n$), $(x_1, \ldots x_N)$ correspond to $(q_1, \ldots q_n, p_1 \ldots p_n)$, $q_i(t+\mathrm{d}t) = q_i(t) + (\partial H/\partial p_i)\mathrm{d}t$, $p_i(t+\mathrm{d}t) = p_i(t) - (\partial H/\partial q_i)\mathrm{d}t$, $\mathbf{A}$ has components $(\partial H/\partial p_1, \ldots \partial H/\partial p_n, -\partial H/\partial q_1, \ldots - \partial H/\partial q_n)$. Thus

$$\mathrm{div}\mathbf{A} = \sum_i \left[\frac{\partial}{\partial q_i}\left(\frac{\partial H}{\partial p_i}\right) + \frac{\partial}{\partial p_i}\left(-\frac{\partial H}{\partial q_i}\right)\right] = 0 \quad .$$

Remark: The Lagrangian for a damped oscillator:

$$L = \mathrm{e}^{\gamma t} m(\dot{x}^2 - \omega^2 x^2)/2 \quad ,$$

$p = \partial L/\partial \dot{x} = m\dot{x}\exp(\gamma t)$, yields the Lagrange equation $\ddot{x} + \gamma\dot{x} + \omega^2 x = 0$. This can be replaced by the system of equations $\dot{x} = A_x$, $\dot{y} = A_y$ with $A_x = y$ and $A_y = -\gamma y - \omega^2 x$.

The area change in the (x, y) plane is determined by

$$\frac{\partial A_x}{\partial x} + \frac{\partial A_y}{\partial y} = -\gamma \quad .$$

On the other hand, working with the canonical variables p and x, since $\dot{x} = A'_x$, $\dot{p} = A'_p$ with $A'_x = (p/m)\exp(-\gamma t)$, $A'_p = -m\omega^2 x\exp(\gamma t)$, the area change in the (p, x) plane is determined by

$$\frac{\partial A'_x}{\partial x} + \frac{\partial A'_p}{\partial p} = 0 + 0 = 0 \quad .$$

There is no area shrinking in the (p, x) plane.

7.5 Cartan's vectors and forms

The purpose of this section is not to give a smattering of modern differential geometry[4], nor to pander to the mathematicians, but rather to introduce the reader to a notation they have invented, which is as useful in Hamiltonian mechanics as vectors are in General Physics.

We define a vector $\bar{U}$ as a set of n numbers, its components U^i, and write

$$\bar{U} = U^i \bar{e}_i \quad ,$$

the $\bar{e}_i$'s being basis vectors.

A 1-form $\tilde{\omega}(\bullet)$ is a linear function on vectors,

$$\tilde{\omega}(\bar{U}) = \text{number} \quad .$$

A 1-form can be expressed as a linear combination of basis 1-forms,

$$\tilde{\omega}(\bullet) = \omega_i \tilde{e}^i(\bullet) \quad .$$

Assuming the orthonormality condition

$$\tilde{e}^i(\bar{e}_j) = \delta^i_j \quad ,$$

and using the linearity of $\tilde{\omega}(\bullet)$, we have

$$\tilde{\omega}(\bar{U}) = \tilde{\omega}(U^i \bar{e}_i) = U^i \tilde{\omega}(\bar{e}_i) = U^i \omega_j \tilde{e}^j(\bar{e}_i) = \omega_i U^i \quad .$$

For example, a force $\tilde{F}$ is a 1-form, a displacement $\bar{D}$ is a vector,

$$\tilde{F}(\bar{D}) = \text{Work (number)} \quad .$$

If $\tilde{F} = F_i \tilde{e}^i$ and $\bar{D} = \delta x^i \, \bar{e}_i$, then

$$\text{Work} = F_i \, \delta x^i \quad .$$

So far so good. The reader may still doubt the usefulness of this notation (why a bar rather than an arrow for vectors?). A mild shock may be experienced when we say that a vector is defined as an operator acting on functions on a manifold, roughly speaking a space, whose points are labeled by coordinates $x^1, \dots x^n$.

---

[4]An excellent introduction for physicists is B.F. Schutz, *Geometrical methods of mathematical physics* (Cambridge University Press, 1980). See also W.L. Burke, *Applied Differential Geometry* (Cambridge University Press,1985), and M. Crampin and F.A.E. Pirani, *Applicable Differential Geometry* (Cambridge University Press,1986).

If $f(x^1, \dots x^n)$ is one such function, then

$$\bar{U} f = U^i \frac{\partial f}{\partial x^i} \text{ , and so } \bar{U} = U^i \frac{\partial}{\partial x^i} \text{ and } \bar{e}_i = \frac{\partial}{\partial x^i} \quad .$$

The following application of this definition of vector is specially useful. If $(x^1(t), \dots x^N(t))$ is a curve in our space, and $U^i = \mathrm{d}x^i/\mathrm{d}t$, then

$$\bar{U} f = \frac{\mathrm{d}x^i}{\mathrm{d}t} \frac{\partial f}{\partial x^i} = \frac{\mathrm{d}f}{\mathrm{d}t} \quad .$$

This is the rate of change of $f(x^1 \dots x^N)$ along the curve, taken at the point of coordinates $x^i(t)$.

The basis 1-form $\tilde{e}^i(\bullet)$ will now be written as

$$\tilde{e}^i(\bullet) = \tilde{d}x^i(\bullet) \quad ,$$

so that the condition $\tilde{e}^i(\bar{e}_j) = \delta^i_j$ will now read (greater shock?)

$$\tilde{d}x^i \left(\frac{\partial}{\partial x^j} \right) = \delta^i_j \quad .$$

A useful 1-form is the gradient

$$\tilde{d}h(\bullet) = \frac{\partial h}{\partial x^i} \, \tilde{d}x^i(\bullet) \quad ,$$

where h is a function on the manifold. The standard example is

h=height above sea level, (x^1, x^2) coordinates on a map etc.

If δx^i are the coordinate changes in an infinitesimal displacement, define the vector

$$\bar{\delta x} = \delta x^j \frac{\partial}{\partial x^j} \quad .$$

Then

$$\tilde{d}h(\bar{\delta x}) = \frac{\partial h}{\partial x^i} \delta x^j \tilde{d}x^i \left(\frac{\partial}{\partial x^j} \right) = \frac{\partial h}{\partial x^i} \delta x^i \quad ,$$

giving the change of h corresponding to the displacement.

What does this have to do with Hamiltonian mechanics? Let our manifold be the phase space with coordinates p_i, q_i.

Define the "Hamiltonian vector field"

$$\bar{U} = U_{p_i} \frac{\partial}{\partial p_i} + U_{q_i} \frac{\partial}{\partial q_i} \quad , \tag{7.32}$$

where

$$U_{p_i} = -\frac{\partial H}{\partial q_i} \quad \text{and} \quad U_{q_i} = \frac{\partial H}{\partial p_i} \quad .$$

Let $f(p,q)$ be a function on phase space. Then

$$\frac{\mathrm{d}f}{\mathrm{d}t} = \frac{\mathrm{d}p_i}{\mathrm{d}t}\frac{\partial f}{\partial p_i} + \frac{\mathrm{d}q_i}{\mathrm{d}t}\frac{\partial f}{\partial q_i}$$

$$= \left(-\frac{\partial H}{\partial q_i}\frac{\partial}{\partial p_i} + \frac{\partial H}{\partial p_i}\frac{\partial}{\partial q_i}\right) f = \bar{U} f \quad .$$

If $A(p,q)$ and $B(p,q)$ are functions in phase space, the expression

$$[A, B]_{p,q} = \sum_i \left(\frac{\partial A}{\partial q_i}\frac{\partial B}{\partial p_i} - \frac{\partial A}{\partial p_i}\frac{\partial B}{\partial q_i}\right)$$

is, as we know, a Poisson bracket. Clearly $\mathrm{d}f/\mathrm{d}t = \bar{U}f = [f, H]$. We already know that a function $f(p,q)$ is a constant of motion if its Poisson bracket with H is zero, $[f, H] = 0$.

We now proceed to define a 2-form as a function on pairs of vectors. For example, the product $(\tilde{dx}^i \otimes \tilde{dx}^j)(\bullet, \bullet)$ is a 2-form which, acting on two vectors $\bar{U}$ and $\bar{V}$ yields

$$(\tilde{dx}^i \otimes \tilde{dx}^j)(\bar{U}, \bar{V}) = (\tilde{dx}^i(\bar{U}))(\tilde{dx}^j(\bar{V})) = U^i V^j \quad .$$

We shall make use of a special kind of 2-forms, the "exterior" 2-forms $\tilde{\omega}^2$, which are antisymmetric,

$$\tilde{\omega}^2(\bar{V}, \bar{U}) = -\tilde{\omega}^2(\bar{U}, \bar{V}) \quad .$$

A typical example is

$$\tilde{\omega}^2 = \tilde{dx}^i \wedge \tilde{dx}^j \equiv \tilde{dx}^i \otimes \tilde{dx}^j - \tilde{dx}^j \otimes \tilde{dx}^i \quad ,$$

so that

$$\tilde{\omega}^2(\bar{U}, \bar{V}) = U^i V^j - U^j V^i \quad .$$

If $\bar{U}$ and $\bar{V}$ are vectors in the $x^1 x^2$ plane (figure 7.1), then

$$(\tilde{dx}^1 \wedge \tilde{dx}^2)(\bar{U}, \bar{V}) = U^1 V^2 - U^2 V^1 = \text{area of parallelogram} \quad .$$

The following in small print will not be needed until later.
Generalizing we define an "exterior" m-form as an antisymmetric function on m vectors,

$$\tilde{\omega}^m(\bar{U}_1 \ldots \bar{U}_j \ldots \bar{U}_i \ldots \bar{U}_m) = -\tilde{\omega}^m(\bar{U}_1 \ldots \bar{U}_i \ldots \bar{U}_j \ldots \bar{U}_m) \quad .$$

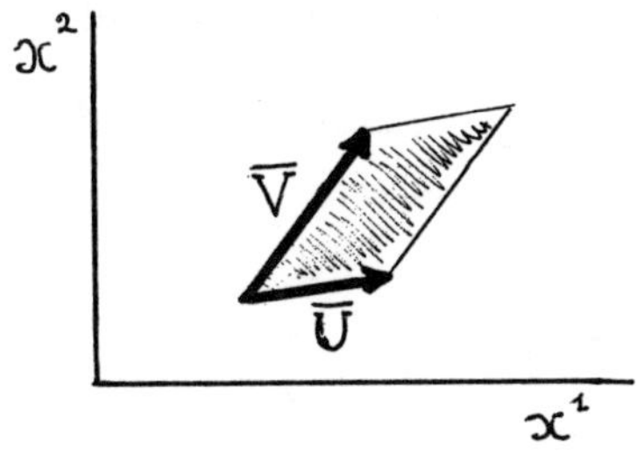

Figure 7.1: Area of parallelogram

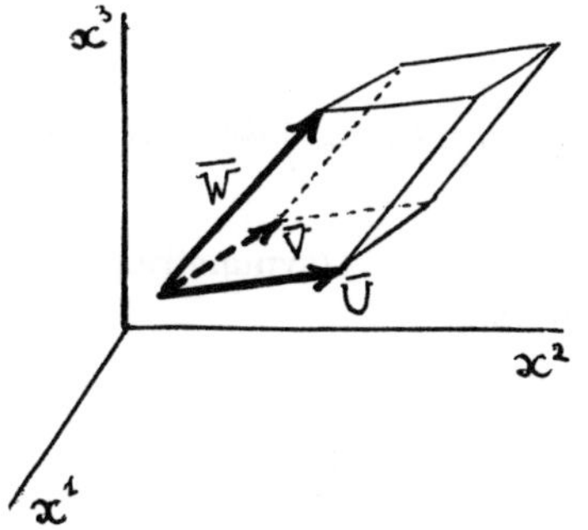

Figure 7.2: Volume of parallelepiped

Basis m-forms are

$$\tilde{d}x^{i_1} \wedge \tilde{d}x^{i_2} \ldots \wedge \tilde{d}x^{i_m} = \sum_P (-)^P \tilde{d}x^{i_1} \otimes \tilde{d}x^{i_2} \ldots \otimes \tilde{d}x^{i_m} \quad ,$$

where P denotes permutation of the indices.

Example: In 3-dimensional space

$$\tilde{d}x^1 \wedge \tilde{d}x^2 \wedge \tilde{d}x^3(\bar{U}, \bar{V}, \bar{W}) = \begin{vmatrix} U^1 & U^2 & U^3 \\ V^1 & V^2 & V^3 \\ W^1 & W^2 & W^3 \end{vmatrix} \quad .$$

This is the volume of the parallelepiped shown in figure 7.2.

In an n-dimensional space, $\tilde{\omega}^m = 0$ $(m > n)$. In fact, any basis form will have two or more indices with the same numerical value, and will vanish because of the antisymmetry.

The wedge product of two exterior forms, $\tilde{\omega}^r \wedge \tilde{\omega}^s$, is an exterior form with $r + s$ slots. For our purpose it is sufficient to say that
$\tilde{\omega}^r \wedge (a\tilde{\omega}^s + b\tilde{\omega}^t) = a\tilde{\omega}^r \wedge \tilde{\omega}^s + b\tilde{\omega}^r \wedge \tilde{\omega}^t$, and that the product of two basis forms is

$$(\tilde{d}x^{i_1} \wedge \ldots \wedge \tilde{d}x^{i_r}) \wedge (\tilde{d}x^{k_1} \wedge \ldots \wedge \tilde{d}x^{k_s}) = \tilde{d}x^{i_1} \wedge \ldots \wedge \tilde{d}x^{i_r} \wedge \tilde{d}x^{k_1} \wedge \ldots \wedge \tilde{d}x^{k_s} \quad .$$

A few words about the operator $\tilde{d}$. This is defined to act on an m-form

$$\tilde{\omega}^m = \sum_{i_1 \dots i_m} f_{i_1 \dots i_m} \tilde{d}x^{i_1} \wedge \dots \wedge \tilde{d}x^{i_m} \tag{7.33}$$

by generating the $(m+1)$-form

$$\tilde{d}\tilde{\omega}^m = \sum_{j i_1 \dots i_m} \left(\frac{\partial}{\partial x^j} f_{i_1 \dots i_m} \right) \tilde{d}x^j \wedge \tilde{d}x^{i_1} \wedge \tilde{d}x^{i_m} \quad . \tag{7.34}$$

Note that

$$\tilde{d}\tilde{d} = 0 \quad . \tag{7.35}$$

In fact

$$\tilde{d}\tilde{d}\tilde{\omega}^m = \sum_{i j i_1 \dots i_m} \left(\frac{\partial^2}{\partial x^i \partial x^j} f_{i_1 \dots i_m} \right) \tilde{d}x^i \wedge \tilde{d}x^j \wedge \tilde{d}x^{i_1} \dots \wedge \tilde{d}x^{i_m} = 0$$

because the double derivative indices (symmetric) are contracted with those of $\tilde{d}x^i \wedge \tilde{d}x^j \wedge \dots$ (antisymmetric).

We now define the 2-form

$$\tilde{\omega}^2 = \sum_i \tilde{d}p_i \wedge \tilde{d}q_i \tag{7.36}$$

operating on pairs of vectors in phase space. If $\bar{V}$ and $\bar{W}$ are two vectors, then

$$\tilde{\omega}^2(\bar{V}, \bar{W}) = \sum_i [(\tilde{d}p_i(\bar{V}))(\tilde{d}q_i(\bar{W})) - (\tilde{d}q_i(\bar{V}))(\tilde{d}p_i(\bar{W}))]$$

is the total area of the projections of the "parallelogram" formed by $\bar{V}$ and $\bar{W}$ on the planes $(p_1, q_1), \dots (p_n, q_n)$.

Let us fill the first of the two slots of $\tilde{\omega}^2(\bullet, \bullet)$ with the Hamiltonian vector field $\bar{U}$, equation (7.32). Thus we obtain the very important equation

$$\tilde{\omega}^2(\bar{U}, \bullet) = -\tilde{d}H(\bullet) \quad . \tag{7.37}$$

Proof for $n = 1$: Remembering that $\tilde{d}p(\partial/\partial q) = \tilde{d}q(\partial/\partial p) = 0$, we have

$$\tilde{\omega}^2(\bar{U}, \bullet) = (\tilde{d}p \wedge \tilde{d}q) \left(-\frac{\partial H}{\partial q}\frac{\partial}{\partial p} + \frac{\partial H}{\partial p}\frac{\partial}{\partial q}, \bullet \right)$$

$$= -\frac{\partial H}{\partial q}\tilde{d}q(\bullet) - \frac{\partial H}{\partial p}\tilde{d}p(\bullet) = -\tilde{d}H(\bullet) \quad .$$

The following section 7.6 is included for completeness. It is not necessary for understanding section 7.7.

7.6 Lie derivatives

The following is a pedestrian definition of Lie derivative.
The Lie derivative $L_{\bar{A}}$ of a geometrical object such as a vector, tensor, or form, is the change undergone by that object when "transported" through $\bar{A}s$ along the vector field $\bar{A}$, divided by the infinitesimal parameter s.

If $f(x)$ is a scalar function,

$$L_{\bar{A}} = [f(x^1 + A^1(x)\, s, \ldots) - f(x^1, \ldots)]/s$$

$$L_{\bar{A}} = A^i \frac{\partial f}{\partial x^i} = \bar{A}\, f(x) \quad . \tag{7.38}$$

If $\bar{B}$ is a vector,

$$L_{\bar{A}}\bar{B} = [\bar{B}(x') - \bar{B}(x)]/s \quad ,$$

where

$$\bar{B}(x') = B^i(x') \frac{\partial}{\partial x'^i} \quad ,$$

$$x'^i = x^i + A^i(x)\, s + O(s^2) \quad , \quad \frac{\partial}{\partial x'^i} = \frac{\partial}{\partial x^i} - s\, \frac{\partial A^j(x)}{\partial x^i}\, \frac{\partial}{\partial x^j} + O(s^2) \quad .$$

Therefore

$$L_{\bar{A}}\bar{B} = \left(\frac{\partial B^i}{\partial x^j}\, A^j - \frac{\partial A^i}{\partial x^j}\, B^j \right) \frac{\partial}{\partial x^i} \quad . \tag{7.39}$$

If $\tilde{\omega}$ is a 1-form,

$$\tilde{\omega}(x') = \omega_i(x')\, \tilde{d}x'^i = \tilde{\omega}(x) + \left(\frac{\partial \omega_i}{\partial x^j}\, A^j s \right) \tilde{d}x^i + \omega_i \left(\frac{\partial A^i}{\partial x^j}\, s\, \tilde{d}x^j \right) \quad ,$$

and so

$$L_{\bar{A}}\tilde{\omega} = \left(\frac{\partial \omega_i}{\partial x^j}\, A^j + \omega_j\, \frac{\partial A^j}{\partial x^i} \right) \tilde{d}x^i \quad . \tag{7.40}$$

It may be interesting to verify equation (7.37). Show first that the Lie derivative $L_{\bar{U}}$ of $\tilde{\omega}^2 = \tilde{d}p \wedge \tilde{d}q$ is zero,

$$L_{\bar{U}}\tilde{\omega}^2 = 0 \quad . \tag{7.41}$$

Then, since $\tilde{d}\tilde{\omega}^2 = \tilde{d}(\tilde{d}p \wedge \tilde{d}q) = 0$ and $L_{\bar{U}}\tilde{\omega}^2 = 0$, using the general formula $L_{\bar{A}}\tilde{\omega} = \tilde{d}[\tilde{\omega}(\bar{A})] + (\tilde{d}\tilde{\omega})(\bar{A})$ (B.F. Schutz, *loc. cit.* eq. (4.67), p. 142), we have

$$\tilde{d}[\tilde{\omega}^2(\bar{U}, \bullet)] = 0 \quad .$$

Since $\tilde{d}\tilde{d} = 0$, this agrees with equation (7.37).

7.7 Time-independent canonical transformations

The "canonical variables" (p_i,q_i) are not unique. A time-independent transformation to new variables ($P_i(p,q)$,$Q_i(p,q)$) is a "canonical transformation", and the new variables are also "canonical", if

$$\sum_i \tilde{d}p_i \wedge \tilde{d}q_i = \sum_i \tilde{d}P_i \wedge \tilde{d}Q_i \quad . \tag{7.42}$$

Filling the slots of equation (7.42) with two phase space displacements $\delta^{(1)}$ and $\delta^{(2)}$, the condition for a transformation to be canonical reads

$$\sum_i (\delta^{(1)} p_i \delta^{(2)} q_i - \delta^{(1)} q_i \delta^{(2)} p_i) = \sum_i (\delta^{(1)} P_i \delta^{(2)} Q_i - \delta^{(1)} Q_i \delta^{(2)} P_i) \quad .$$

It is given in this form in several textbooks.

Hamilton's equations are invariant under time-independent canonical transformations,

$$\dot{p}_i = -\frac{\partial H(p,q)}{\partial q_i} \ , \ \dot{q}_i = \frac{\partial H(p,q)}{\partial p_i} \rightarrow \dot{P}_i = -\frac{\partial K(P,Q)}{\partial Q_i} \ , \dot{Q}_i = \frac{\partial K(P,Q)}{\partial P_i} \ ,$$

with $K(P,Q) = H(p(P,Q), q(P,Q))$.

Proof: We know that

$$\sum_i (\tilde{d}p_i \wedge \tilde{d}q_i)(\bar{U}, \bullet) = -\tilde{d}H(p,q)(\bullet)$$

with

$$\bar{U} = \sum_j \left(-\frac{\partial H(p,q)}{\partial q_j} \frac{\partial}{\partial p_j} + \frac{\partial H(p,q)}{\partial p_j} \frac{\partial}{\partial q_j} \right) \quad .$$

Expressing $\tilde{d}H$ in terms of the new variables,

$$\tilde{d}H = \sum_j \left(\frac{\partial K(P,Q)}{\partial P_j} \tilde{d}P_j + \frac{\partial K(P,Q)}{\partial Q_j} \tilde{d}Q_j \right) \quad ,$$

and using equation (7.37), we have

$$\sum_i (\tilde{d}P_i \wedge \tilde{d}Q_i)(\bar{U}, \bullet) = -\tilde{d}K(P,Q)(\bullet) \quad .$$

Hence

$$U_{P_i} = -\frac{\partial K}{\partial Q_i} \ , \ U_{Q_i} = \frac{\partial K}{\partial P_i} \quad .$$

Example: The transformation $P_i = -q_i$, $Q_i = p_i$ is obviously canonical. Note the minus sign!

Example: Coordinates in a plane, $(p_1, x_1)(p_2, x_2) \to (p_r, r)(p_\theta, \theta)$.
We have

$$r = \sqrt{x_1^2 + x_2^2}\,,\ \theta = \tan^{-1}(x_2/x_1),\ x_1 = r\ \cos\theta,\ x_2 = r\ \sin\theta,$$
$$p_r = (p_1 x_1 + p_2 x_2)/\sqrt{x_1^2 + x_2^2}\,,\ p_\theta = x_1 p_2 - x_2 p_1,$$
$$p_1 = p_r \cos\theta - (p_\theta/r)\sin\theta,\ p_2 = p_r \sin\theta + (p_\theta/r)\cos\theta.$$

Using these formulae, and performing a few differentiations, it is easy to verify that

$$\tilde{d}p_1 \wedge \tilde{d}x_1 + \tilde{d}p_2 \wedge \tilde{d}x_2 = \tilde{d}p_r \wedge \tilde{d}r + \tilde{d}p_\theta \wedge \tilde{d}\theta \quad .$$

7.8 Generating functions

In the above two examples we expressed the new canonical variables in terms of the old, and proceeded to verify that $\sum_i \tilde{d}P_i \wedge \tilde{d}Q_i$ was indeed equal to $\sum_i \tilde{d}p_i \wedge \tilde{d}q_i$.

Suppose now that we have formulae expressing p and P in terms of q and Q, $p = p(q, Q)$, $P = P(q, Q)$. (For simplicity's sake we confine ourselves to one dimension.) Then, substituting

$$\tilde{d}p = \frac{\partial p}{\partial q}\tilde{d}q + \frac{\partial p}{\partial Q}\tilde{d}Q \quad , \quad \tilde{d}P = \frac{\partial P}{\partial q}\tilde{d}q + \frac{\partial P}{\partial Q}\tilde{d}Q$$

in $\tilde{d}p \wedge \tilde{d}q = \tilde{d}P \wedge \tilde{d}Q$, we find

$$\frac{\partial p}{\partial Q}\,\tilde{d}Q \wedge \tilde{d}q = \frac{\partial P}{\partial q}\,\tilde{d}q \wedge \tilde{d}Q \quad , \quad \frac{\partial p}{\partial Q} = -\frac{\partial P}{\partial q} \quad .$$

If the functions $p = p(q, Q)$ and $P = P(q, Q)$ can be expressed in the form

$$p = \frac{\partial G_1}{\partial q} \quad , \quad P = -\frac{\partial G_1}{\partial Q} \quad , \tag{7.43}$$

where the "generating function" G_1 is a function of q and Q, the relation $\partial p/\partial Q = -\partial P/\partial q$ is satisfied, and the transformation is canonical.

For the canonical transformation $Q = p$, $P = -q$, the generating function is $G_1 = qQ$.

Example: The generating function

$$G_1 = (q^2 \cos\alpha - 2qQ + Q^2 \cos\alpha)/2\sin\alpha$$

yields

$$p = \frac{\partial G_1}{\partial q} = \frac{q\cos\alpha - Q}{\sin\alpha} \quad , \quad P = -\frac{\partial G_1}{\partial Q} = \frac{q - Q\cos\alpha}{\sin\alpha} \quad ,$$

namely the rotation

$$Q = q\cos\alpha - p\sin\alpha \quad , \quad P = q\sin\alpha + p\cos\alpha$$

in the (p, q)-plane. For $\alpha = -\pi/2$, $G_1 = qQ$, $Q = p$, $P = -q$. There are altogether four types of generating functions as listed below:

$G_1(q,Q)$	$p = \partial G_1/\partial q\ ,\ P = -\partial G_1/\partial Q$
$G_2(q,P)$	$p = \partial G_2/\partial q\ ,\ Q = \partial G_2/\partial P$
$G_3(p,Q)$	$q = -\partial G_3/\partial p\ ,\ P = -\partial G_3/\partial Q$
$G_4(p,P)$	$q = -\partial G_4/\partial p\ ,\ Q = \partial G_4/\partial P$

Note the following relations, which can be easily verified:

$$p\,\tilde{d}q - P\,\tilde{d}Q = \tilde{d}G_1 = \tilde{d}(G_2 - PQ)$$

$$= \tilde{d}(G_3 + pq) = \tilde{d}(G_4 - PQ + pq) \quad .$$

This string of Legendre transformations is analogous to the well known Thermodynamics relations

$$T\ \mathrm{d}s - p\ \mathrm{d}v = \mathrm{d}u(s,v) = \mathrm{d}(h(s,p) - pv)$$

$$= \mathrm{d}(f(T,v) + Ts) = \mathrm{d}(g(T,p) + Ts - pv) \quad ,$$

where v is the volume of the system.

The generating functions G_1 and G_2 are rather special. What is special about G_1? Just that

$$\tilde{d}G_1 = \frac{\partial G_1}{\partial q}\,\tilde{d}q + \frac{\partial G_1}{\partial Q}\,\tilde{d}Q = p\,\tilde{d}q - P\,\tilde{d}Q \quad . \tag{7.44}$$

A transformation is canonical if the difference between the action elements $\sum_i p_i\tilde{d}q_i(\bullet)$ and $\sum_i P_i\tilde{d}Q_i(\bullet)$ is a gradient $\tilde{d}G_1(\bullet)$.

The special role played by $G_2(q,P)$ is due to the fact that the identity transformation $P_i = p_i$, $Q_i = q_i$ $(i = 1, \ldots n)$ is induced by $G_2 = q_iP_i$, where we start using the sum convention of equal indices. Therefore a G_2 generating function is convenient to describe infinitesimal transformations.

For the moment, consider

$$G_2 = (q_i + \delta q_i)(P_i - \delta p_i)$$

with δq_i and δp_i not necessarily infinitesimal. This yields

$$p_i = \frac{\partial G_2}{\partial q_i} = P_i - \delta p_i \quad , \quad Q_i = \frac{\partial G_2}{\partial P_i} = q_i + \delta q_i \quad .$$

For a particle, $G_2 = x_i P_i + \delta G_2$ with $\delta G_2 = P_i\ \delta a_i$ and $\delta G_2 = -x_i\ \delta b_i$ generate the translation $X_i = x_i + \delta a_i$ and the boost $P_i = p_i + \delta b_i$, respectively.

For both coordinates and momentum components, an orthogonal transformation with coefficients a_{ij} satisfying $a_{ki}a_{kj} = \delta_{ij}$ is induced by

$$G_2 = a_{kj} q_j P_k \quad .$$

In fact, $Q_i = \partial G_2/\partial P_i = a_{ij} q_j$ and $p_i = \partial G_2/\partial q_i = a_{ki} P_k$, or, inverting by use of the relations for the a_{ij}'s, $P_j = a_{ji} p_i$.
The inversion $Q_i = -q_i$, $P_i = -p_i$ is given by the above with $a_{ij} = -\delta_{ij}$, so that $G_2^{(\mathrm{inv})} = -G_2^{(\mathrm{id})}$.

It is easy to show that $G_2 = x_i P_i + \delta G_2$ with $\delta G_2 = \delta\alpha\ \epsilon_{ijk} x_i P_j n_k$ generates the infinitesimal rotation

$$X_i = x_i - \delta\alpha\ \epsilon_{ijk} x_j n_k,\ P_i = p_i - \delta\alpha\ \epsilon_{ijk} p_j n_k.$$

In general, infinitesimal transformations about the identity are induced by a generating function of the form

$$G_2 = q_i P_i + \delta G(q, P) \quad .$$

Then

$$p_i = \frac{\partial G_2}{\partial q_i} = P_i + \frac{\partial \delta G}{\partial q_i} \quad , \quad Q_i = \frac{\partial G_2}{\partial P_i} = q_i + \frac{\partial \delta G}{\partial P_i} \quad ,$$

and so

$$P_i = p_i + \delta p_i \quad , \quad Q_i = q_i + \delta q_i$$

with $\delta p_i = -\partial \delta G/\partial q_i$ and $\delta q_i = \partial \delta G/\partial P_i$.

The Hamiltonian flow $(p(t), q(t)) \rightarrow (p(t + \delta t), q(t + \delta t))$ can be regarded as an infinitesimal canonical transformation with generating function

$$G_2 = q(t) p(t + \delta t) + H(p(t + \delta t), q(t)) \delta t \quad ,$$

where $H(p(t+\delta t), q(t))$ is obtained from $H(p(t), q(t))$ by replacing $p(t)$ by $p(t+\delta t)$. In fact,

$$\begin{aligned} p(t + \delta t) = \frac{\partial G_2}{\partial q(t)} &= p(t) - \frac{\partial H(p(t + \delta t), q(t))}{\partial q(t)} \delta t \\ &\simeq p(t) - \frac{\partial H(p(t), q(t))}{\partial q(t)} \delta t \end{aligned}$$

and similarly for $q(t + \delta t)$.

Consider now a function $f(p, q)$. Its change if p and q are replaced by $P = p + \delta p$ and $Q = q + \delta q$ with $\delta p = -\partial \delta G/\partial q$ and $\delta Q = \partial \delta G/\partial P$ is

$$\delta f = f(p + \delta p, q + \delta q) - f(p, q) = -\frac{\partial f}{\partial p} \frac{\partial \delta G(q, P)}{\partial q} + \frac{\partial f}{\partial q} \frac{\partial \delta G(q, P)}{\partial P}$$

$$\simeq -\frac{\partial f}{\partial p}\frac{\partial \delta G(q,p)}{\partial q} + \frac{\partial f}{\partial q}\frac{\partial \delta G(q,p)}{\partial p} = [f, \delta G(q,p)] \quad .$$

Here $[f, \delta G(q,p)]$ is the Poisson bracket of $f(p,q)$ and $\delta G(q,p) = \delta G(q,P)|_{P\to p}$. In particular, $\delta p = [p, \delta G]$ and $\delta q = [q, \delta G]$.

It is interesting that $\delta G_2 = \delta\alpha\ l$, with $l = |\mathbf{l}| = \sqrt{l_1^2 + l_2^2 + l_3^2}$, generates infinitesimal rotations about $\mathbf{l}$. In fact,

$$[x_i, l] = \sum_k \left(\frac{\partial x_i}{\partial x_k}\frac{\partial l}{\partial p_k} - \frac{\partial x_i}{\partial p_k}\frac{\partial l}{\partial x_k}\right) = \frac{\partial l}{\partial p_i} = \frac{\partial l_k}{\partial p_i}\frac{\partial l}{\partial l_k} = \frac{l_k}{l}\,\epsilon_{ijk}x_j \quad ,$$

$$[p_i, l] = \frac{l_k}{l}\,\epsilon_{ijk}p_j \ , \ [l_i, l] = 0 \quad .$$

Hence $X_i = x_i + \delta\alpha\,[x_i, l] = x_i - \delta\alpha\,\epsilon_{ijk}x_j l_k/l$, $P_i = p_i + \delta\alpha\,[p_i, l]$ $= p_i - \delta\alpha\,\epsilon_{ijk}p_j l_k/l$, $L_i = l_i$.

If the Hamiltonian is invariant under an infinitesimal transformation, then the corresponding $\delta G(q,p)$ is a constant of motion. In fact,

$$\mathrm{d}\delta G/\mathrm{d}t = [\delta G, H] = -\delta H = 0 \quad .$$

Therefore, if the Hamiltonian of a particle is invariant under a space translation in the direction of the unit vector $\hat{\mathbf{n}}$, the component of the momentum in that direction is a constant of motion; if it is invariant under rotations about $\hat{\mathbf{n}}$, then the component of the angular momentum $\hat{\mathbf{n}}\cdot\mathbf{l}$ is a constant of motion.

7.9 Lagrange and Poisson brackets

In the condition $\sum_i \tilde{d}p_i \wedge \tilde{d}q_i = \sum_i \tilde{d}P_i \wedge \tilde{d}Q_i$ for time-independent canonical transformations, let us express p_i and q_i in terms of the new variables (P, Q). Defining a Lagrange bracket as

$$\{A, B\}_{p,q} = -\{B, A\}_{p,q} = \sum_i \left(\frac{\partial q_i}{\partial A}\frac{\partial p_i}{\partial B} - \frac{\partial q_i}{\partial B}\frac{\partial p_i}{\partial A}\right) \quad , \tag{7.45}$$

we find

$$\sum_{ijk}[\{Q_j, Q_k\}\tilde{d}Q_k\wedge\tilde{d}Q_j + \{P_j, P_k\}\tilde{d}P_k\wedge\tilde{d}P_j + 2\{Q_j, P_k\}\tilde{d}P_k\wedge\tilde{d}Q_j] = 2\sum_i \tilde{d}P_i\wedge\tilde{d}Q_i \quad .$$

Hence the conditions for a transformation to be canonical

$$\{Q_j, Q_k\}_{p,q} = 0 \ , \ \{P_j, P_k\}_{p,q} = 0 \ , \ \{Q_j, P_k\}_{p,q} = \delta_{jk} \quad . \tag{7.46}$$

The identity transformation $Q_i = q_i$, $P_i = p_i$ $(i = 1, \dots n)$ satisfies these relations.

We shall show that the above conditions for Lagrange brackets are equivalent to the conditions

$$[Q_j, Q_k]_{p,q} = 0 \ , \ [P_j, P_k]_{p,q} = 0 \ , \ [Q_j, P_k]_{p,q} = \delta_{jk} \tag{7.47}$$

for Poisson brackets, also satisfied by the identity transformation.

Proof: Put $x_{2i-1} = q_i$, $x_{2i} = p_i$ $(i = 1, \dots n)$ $(x_1 = q_1, x_2 = p_1, \dots x_{2n-1} = q_n$, $x_{2n} = p_n)$ and define the symplectic metric [5] $g_{2i-1,2i} = +1$, $g_{2i,2i-1} = -1$ $(g_{12} = 1, g_{21} = -1, g_{34} = 1, g_{43} = -1, \dots$, all other g_{ij} being zero),

$$\mathsf{G} = \|g_{ij}\| = \begin{pmatrix} 0 & 1 & 0 & 0 & \dots \\ -1 & 0 & 0 & 0 & \dots \\ 0 & 0 & 0 & 1 & \dots \\ 0 & 0 & -1 & 0 & \dots \\ 0 & 0 & 0 & 0 & \dots \\ \vdots & \vdots & \vdots & \vdots & \vdots \end{pmatrix} \ . \tag{7.48}$$

The Lagrange bracket $\{A, B\}_{p,q}$ can be written as

$$\{A, B\}_{p,q} = \sum_{ij} g_{ij} \frac{\partial x_i}{\partial A} \frac{\partial x_j}{\partial B} \ , \tag{7.49}$$

and the Poisson bracket $[A, B]_{p,q}$ as

$$[A, B]_{p,q} = \sum_{ij} g_{ij} \frac{\partial A}{\partial x_i} \frac{\partial B}{\partial x_j} \ . \tag{7.50}$$

Put now $X_{2k-1} = Q_k$, $X_{2k} = P_k$. Then

$$\sum_l \{X_l, X_i\}_{p,q} \{X_l, X_j\}_{p,q} = \delta_{ij} \ . \tag{7.51}$$

---

[5] The condition

$$\sum_i (\delta^{(1)} p_i \delta^{(2)} q_i - \delta^{(1)} q_i \delta^{(2)} p_i) = \sum_i (\delta^{(1)} P_i \delta^{(2)} Q_i - \delta^{(1)} Q_i \delta^{(2)} P_i) \ .$$

for a transformation to be canonical, can be written concisely as

$$\sum_{ij} g_{ij} \delta^{(2)} x_i \delta^{(1)} x_j = \sum_{ij} g_{ij} \delta^{(2)} X_i \delta^{(1)} X_j \ .$$

In one dimension

$$\{q,p\}_{P,Q} = \frac{\partial Q}{\partial q}\frac{\partial P}{\partial p} - \frac{\partial P}{\partial q}\frac{\partial Q}{\partial p} = \frac{\partial(Q,P)}{\partial(q,p)} = ([Q,P]_{p,q})^{-1} \ ,$$

and so

$$\{q,p\}_{P,Q}[Q,P]_{p,q} = 1 \ .$$

The proof is straightforward, we only have to use $\mathsf{G}^T\mathsf{G} = \mathsf{I}$ ($\sum_a g_{ab}g_{ac} = \delta_{bc}$) and $\sum_a(\partial x_a/\partial X_i)(\partial X_j/\partial x_a) = \delta_{ij}$.
Using (7.51), we find

$$\sum_l (\{Q_l,Q_i\}[Q_l,Q_j] + \{P_l.Q_i\}[P_l,Q_j]) = \delta_{ij} \ ,$$

$$\text{zero} \ - \sum_l \delta_{il}[P_l,Q_j] = \delta_{ij} \ , \quad [Q_j,P_i] = \delta_{ij} \ .$$

Similarly we find the other Poisson brackets.

The reader can easily show that the Lagrange bracket of two functions (A,B) on phase space is invariant under canonical transformations, namely

$$\{A,B\}_{p,q} = \{A,B\}_{P,Q} \ ,$$

and so is also the Poisson bracket

$$[A,B]_{p,q} = [A.B]_{P,Q} \ .$$

Therefore from now on we omit the subscripts $(p.q)$, and write simply $[A,B]$ for the Poisson bracket of A and B.

7.10 Hamilton-Jacobi equation revisited

Let $H(p,q)$ be the (not explicitly time-dependent) Hamiltonian of a system. Perform a canonical transformation from (p_i,q_i) to (P_i,Q_i) with generating function $G_2(q,P) \Rightarrow S(q,P)$ such that the new momenta are constants of motion, $\dot{P}_i = 0$ $(i = 1,\ldots n)$.

Since $\dot{P}_i = -\partial K/\partial Q_i$, where $K(P,Q)$ is the transformed Hamiltonian, $\dot{P}_i = 0$ requires that K be a function of the P_i's, but not of the Q_i's, $K = K(P)$. We have also $K = H(p,q) = E$.

Let $P_1 = E$, and P_i $(i > 1)$ $(n-1)$ more constants of motion. Then

$$G_2(q,P) = S(q,P_1 = E, P_2 \ldots P_n) \ . \tag{7.52}$$

Since $p_i = \partial G_2/\partial q_i = \partial S/\partial q_i$, substituting in $H(p,q) = E$ we obtain the Hamilton-Jacobi equation

$$H\left(\frac{\partial S}{\partial q_i}, q_i\right) = E \ . \tag{7.53}$$

What about the coordinates Q_i conjugate to the P_i? From the equations $\dot{Q}_i = \partial K/\partial P_i$, since K depends only on the P_i's, and these are constants of motion, we have that all the Q_i's are constants of motion. Note that $\dot{Q}_1 = \partial K/\partial P_1 = \partial P_1/\partial P_1 = 1$. For $i > 1$ we have $\dot{Q}_i = \partial P_1/\partial P_i = 0$. Therefore $Q_1 = t - t_0$, while $Q_2, \ldots Q_n$ are $(n-1)$ constants.
We have collected the right number $(2n)$ of constants of motion, (n) P_i's (including the energy), $\dot{Q}_1 = 1$, and $(n-1)$ Q_i's $(i > 1)$.

We have the equations $\partial S/\partial E = \partial S/\partial P_1 = t - t_0$ and, for $i > 1$, $\partial S/\partial P_i = Q_i$.

Going back to chapter 1, equation (1.21) $(t_2 - t_1 = \partial S(q_2, q_1; E)/\partial E)$ agrees with $t - t_0 = \partial S/\partial E$, and there is only one P_i, $P_1 = E$.

In the two-dimensional example in section 1.6, $P_1 = E$ and P_2 was denoted by α. (In most books the P_i's are denoted by α_i.) We have also

$$Q_2 = \partial S/\partial P_2 = \partial S/\partial p_x ,$$

$$Q_2 = x - p_x \left[\sqrt{2m(E - E_x)} - \sqrt{2m(E - E_x - mgy)}\right] /m^2 g$$
$$= x - v_{0x}[v_{0y} + gt - v_{0y}]/g = v_{0x}t - v_{0x}t = 0 \quad .$$

Q_2 is indeed a constant, which in this case happens to be zero.

Returning to the general case, we now state that

$$S = \int \sum_i p_i \, \tilde{d}q_i \quad , \tag{7.54}$$

where the integral of the 1-form $\sum_i p_i \, \tilde{d}q_i$ is along a trajectory "c" in phase space from (p_0, q_0) to (p, q).

We must first define the notation, which is also necessary for later use. Divide "c" into intervals, each of which is represented by a vector

$$\bar{\mathrm{d}s} = \sum_j \left(\mathrm{d}q_j \frac{\partial}{\partial q_j} + \mathrm{d}p_j \frac{\partial}{\partial p_j} \right) \quad .$$

Do not confuse the numbers $(\mathrm{d}q_j, \mathrm{d}p_j)$ with the 1-forms $(\tilde{d}q_j, \tilde{d}p_j)$.
In (7.54) $\tilde{d}q_i$ stands for

$$\tilde{d}q_i(\bar{\mathrm{d}s}) = \sum_j \left[\mathrm{d}q_j \tilde{d}q_i \left(\frac{\partial}{\partial q_j} \right) + \mathrm{d}p_j \tilde{d}q_i \left(\frac{\partial}{\partial p_j} \right) \right]$$
$$= \sum_j \left[\mathrm{d}q_j \, \delta_{ij} + \text{zero} \right] = \mathrm{d}q_i \quad .$$

Thus

$$S(p, q, p_0, q_0) = \int_{p_0, q_0}^{p, q} \sum_i p_i \, \mathrm{d}q_i = \int_{p_0, q_0}^{p, q} \sum_i p_i \dot{q}_i \, \mathrm{d}t$$

$$= \int_{t_0}^{t} (L+H)\mathrm{d}t = \int_{t_0}^{t} L\mathrm{d}t + E(t-t_0) \quad .$$

Then for an isochronous variation

$$\delta S = \int_{t_0}^{t} \delta L \, \mathrm{d}t + (t-t_0)\delta E = \left[\sum_i \frac{\partial L}{\partial \dot{q}_i} \delta q_i \right]_{t_0}^{t} + (t-t_0)\delta E$$

$$= \sum_i p_i(t)\delta q_i(t) - \sum_i p_i(t_0)\delta q_i(t_0) + (t-t_0)\delta E \quad .$$

This tells us that

$$\frac{\partial S}{\partial q_i} = p_i \quad , \quad \frac{\partial S}{\partial q_{0i}} = -p_{0i} \quad , \quad \frac{\partial S}{\partial t} = E \quad .$$

7.11 Time-dependent canonical transformations

They leave invariant the 2-form $\tilde{\omega}^2 = \sum_i \tilde{d}p_i \wedge \tilde{d}q_i - \tilde{d}H \wedge \tilde{d}t$,

$$\tilde{\omega}^2(P,Q,K,T) = \tilde{\omega}^2(p,q,H,t) \quad , \tag{7.55}$$

$$\sum_i \left(\tilde{d}P_i \wedge \tilde{d}Q_i - \frac{\partial K}{\partial P_i} \tilde{d}P_i \wedge \tilde{d}T - \frac{\partial K}{\partial Q_i} \tilde{d}Q_i \wedge \tilde{d}T \right)$$

$$= \sum_i \left(\tilde{d}p_i \wedge \tilde{d}q_i - \frac{\partial H}{\partial p_i} \tilde{d}p_i \wedge \tilde{d}t - \frac{\partial H}{\partial q_i} \tilde{d}q_i \wedge \tilde{d}t \right) \quad ,$$

where $H(p,q,t)$ and $K(P,Q,T)$ are the Hamiltonians.

This is equivalent to the condition

$$\sum_i (p_i \tilde{d}q_i - P_i \tilde{d}Q_i) - H\tilde{d}t + K\tilde{d}T = \tilde{d}G_1 \quad , \tag{7.56}$$

which is an extension of equation (7.44) to the $(2n+1)$-dimensional phase space (p_i, q_i, t). If $T = t$, then

$$K = H + \frac{\partial G_1}{\partial t} \quad . \tag{7.57}$$

From Hamilton's equations in terms of the old variables

$$\frac{\mathrm{d}p_i}{\mathrm{d}t} = -\frac{\partial H}{\partial q_i} \quad , \quad \frac{\mathrm{d}q_i}{\mathrm{d}t} = \frac{\partial H}{\partial p_i} \quad ,$$

follow those in the new variables

$$\frac{\mathrm{d}P_i}{\mathrm{d}T} = -\frac{\partial K}{\partial Q_i} \ , \ \frac{\mathrm{d}Q_i}{\mathrm{d}T} = \frac{\partial K}{\partial P_i} \ .$$

Formal proofs will be given shortly (see section 7.13).

Example:

Galilean transformation for a free particle, $q_i = x_i$, $Q_i = x'_i$ $(i = 1, 2, 3)$ etc. ,

$$\mathbf{r}' = \mathbf{r} - \mathbf{u}t \ , \ \mathbf{p}' = \mathbf{p} - m\mathbf{u} \ , \ t' = t \ , \ H = \mathbf{p}^2/2m \ , \ K = \mathbf{p}'^2/2m \ .$$

Note first that

$$H = \mathbf{p}^2/2m = (\mathbf{p}' + m\mathbf{u})^2/2m \neq K = \mathbf{p}'^2/2m \quad .$$

We verify that equation (7.55) is satisfied. Since H and K depend only on the momenta, we have

$$\sum_i (\tilde{d}p'_i \wedge \tilde{d}x'_i - (\partial K/\partial p'_i)\tilde{d}p'_i \wedge \tilde{d}t')$$

$$= \sum_i (\tilde{d}p_i \wedge (\tilde{d}x_i - u_i\tilde{d}t) - m^{-1}(p_i - mu_i)\tilde{d}p_i \wedge \tilde{d}t)$$

$$= \sum_i (\tilde{d}p_i \wedge \tilde{d}x_i - m^{-1}p_i\tilde{d}p_i \wedge \tilde{d}t) = \sum_i (\tilde{d}p_i \wedge \tilde{d}x_i - (\partial H/\partial p_i)\tilde{d}p_i \wedge \tilde{d}t) \quad .$$

Equation (7.56) is satisfied if we take

$$G_1(\mathbf{r}, \mathbf{r}', t) = (\mathbf{r} - \mathbf{r}') \cdot \mathbf{p}' + m\,\mathbf{u} \cdot \mathbf{r} - m\,\mathbf{u}^2 t/2 - \mathbf{u} \cdot \mathbf{p}' t \quad .$$

In fact

$$\frac{\partial G_1}{\partial x_i} = p'_i + mu_i = p_i \ , \ \frac{\partial G_1}{\partial x'_i} = -p'_i \ ,$$

$$\frac{\partial G_1}{\partial t} = -m\mathbf{u}^2/2 - \mathbf{u} \cdot \mathbf{p}' = K - H \quad .$$

7.12 Time-dependent Hamilton-Jacobi equation

If the Hamiltonian H depends explicitly on the time, we perform a canonical transformation with generating function $G_2(q, P, t) \Rightarrow S^\star(q, P, t)$ chosen in such a way that the new Hamiltonian K vanishes.

This means $H + \partial G_2/\partial t = K = 0$. We still have $p_i = \partial G_2/\partial q_i$. Hence the Hamilton-Jacobi equation

$$H\left(\frac{\partial S^\star}{\partial q_i}, q_i, t\right) + \frac{\partial S^\star}{\partial t} = 0 \quad . \tag{7.58}$$

Since $K = 0$, $\dot{P}_i = -\partial K/\partial Q_i = 0$ and $\dot{Q}_i = \partial K/\partial P_i = 0$, and so the new coordinates and momenta are constants of motion.

Equation (7.58) is a first order differential equation for a function of $n+1$ variables, the q_i's and t. Therefore we would expect $S^\star$ to depend on $n+1$ constants of motion. Of these, n are provided by the P_i's, while another can be the value of $S^\star$ at some particular time, $S^\star(q_{0i}, P_i, t_0)$.

If H does not depend explicitly on t, the relation between Hamilton's "principal function" $S^\star$ and Hamilton's "characteristic function" S is

$$S^\star = S - Et \quad . \tag{7.59}$$

Like S, $S^\star$ is related to the action integral,

$$S^\star(q, q_0, t, t_0) = \int_c L\mathrm{d}t \quad ,$$

where "c" is the trajectory. In fact

$$\frac{\mathrm{d}S^\star}{\mathrm{d}t} = \sum_i \frac{\partial S^\star}{\partial q_i}\dot{q}_i + \frac{\partial S^\star}{\partial t} = \sum_i p_i\dot{q}_i + \frac{\partial S^\star}{\partial t} = \sum_i p_i\dot{q}_i - H = L \quad .$$

Example: Suppose for a particle of mass m we are given

$$S^\star(q, P, t) = (p_0 q - p_0^2 t/2m)\theta(-t) + [(p_0 + P)q - (p_0 + P)^2 t/2m]\theta(t) \quad ,$$

where p_0 is a constant, and $\theta(\xi) = 1$ for $\xi > 0$, $= 0$ for $\xi < 0$. We have
$p = \partial S^\star/\partial q = p_0 + P\theta(t)$ and $Q = \partial S^\star/\partial P = [q - (p_0 + P)t/m]\theta(t)$.
Thus $p = p_0$ for $t < 0$ and $p = p_0 + P$ for $t > 0$.
On the other hand, $Q = 0$ for $t < 0$ because $\theta(t) = 0$, and so, being a constant of motion, Q must be zero also for $t > 0$. Hence $q - (p_0 + P)t/m = 0$ for $t > 0$.
Here is clearly a particle of mass m with momentum p_0 for $t < 0$, receiving an impulse P at $t = 0$, after which it moves with uniform velocity $(p_0 + P)/m$.

Now

$$\frac{1}{2m}\left(\frac{\partial S^\star}{\partial q}\right)^2 + \frac{\partial S^\star}{\partial t}$$

$$= ([p_0\theta(-t) + (p_0 + P)\theta(t)]^2 - p_0^2\theta(-t) - (p_0 + P)^2\theta(t))/2m$$
$$-(p_o q - p_0^2 t/2)\delta(t) + [(p_0 + P)q - (p_0 + P)^2 t/2m]\delta(t) \quad .$$

Since $\theta(-t)\theta(t) = 0$ and $t\,\delta(t) = 0$, we find

$$\frac{1}{2m}\left(\frac{\partial S^\star}{\partial q}\right)^2 + U(q, t) + \frac{\partial S^\star}{\partial t} = 0 \quad ,$$

where $U(q, t) = -Pq\,\delta(t)$, the potential energy for an impulsive force $P\delta(t)$.

We want also to check the invariance of $\tilde{\omega}^2$:

$$\tilde{d}p \wedge \tilde{d}q - \tilde{d}H \wedge \tilde{d}t = \tilde{d}P \wedge \tilde{d}Q - \tilde{d}K \wedge \tilde{d}t$$

The right member is zero, because P and Q are constants and $K = 0$. Substituting $p = p_0 + P\theta(t)$ and $H = p^2/2m - Pq\delta(t)$ in the left member, since $\tilde{d}p = P\delta(t)\tilde{d}t$ and $\tilde{d}t \wedge \tilde{d}t = 0$, we find $P\delta(t)\tilde{d}t \wedge \tilde{d}q + \text{zero} + P\delta(t)\tilde{d}q \wedge \tilde{d}t = 0$.

7.13 Stokes' theorem and some proofs

First a bit of mathematics. In section 7.5 we saw that the operator $\tilde{d}$ acting on a 0-form (a function) h yields the 1-form $\tilde{d}h(\bullet) = \sum_i (\partial h/\partial x^i)\tilde{d}x^i(\bullet)$. The action of $\tilde{d}$ on a 1-form is defined as follows (see equation (7.34)). If

$$\tilde{\omega}^1(\bullet) = \sum_i X_i(x^1, \ldots x^n)\tilde{d}x^i(\bullet)$$

is a 1-form, then

$$\tilde{d}\tilde{\omega}^1(\bullet, \bullet) = \sum_{i,j} \frac{\partial X_i}{\partial x^j}\tilde{d}x^j \wedge \tilde{d}x^i(\bullet, \bullet) \quad .$$

For instance, in three dimensions

$$\tilde{d}(X\tilde{d}x + Y\tilde{d}y + Z\tilde{d}z)$$

$$= \left(\frac{\partial Z}{\partial y} - \frac{\partial Y}{\partial z}\right)\tilde{d}y \wedge \tilde{d}z + \left(\frac{\partial X}{\partial z} - \frac{\partial Z}{\partial x}\right)\tilde{d}z \wedge \tilde{d}x + \left(\frac{\partial Y}{\partial x} - \frac{\partial X}{\partial y}\right)\tilde{d}x \wedge \tilde{d}y \quad ,$$

which corresponds to the elementary vector calculus operation

$$\text{curl}\mathbf{V} = \left(\frac{\partial Z}{\partial y} - \frac{\partial Y}{\partial z}, \ldots, \ldots\right) \quad \text{if} \quad \mathbf{V} = (X, Y, Z) \quad .$$

It is apparent that if $\tilde{\omega}^1$ is a gradient, $X_i = \partial h/\partial x^i$, then, combining in $\tilde{d}\tilde{\omega}^1$ a pair of terms $(\partial X_i/\partial x^j)\tilde{d}x^j \wedge \tilde{x}^i$ and $(\partial X_j/\partial x^i)\tilde{d}x^i \wedge \tilde{x}^j$ we have $[(\partial^2 h/\partial x^j \partial x^i) - (\partial^2 h/\partial x^i \partial x^j)]\tilde{d}x^j \wedge \tilde{d}x^i = 0$. Thus,

$$\text{if} \quad \tilde{\omega}^1 = \tilde{d}h \ , \quad \text{then} \quad \tilde{d}\tilde{\omega}^1 = \tilde{d}^2 h = 0 \quad ,$$

which corresponds to curlgrad$h = 0$. We already know from (7.35), in general $\tilde{d}^2 = 0$, the double application of the operator $\tilde{d}$ gives zero. We shall encounter later n-forms $\tilde{\omega}^n$. These will be defined so that $\tilde{d}^2\tilde{\omega}^n = 0$.

A special case of Stokes' integral theorem can now be expressed in the following concise form:

$$\oint_{\mathrm{c}} \tilde{\omega}^1 = \int_{\mathrm{S}} \tilde{d}\tilde{\omega}^1 \quad , \tag{7.60}$$

where "c" is a closed curve, and "S" a surface having "c" as boundary.

The left side simply means that we must fill the single slot of $\tilde{\omega}^1(\bullet)$ with the vectors $\bar{\mathrm{d}}s$ defined in section 7.10. The right side means that we must fill the two slots of $\tilde{d}\tilde{\omega}^1(\bullet, \bullet)$ with pairs of vectors $(\bar{\mathrm{d}}s, \bar{\delta}s)$, each pair forming two sides of an infinitesimal parallelogram of a set covering all "S".
In three dimensions, with $\tilde{\omega}^1 = X\tilde{d}x + Y\tilde{d}y + Z\tilde{d}z$,

$$\oint_{\mathrm{c}} \tilde{\omega}^1 = \oint_{\mathrm{c}} (X\tilde{d}x + Y\tilde{d}y + Z\tilde{d}z)(\bar{\mathrm{d}}s) = \oint_{\mathrm{c}} (X\,\mathrm{d}x + Y\,\mathrm{d}y + Z\,\mathrm{d}z) \Leftrightarrow \oint_{\mathrm{c}} \mathbf{V}\cdot\mathrm{d}\mathbf{r} \quad ,$$

while

$$\int_{\mathrm{S}} \tilde{d}\tilde{\omega}^1 = \int_{\mathrm{S}} \left[\left(\frac{\partial Z}{\partial y} - \frac{\partial Y}{\partial z} \right) \tilde{d}y \wedge \tilde{d}z(\bar{\mathrm{d}}s, \bar{\delta}s) + \ldots + \ldots \right]$$

$$= \int_{\mathrm{S}} \left[\left(\frac{\partial Z}{\partial y} - \frac{\partial Y}{\partial z} \right) (\mathrm{d}y\delta z - \mathrm{d}z\delta y) + \ldots + \ldots \right] \Leftrightarrow \int_{\mathrm{S}} \mathrm{curl}\mathbf{V}\cdot\hat{\mathbf{n}}\,\mathrm{d}A \quad .$$

In our extended $(2n+1)$-dimensional phase space, if

$$\tilde{\omega}^1 = \sum_i p_i \tilde{d}q_i - H\tilde{d}t \quad ,$$

then

$$\tilde{d}\tilde{\omega}^1 = \sum_i \tilde{d}p_i \wedge \tilde{d}q_i - \tilde{d}H \wedge \tilde{d}t = \tilde{\omega}^2(p, q, H, t) \quad .$$

Note that

$$\tilde{d}\textstyle\sum_i p_i\tilde{d}q_i = \sum_{ij} \left[(\partial p_i/\partial p_j)\tilde{d}p_j \wedge \tilde{d}q_i + (\partial p_i/\partial q_j)\tilde{d}q_j \wedge \tilde{d}p_i \right]$$
$$= \textstyle\sum_{ij} \delta_{ij}\tilde{d}p_j \wedge \tilde{d}q_i + \text{zero} = \sum_i \tilde{d}p_i \wedge \tilde{d}q_i \quad .$$

For an arbitrary closed curve "c" we have

$$\oint_{\mathrm{c}} \tilde{\omega}^1(p, q, H, t) = \int_{\mathrm{S}} \tilde{\omega}^2(p, q, H, t)$$

$$= \int_{\mathrm{S}} \tilde{\omega}^2(P, Q, K, T) = \oint_{\mathrm{c}} \tilde{\omega}^1(P, Q, K, T) \quad ,$$

where we have used the definition (7.55) of canonical transformations.
Hence

$$\oint_{\mathrm{c}} [\tilde{\omega}^1(p, q, H, t) - \tilde{\omega}^1(P, Q, K, T)] = 0$$

for an arbitrary "c", implying that

$$\tilde{\omega}^1(p,q,H,t) - \tilde{\omega}^1(P,Q,K,T) = \tilde{d}G_1 \quad ,$$

namely (7.56).

In order to prove the invariance of Hamilton's equations we use Hamilton's principle for isochronous variations vanishing at t_1 and t_2, namely

$$\delta \int_{t_1}^{t_2} (\sum_i p_i \dot{q}_i - H)\mathrm{d}t = 0 \quad \text{or} \quad \delta \int_1^2 \tilde{\omega}^1(p,q,H,t) = 0 \quad .$$

Since by assumption $\delta[G_1(2) - G_1(1)] = 0$, Hamilton's equations for (K,P,Q,T) follow from the following steps:

$$\delta \int_1^2 \tilde{\omega}^1(P,Q,K,T) = \delta \int_1^2 [\tilde{\omega}^1(P,Q,K,T) - \tilde{d}G_1]$$

$$= \delta \int_1^2 \tilde{\omega}^1(p,q,H,t) = 0 \quad .$$

7.14 Hamiltonian flow as a Lie-Cartan group

Let

$$\frac{\mathrm{d}x_i}{\mathrm{d}t} = A_i(\mathbf{x}) \quad , \quad \mathbf{x} = (x_1, \ldots x_N)$$

be a system of first order differential equations. In an <u>infinitesimal</u> interval t, a point $\mathbf{x} = (x_1, \ldots x_N)$ is shifted to the point

$$A^t\mathbf{x} = (x_1 + A_1(\mathbf{x})t, \ldots x_N + A_N(\mathbf{x})t).$$

The corresponding change of a function $f(\mathbf{x})$ of the coordinates is

$$\mathrm{d}f(\mathbf{x}) = f(A^t\mathbf{x}) - f(\mathbf{x}) = t\bar{A}(\mathbf{x})f(\mathbf{x}) \quad ,$$

where $\bar{A}(\mathbf{x})$ is the vector field

$$\bar{A}(\mathbf{x}) = \sum_i A_i(\mathbf{x}) \frac{\partial}{\partial x_i} \quad .$$

Consider now another system

$$\frac{\mathrm{d}x_i}{\mathrm{d}t} = B_i(\mathbf{x})$$

and rewrite the above formulae with $A \to B$ and $t \to s$.

Let the point $\mathbf{x}$ be shifted first by the vector field $\bar{A}$ acting for the infinitesimal parameter interval t, and then by $\bar{B}$ for the infinitesimal interval s,

$$x_i \to x_i + tA_i(\mathbf{x}) \to x_i + tA_i(\mathbf{x}) + sB_i(x_1 + tA_1(\mathbf{x}), \ldots)$$
$$= x_i + (tA_i(\mathbf{x}) + sB_i(\mathbf{x})) + st\sum_j \frac{\partial B_i(\mathbf{x})}{\partial x_j} A_j(\mathbf{x}) + \ldots$$

If $\mathbf{x}$ is shifted first by $\bar{B}$ through s and then by $\bar{A}$ through t, we have

$$x_i \to \ldots \to x_i + (sB_i(\mathbf{x}) + tA_i(\mathbf{x})) + ts\sum_j \frac{\partial A_i(\mathbf{x})}{\partial x_j} B_j(\mathbf{x}) + \ldots$$

The difference of the values of $f(\mathbf{x})$ at $B^s A^t(\mathbf{x})$ and $A^t B^s(\mathbf{x})$ is

$$f(B^s A^t \mathbf{x}) - f(A^t B^s \mathbf{x}) = st\sum_{ij} \left(\frac{\partial B_i}{\partial x_j} A_j - \frac{\partial A_i}{\partial x_j} B_j \right) \frac{\partial f(\mathbf{x})}{\partial x_i} \quad ,$$

the error being of the third order in the infinitesimals.

This difference has the form of a vector field acting on $f(\mathbf{x})$. This vector field, divided by st, is denoted by $[\bar{A}, \bar{B}]$. Its components are

$$[\bar{A}, \bar{B}]_i = \sum_j \left(\frac{\partial B_i}{\partial x_j} A_j - \frac{\partial A_i}{\partial x_j} B_j \right) \quad .$$

It is called the "commutator" of $\bar{A}$ and $\bar{B}$. It is skewsymmetric in $\bar{A}$ and $\bar{B}$, $[\bar{B}, \bar{A}] = -[\bar{A}, \bar{B}]$.
If $\bar{A}$ and $\bar{B}$ commute ($[\bar{A}, \bar{B}] = 0$), it does not make any difference in what order they act on a function.

Example:
$A_1 = 0$, $A_2 = -x_3$, $A_3 = x_2$ ($\bar{A}t$ induces a rotation through t about x_1),
$B_1 = x_3$, $B_2 = 0$, $B_3 = -x_1$ ($\bar{B}s$ induces a rotation through s about x_2)

$$\bar{A} = -x_3 \frac{\partial}{\partial x_2} + x_2 \frac{\partial}{\partial x_3} = \mathrm{i}\bar{L}_1 \ , \ \bar{B} = x_3 \frac{\partial}{\partial x_1} - x_1 \frac{\partial}{\partial x_3} = \mathrm{i}\bar{L}_2 \ ,$$

$$[\bar{A}, \bar{B}] = x_2 \frac{\partial}{\partial x_1} - x_1 \frac{\partial}{\partial x_2} = -\mathrm{i}\bar{L}_3$$

Hence $[\bar{L}_1, \bar{L}_2] = \mathrm{i}\bar{L}_3$ and cyclic permutations. The imaginary unit has been introduced so that ($\bar{L}_1$,$\bar{L}_2$,$\bar{L}_3$) are the components of the quantum mechanics angular momentum operator in the coordinate representation ($\hbar = 1$).

Consider now two Hamiltonian fields $\bar{U}_1$ and $\bar{U}_2$,

$$\bar{U}_\alpha = \sum_i \left(-\frac{\partial H_\alpha}{\partial q_i} \frac{\partial}{\partial p_i} + \frac{\partial H_\alpha}{\partial p_i} \frac{\partial}{\partial q_i} \right) \quad (\alpha = 1, 2)$$

and evaluate their commutator. A not-too-long calculation gives

$$[\bar{U}_1, \bar{U}_2]_{p_i} = \frac{\partial}{\partial q_i}[H_1, H_2] \quad , \quad [\bar{U}_1, \bar{U}_2]_{q_i} = -\frac{\partial}{\partial p_i}[H_1, H_2] \quad ,$$

where $[H_1, H_2]$ is the Poisson bracket of H_1 and H_2. Hence

$$[\bar{U}_1, \bar{U}_2] = \sum_i \left(-\frac{\partial H}{\partial q_i}\frac{\partial}{\partial p_i} + \frac{\partial H}{\partial p_i}\frac{\partial}{\partial q_i} \right) \quad ,$$

where

$$H = [H_2, H_1] \quad .$$

Summarizing, the commutator of two Hamiltonian fields is also a Hamiltonian field with a Hamiltonian which is the commutator of the Hamiltonians in reverse order.

In light of the above, the condition $\mathrm{d}f(p,q)/\mathrm{d}t = [f, H] = 0$ for a quantity $f(p,q)$ to be a constant of motion, can be expressed by stating that $f(p,q)$ is a constant of motion if the Hamiltonian field with f as a Hamiltonian commutes with that with Hamiltonian H.

Finally here is the neat proof of the Jacobi identity for Poisson brackets learned from Arnold and promised in section 7.2. The expression

$$[[A, B], C] + [[B, C], A] + [[C, A], B]$$

is clearly a sum of terms each containing a second derivative. The terms containing second derivatives of A are the first and the last. Denoting by $(\mathrm{d}A/\mathrm{d}t)_B$ and $(\mathrm{d}A/\mathrm{d}t)_C$ the rate of change of A under a flow with Hamiltonians B and C, respectively, we have

$$[[A, B], C] + [[C, A], B] = \left[\left(\frac{\mathrm{d}A}{\mathrm{d}t} \right)_B, C \right] - \left[\left(\frac{\mathrm{d}A}{\mathrm{d}t} \right)_C, B \right]$$

$$= [\bar{B}A, C] - [\bar{C}A, B] = (\bar{C}\bar{B} - \bar{B}\bar{C})A = [\bar{C}, \bar{B}]A \quad ,$$

where $\bar{B}$ and $\bar{C}$ are the Hamiltonian fields with Hamiltonians B and C, respectively. But $[\bar{C}, \bar{B}]$ is a <u>first</u> order differential operator. Therefore our original expression cannot contain second derivatives of A. Similarly we show that it cannot contain second derivatives of B and C. Therefore it must be zero.

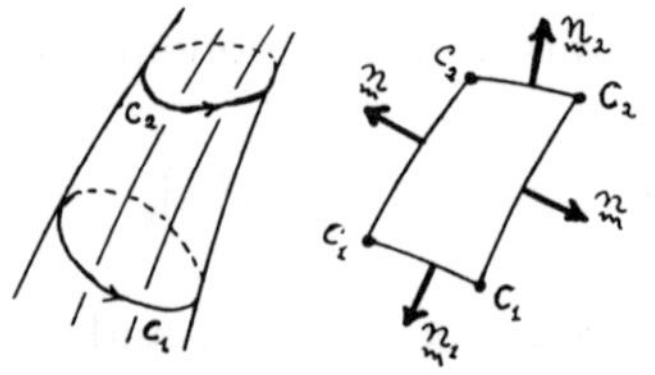

Figure 7.3: Magnetostatic field example

7.15 Poincaré-Cartan integral invariant

In phase space consider a tube formed by flow lines, and two arbitrary closed curves c_1 and c_2 encircling it.

We shall show that

$$\oint_{c_1} \tilde{\omega}^1 = \oint_{c_2} \tilde{\omega}^1 \quad , \tag{7.61}$$

where $\tilde{\omega}^1 = \sum_i p_i \tilde{d}q_i$.

Proof: Following Arnold we begin with a similar elementary problem. Consider the curves c_1 and c_2 encircling a tube of lines of force of a magnetostatic field $B = \nabla \times \mathbf{A}$ (see figure 7.3).

Let S_1 and S_2 be two surfaces having c_1 and c_2 as rims, respectively. Since $\nabla \cdot \mathbf{B} = 0$, Gauss' theorem applied to the closed surface consisting of S_1, S_2, and the portion of the tube comprised between c_1 and c_2, gives

$$\left(\int_{S_1} + \int_{S_2} + \int_{\text{tube}}\right) \mathbf{B} \cdot \hat{\mathbf{n}}\, dA = 0 \quad . \tag{7.62}$$

The integral over the tube is zero and so we find (see figure 7.5)

$$\int_{S_1} \mathbf{B} \cdot \hat{\mathbf{n}}_1\, dA_1 = \int_{S_2} \mathbf{B} \cdot (-\hat{\mathbf{n}}_2)\, dA_2 \quad . \tag{7.63}$$

Then Stokes' theorem yields

$$\int_{c_1} \mathbf{A} \cdot d\mathbf{r} = \int_{c_2} \mathbf{A} \cdot d\mathbf{r} \quad .$$

We want to cast the above into a new form. Define

$$\tilde{\omega}^1 = A_x \tilde{d}x + A_y \tilde{d}y + A_z \tilde{d}z,$$
$$\tilde{\omega}^2 = \tilde{d}\tilde{\omega}^1 = B_x \tilde{d}y \wedge \tilde{d}z + B_y \tilde{d}z \wedge \tilde{d}x + B_z \tilde{d}x \wedge \tilde{d}y,$$

where $B_x = \partial A_z/\partial y - \partial A_y/\partial z$ etc.

Filling the slots of $\tilde{\omega}^2$ with two vectors $\bar{U}$ and $\bar{V}$, we have

$$\tilde{\omega}^2(\bar{U}, \bar{V}) = \mathbf{B} \cdot (\mathbf{U} \times \mathbf{V}) = (\mathbf{B} \times \mathbf{U}) \cdot \mathbf{V} = (\mathsf{A}\bar{U}, \bar{V}) \quad ,$$

where

$$\mathsf{A} = \begin{pmatrix} 0 & -B_z & B_y \\ B_z & 0 & -B_x \\ -B_y & B_x & 0 \end{pmatrix} \quad ,$$

and $(\mathsf{A}\bar{U}, \bar{V}) = (A_{ij}U_j)V_i$.

The matrix A has zero determinant. Therefore the secular equation $\det(\mathsf{A} - \lambda\mathsf{I}) = 0$ has a solution $\lambda = 0$, and so A has an eigenvalue zero. This zero eigenvalue corresponds to an eigenvector with components (B_x, B_y, B_z),

$$\mathsf{A} \begin{pmatrix} B_x \\ B_y \\ B_z \end{pmatrix} = 0 \quad .$$

In our elementary treatment we applied Gauss' theorem to a closed surface consisting of S_1, S_2, and a surface spanned by vectors $\bar{U}$ which are eigenvectors of A belonging to the eigenvalue zero (the "flow lines" of **B**), so that

$$\int_{\text{tube}} \tilde{\omega}^2 = 0 \quad \text{because} \quad \tilde{\omega}^2(\bar{U}, \bullet) = 0 \quad .$$

The above method can be extended to any <u>odd-dimensional</u> space, in particular to the $(2n+1)$-dimensional space $(p_1, \ldots p_n, q_1, \ldots q_n, t)$. We show this for $n = 1$, the extension to $n > 1$ is trivial.

We have $\tilde{\omega}^1 = p\tilde{d}q - H\tilde{d}t$,

$$\tilde{\omega}^2 = \tilde{d}\tilde{\omega}^1 = \tilde{d}p \wedge \tilde{d}q - (\partial H/\partial p)\tilde{d}p \wedge \tilde{d}t - (\partial H/\partial q)\tilde{d}q \wedge \tilde{d}t \quad ,$$

$$\begin{aligned} \tilde{\omega}^2(\bar{U}, \bar{V}) = U_pV_q - U_qV_p - (\partial H/\partial p)(U_pV_t - U_tV_p) - (\partial H/\partial q)(U_qV_t - U_tV_q) \\ = (\mathsf{A}\bar{U}, \bar{V}) \end{aligned}$$

with

$$\bar{U} = \begin{pmatrix} U_p \\ U_q \\ U_t \end{pmatrix} \quad , \quad \bar{V} = \begin{pmatrix} V_p \\ V_q \\ V_t \end{pmatrix} \quad ,$$

and

$$\mathsf{A} = \begin{pmatrix} 0 & -1 & \partial H/\partial p \\ 1 & 0 & \partial H/\partial q \\ -\partial H/\partial p & -\partial H/\partial q & 0 \end{pmatrix} \quad .$$

Now $\det\mathsf{A} = 0$. In fact, the determinant of any skewsymmetric $(2n+1) \times (2n+1)$ matrix is zero:

$$\det\mathsf{A} = \det\mathsf{A}^{\mathrm{T}} = \det(-\mathsf{A}) = (-)^{2n+1}\det\mathsf{A}.$$

Therefore A has a zero eigenvalue, corresponding to the eigenvector

$$\begin{pmatrix} -\partial H/\partial q \\ \partial H/\partial p \\ 1 \end{pmatrix} = \begin{pmatrix} \dot{p} \\ \dot{q} \\ 1 \end{pmatrix} \quad .$$

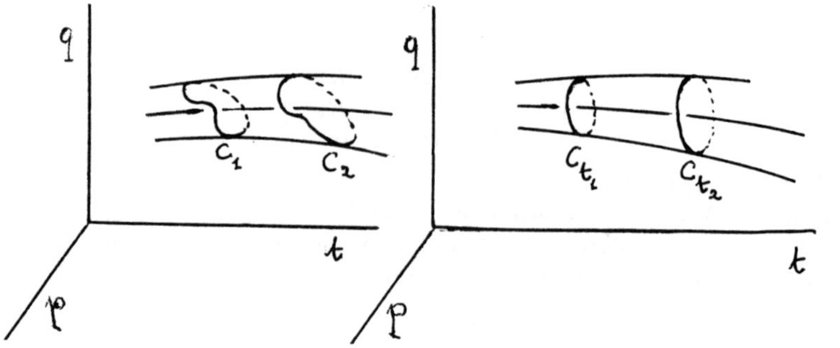

Figure 7.4: Equation (7.63)

Hence (see figure 7.4)

$$\int_{S_2} \tilde{\omega}^2 = \int_{S_1} \tilde{\omega}^2 \overset{\text{Stokes}}{\rightarrow} \oint_{c_2} \tilde{\omega}^1 = \oint_{c_1} \tilde{\omega}^1 \quad . \tag{7.64}$$

If c_1 and c_2 are the intersections of the planes $t = t_1$ and $t = t_2$ with the tube of flow, then we have, generalizing to $n > 1$,

$$\oint_{c_{t_2}} \sum_i p_i \tilde{d}q_i = \oint_{c_{t_1}} \sum_i p_i \tilde{d}q_i$$

and [6]

$$\int_{S_{t_2}} \tilde{\omega}^2 = \int_{S_{t_1}} \tilde{\omega}^2 \quad ,$$

where $\tilde{\omega}^2 = \sum_i \tilde{d}p_i \wedge \tilde{d}Q_i$ (no $\tilde{d}t$!), and S_{t_2} and S_{t_1} are the portions of the planes $t = t_2$ and $t = t_1$ having as boundaries c_{t_2} and c_{t_1}. Since the tube can be as thin as we want, we have also that $\tilde{\omega}^2 = \sum_i \tilde{d}p_i \wedge \tilde{d}q_i$ is invariant under the Hamiltonian flow, as it is under canonical transformations. We already showed in section 7.5 that the Hamiltonian flow is a canonical transformation. This is a mere confirmation of something we already know.

[6]For $n = 1$, this is Liouville's theorem.

7.16 Poincaré's invariants

The invariance of $\tilde{\omega}^2 = \sum_i \tilde{d}p_i \wedge \tilde{d}q_i$ under canonical transformations $(p, q) \to (P, Q)$ implies that

$$\int_S \sum_i \tilde{d}p_i \wedge \tilde{d}q_i = \int_S \sum_i \tilde{d}P_i \wedge \tilde{d}Q_i \quad ,$$

where S is a two-dimensional surface. This is Poincaré's first integral invariant

$$I_1 = \int_S \sum_i \mathrm{d}p_i \mathrm{d}q_i = \int_S \sum_i \mathrm{d}P_i \mathrm{d}Q_i \quad . \tag{7.65}$$

If the surface S is parametrized, $p_i = p_i(u, v)$, $q_i = q_i(u, v)$, $P_i = P_i(p(u, v), q(u, v))$, $Q_i = Q_i(p(u, v), q(u, v))$, this implies

$$\int_S \sum_i \frac{\partial(p_i, q_i)}{\partial(u, v)} \mathrm{d}u\mathrm{d}v = \int_S \sum_i \frac{\partial(P_i, Q_i)}{\partial(u, v)} \mathrm{d}u\mathrm{d}v \quad ,$$

namely

$$\sum_i \frac{\partial(p_i, q_i)}{\partial(u, v)} = \sum_i \frac{\partial(P_i, Q_i)}{\partial(u, v)} \quad .$$

Here

$$\frac{\partial(p_i, q_i)}{\partial(u, v)} = \begin{vmatrix} \partial p_i/\partial u & \partial p_i/\partial v \\ \partial q_i/\partial u & \partial q_i/\partial v \end{vmatrix} \quad .$$

The powers $(\tilde{\omega}^2)^i$ are also invariant. Note that the highest non-vanishing power is

$$(\tilde{\omega}^2)^n = \tilde{d}p_1 \wedge \ldots \tilde{d}p_n \wedge \tilde{d}q_1 \wedge \ldots \tilde{d}q_n$$

up to a numerical factor depending on the definition of a product of exterior forms.

Integrating over a region V of phase space, we have

$$I_n = \int_V [(\tilde{\omega}^2)^n]_{p,q} = \int_V [(\tilde{\omega}^2)^n]_{P,Q} \tag{7.66}$$

expressing the invariance of phase space volumes under canonical transformations.

For $p \to p(t)$, $q \to q(t)$, $P \to p(t')$, $Q \to q(t')$ we have Liouville's theorem for finite time intervals [7], while the proof in section 7.4 was for the interval $(t, t + \mathrm{d}t)$.

---

[7] In his *Classical Dynamical Systems* vol.1 of *A Course in Mathematical Physics*, (Springer,1978) p. 84, W. Thirring writes:
"In the framework of [old] classical mechanics the proof of [Liouville's] theorem requires some effort. But modern concepts are so formulated that there is really nothing to prove."

7.17 Chapter 7 problems

7.1 (i) Write the non-relativistic Hamiltonian and Hamilton's equations for a particle of mass m and charge q in the uniform static electromagnetic field $\mathbf{E} = (0, E, 0)$, $\mathbf{B} = (0, B, 0)$.
(ii) Verify that $C_x = p_x$, $C_y = p_y + (qB/c)x - qEt$, and $C_z = p_z$ are constants of motion.
(iii) Assume that at $t = 0$ the particle is at rest at $x = y = z = 0$. Expressing the Hamiltonian in terms of C_x, C_y, and C_z, find the trajectory of the particle.

7.2 (i) What is the Hamiltonian H_c derived from the complementary Lagrangian L_c in problem 6.7?

Comment on the invariance of Hamilton's equations under the transformation $(p, q, H) \to (P, Q, H_c)$.

7.3 Using the canonical transformation $\mathbf{Q} = \mathbf{p}$, $\mathbf{P} = -\mathbf{q}$, write the Hamiltonian $H = \mathbf{p}^2/2m - k/|\mathbf{q}|$ and Hamilton's equations in "momentum space", and show that the Laplace-Runge-Lenz vector is constant.

7.4 Study the canonical transformation generated by

$$G_2 = \frac{2qP - \sin\alpha\,(q^2 + P^2)}{2\cos\alpha} \quad ,$$

containing a parameter α, and the case of α infinitesimal. What happens for $\alpha = -\pi/2$? Use the corresponding G_1 for transformations differing infinitesimally from $p = Q$, $P = -q$.

7.5 (i) Show that the generating function

$$G_2 = (P_1 q_1 + P_2 q_2)\cos\alpha + (P_2 q_1 - P_1 q_2)\sin\alpha$$

induces the canonical transformation

$$(p_1, q_1, p_2, q_2) \to (P_1, Q_1, P_2, Q_2)$$

with

$$P_1 = p_1\cos\alpha - p_2\sin\alpha \;, P_2 = p_1\sin\alpha + p_2\cos\alpha \quad ,$$
$$Q_1 = q_1\cos\alpha - q_2\sin\alpha \;, Q_2 = q_1\sin\alpha + q_2\cos\alpha \quad .$$

(ii) Show that this transformation with

$$\tan(2\alpha) = -\frac{2f}{\omega_1^2 - \omega_2^2}$$

uncouples the two harmonic oscillators with Hamiltonian ($\omega_1 > \omega_2$)

$$H = (p_1^2 + \omega_1^2 q_1^2)/2 + (p_2^2 + \omega_2^2 q_2^2)/2 + f q_1 q_2 \quad .$$

Expressing H in terms of the new variables, one has

$$H = (P_1^2 + \Omega_1^2 Q_1^2)/2 + (P_2^2 + \Omega_2^2 Q_2^2)/2$$

with

$$\Omega_1^2 = \left(\omega_1^2 + \omega_2^2 + \sqrt{(\omega_1^2 - \omega_2^2)^2 + 4f^2}\right)/2 \quad ,$$

$$\Omega_2^2 = \left(\omega_1^2 + \omega_2^2 - \sqrt{(\omega_1^2 - \omega_2^2)^2 + 4f^2}\right)/2 \quad .$$

(iii) What happens is $\omega_1 = \omega_2 \to \omega$? Nothing bad, $\alpha = -\pi/4$, $p_1 = (P_1 - P_2)/\sqrt{2}$, $q_1 = (Q_1 - Q_2)/\sqrt{2}$, $p_2 = (P_1 + P_2)/\sqrt{2}$, $q_2 = (Q_1 + Q_2)/\sqrt{2}$, $\Omega_1^2 = \omega^2 + f$, $\Omega_2^2 = \omega^2 - f$.

7.6 Find the generating function for the canonical transformation $(r, \theta, \phi, p_r, p_\theta, p_\phi) \to (x, y, z, p_x, p_y, p_z)$.

7.7 (i) Consider the canonical transformation induced by the generating function

$$G_2(\mathbf{r}, \mathbf{P}) = x_3 P_3 + P_1 P_2 + \sqrt{\lambda}\,(x_1 P_2 + x_2 P_1) + \lambda x_1 x_2$$

and use it to transform the Hamiltonian

$$H = [p_1^2 + (p_2 - \lambda x_1)^2 + p_3^2]/2m \quad .$$

(ii) Write and solve Hamilton's equations in the new variables.

(iii) Use the result to study the motion of a positron (charge $+e$) in a uniform magnetic field of intensity B in the x_3 direction ($A_1 = A_3 = 0$, $A_2 = Bx_1$).

7.8 (i) Verify that the Poisson bracket of the Toda Hamiltonian

$$H = \frac{1}{2}\,(p_x^2 + p_y^2) + \frac{1}{24}[e^{2(x - y\sqrt{3})} + e^{2(x + y\sqrt{3})} + e^{-4x}] - \frac{1}{8}$$

with the Hénon integral of motion

$$I = 8p_y(p_y^2 - 3p_x^2) + (p_y + p_x\sqrt{3}\,)e^{2(x - y\sqrt{3})} + (p_y - p_x\sqrt{3}\,)e^{2(x + y\sqrt{3})} - 2p_y e^{-4x}$$

is zero.

(ii) Since I does not explicitly depend on t, its conservation must correspond to the invariance of the Hamiltonian under a transformation $\mathbf{p} \to \mathbf{p} + \delta\mathbf{p}$, $\mathbf{x} \to \mathbf{x} + \delta\mathbf{x}$, with $\delta\mathbf{p}$ and $\delta\mathbf{x}$ of the order of an infinitesimal parameter ϵ, $\delta H = O(\epsilon^2)$. Find $\delta\mathbf{p}$ and $\delta\mathbf{x}$.

7.9 (i) What are the infinitesimal transformations for x_k and p_k generated by ηA_i, where η is an infinitesimal parameter and

$$A_i = -\frac{x_i}{r} + \frac{1}{km}(|\mathbf{p}|^2 x_i - (\mathbf{r} \cdot \mathbf{p})p_i)$$

is the i-th component of the Laplace-Runge-Lenz vector.

(ii) Show that the Hamiltonian $H = (|\mathbf{p}|^2/2m \ - \ k/r)$ is invariant in the first order in η.

7.10 Let p and q be canonical variables for a particle with Hamiltonian $H = p^2/2m$.

(i) Perform the canonical transformation $(p, q) \to (P, Q)$ with

$$G_2(q, P) = (q - at)(P + b) \quad ,$$

where a and b are constants. Find P and Q.

(ii) We calculate $\dot{Q} = \partial H(p(P,Q), q(P,Q))/\partial P$, then we express the resuly in terms of q and p finding $\dot{q} = p/m \ + a$ (strange!). What have we done wrong? What should we have done?

7.11 In a frame XYZ an electron is subject to the uniform magnetic field $\mathbf{B} = (0, 0, B)$, $\mathbf{A} = (-By, 0, 0)$. In a frame $X'Y'Z'$ moving in the x direction withe velocity u ($x' = x - ut, y' = y, z' = z$), it is subject to the magnetic field $\mathbf{B}' = \mathbf{B}$ and the electric field $\mathbf{E}' = (0, -Bu/c, 0)$.

(i) Write the Hamiltonians H and K for the electron in XYZ and $X'Y'Z'$, respectively.

(ii) Find $\mathbf{p}$ and $\mathbf{p}'$, and Hamilton's equations in the two frames.

(iii) Verify that

$$G_2(\mathbf{r}, \mathbf{p}', t) = -up'_x t - mu^2 t/2 \ + \mathbf{r} \cdot \mathbf{p}' + mux$$

induces the Galilean transformation from $(\mathbf{r}, \mathbf{p})$ to $(\mathbf{r}', \mathbf{p}')$. Note that this corresponds to

$$G_1(\mathbf{r}, \mathbf{r}', t) = G_2(\mathbf{r}, \mathbf{p}', t) - \mathbf{r}' \cdot \mathbf{p}'$$

as we had in section 7.11.

7.12 Suppose we have the integral of motion $u(p, q, t)$ explicitly dependent on t

$$u(p, q, t) = \ln(p + im\omega q) - i\omega t$$

for the harmonic oscillator with Hamiltonian $H = p^2/2m + m\omega^2 q^2/2$. (For $q = A\cos(\omega t + \alpha)$, $p = -m\omega A\sin(\omega t + \alpha)$, $u = \ln(m\omega A) + \mathrm{i}(\alpha + \pi/2)$.) Then

$$0 = \frac{\mathrm{d}u}{\mathrm{d}t} = [u, H] + \frac{\partial u}{\partial t} \quad ,$$

which can also be written in the form

$$\frac{\partial u}{\partial q}\frac{\partial H}{\partial p} - \frac{\partial u}{\partial p}\frac{\partial H}{\partial q} + \frac{\partial u}{\partial t} = 0 \quad ,$$

$$\frac{\partial H}{\partial p}[u, p] + \frac{\partial H}{\partial q}[u, q] + \frac{\partial u}{\partial t} = 0 \quad .$$

Find the corresponding infinitesimal transformation under which

$$\delta H = -\epsilon\,\frac{\partial u}{\partial t} \quad .$$

7.13 (i) Find the infinitesimal transformations induced in p and q by the integral of motion $I = up - H$ for a particle in one dimension with Hamiltonian $H = p^2/2m + U(x - ut)$ (u =constant).

(ii) Show that under this transformation $\delta H = -\epsilon\ \partial I/\partial t$.

7.14 For $\tilde{\omega}^2 = \tilde{d}p \wedge \tilde{d}q$ and $\bar{U} = -(\partial H/\partial q)\partial/\partial p + (\partial H/\partial p)\partial/\partial q$, show that $L_{\bar{U}}\tilde{\omega}^2 = 0$.

7.15 Verify the formula

$$L_{\bar{A}}\tilde{\omega} = \tilde{d}[\tilde{\omega}(\bar{A})] + (\tilde{d}\tilde{\omega})(\bar{A})$$

in the case where $\tilde{\omega}$ is a 1-form.

7.16 Show that the operators $\tilde{d}$ and $L_{\bar{A}}$ acting on a form commute, $\tilde{d}L_{\bar{A}}\tilde{\omega} = L_{\bar{A}}\tilde{d}\tilde{\omega}$.

7.17 Consider a flow generated by the vector field $\bar{V}$ and a closed curve C_t whose points are transported by the flow, so that at the time $t + \mathrm{d}t$ they cover the curve $C_{t+\mathrm{d}t}$.
If $\tilde{\omega}^1 = A_j\tilde{d}x^j$ is a 1-form, one has

$$\oint_{C_{t+\mathrm{d}t}} \tilde{\omega}^1 = \oint_{C_t} \tilde{\omega}^1 + \mathrm{d}t \oint_{C_t} L_{\bar{V}}\tilde{\omega}^1 \quad .$$

Interpret this formula in three dimensions.

7.18 Use the result of problem 7.17 to show that

$$\oint_{C_{t+dt}} p\,\tilde{d}q = \oint_{C_t} p\,\tilde{d}q \quad .$$

Solutions to ch. 7 problems

S7.1 (i)

$$H = (\mathbf{p} - q\mathbf{A}/c)^2/2m \; + \; qV, \; V = -Ey, \; A_x = -By, \; A_y = A_z = 0,$$

$$H = [(p_x + qBy/c)^2 + p_y^2 + p_z^2]/2m \; - \; qEy$$

$$p_x = m\dot{x} + qA_x/c = m\dot{x} - qBy/c \; , \; p_y = m\dot{y} \; , \; p_z = m\dot{z}$$

At $t = 0$, $p_x = p_y = p_z = 0$, $x = y = z = 0$.
(ii) $C_x = p_x$ and $C_z = p_z$ are constants of motion because x and z do not occur in the Hamiltonian. It is easily verified that $C_y = p_y + (qB/c)x$ is a constant of motion. In fact $\dot{C}_y = -(\partial H/\partial y) + (qB/c)\dot{x} - qE$,

$$\dot{C}_y = -(qB/mc)(p_x + qBy/c) + qE + qB\dot{x}/c - qE = 0 \; .$$

(iii) Since $C_x = C_y = C_z = 0$ at $t = 0$, we have

$$H = [(qB/c)^2 y^2 + (qBx/c - qEt)^2]/2m \; - \; qEy \; .$$

From this we see that $H = 0$.
The trajectory is the cycloid

$$\left(x - \frac{Ect}{B}\right)^2 + \left(y - \frac{Emc^2}{qB^2}\right)^2 = r^2$$

with $r = Emc^2/qB^2$.

S7.2 (i)

$$H_c = P\dot{Q} - (P\dot{Q} + \alpha p\dot{q} - \alpha L) = -\alpha H(p, q, t)$$

(ii) The equations

$$\dot{Q} = \frac{\partial H_c}{\partial P} \quad , \quad \dot{P} = -\frac{\partial H_c}{\partial Q}$$

give

$$\dot{p} = \frac{\partial(-\alpha H)}{\partial(\alpha q)} = -\frac{\partial H}{\partial q} \quad , \quad \alpha\dot{q} = -\frac{\partial(-\alpha H)}{\partial p} \quad , \quad \dot{q} = \frac{\partial H}{\partial p} \; .$$

Hamilton's equations are invariant under the canonical transformation $Q = p$, $P = -q$ without transforming the Hamiltonian, while their invariance under the (non-canonical) transformation $Q = p$, $P = q$ requires a change of sign for the Hamiltonian.

S7.3

$$H = \mathbf{Q}^2/2m \; - k/|\mathbf{P}| \, , \; \dot{Q}_i = \partial H/\partial P_i = kP_i/|\mathbf{P}|^3, \; \dot{P}_i = -\partial H/\partial Q_i = -Q_i/m$$

$$\mathbf{l} = \mathbf{q} \times \mathbf{p} = \mathbf{Q} \times \mathbf{P} \, , \; \dot{l}_i = \epsilon_{ijk}(kP_jP_k/|\mathbf{P}|^3 \; - Q_jQ_k/m) = 0$$

$$\mathbf{A} = -\mathbf{q}/|\mathbf{q}| \; + \mathbf{p} \times \mathbf{l}/km = \mathbf{P}/|\mathbf{P}| \; - \mathbf{l} \times \mathbf{Q}/km$$

$$\mathrm{d}(\mathbf{P}/|\mathbf{P}|)/\mathrm{d}t = \dot{\mathbf{P}}/|\mathbf{P}| - (\mathbf{P} \cdot \dot{\mathbf{P}})\mathbf{P}/|\mathbf{P}|^3 = [-|\mathbf{P}|^2\mathbf{Q} + (\mathbf{P} \cdot \mathbf{Q})\mathbf{P}]/m|\mathbf{P}|^3$$

$$\mathrm{d}(\mathbf{l} \times \mathbf{Q})/\mathrm{d}t = \mathbf{l} \times \dot{\mathbf{Q}} = k\mathbf{l} \times \mathbf{P}/|\mathbf{P}|^3 = k[-|\mathbf{P}|^2\mathbf{Q} + (\mathbf{P} \cdot \mathbf{Q})\mathbf{P}]/|\mathbf{P}|^3$$

S7.4

$p = \partial G_2/\partial q = (P - q\ \sin\alpha)/\cos\alpha$, $Q = \partial G_2/\partial P = (q - P\ \sin\alpha)/\cos\alpha$, namely $P = p\ \cos\alpha + q\ \sin\alpha$, $q = Q\ \cos\alpha + P\ \sin\alpha$. Note that $G_2 = G_1 + PQ$, where

$$G_1 = \frac{1}{2\ \sin\alpha}(q^2\ \cos\alpha - 2qQ + Q^2\ \cos\alpha) \quad .$$

While G_1 is singular for $\alpha = 0$, G_2 is singular for $\alpha = -\pi/2$. Therefore G_2 with $\alpha \to \delta\alpha$ (infinitesimal) can be used for a transformation differing infinitesimally from the identity $P = p$, $Q = q$, in which case $P \simeq p + q\ \delta\alpha$, $Q \simeq q - p\ \delta\alpha$. On the other hand, G_1 with $\alpha = -\pi/2 + \delta\alpha$ could be used for a transformation differing infinitesimally from $p = Q$, $P = -q$.

S7.6 A simple extension of the two-dimensional case, gives

$$G_2(r,\theta,\phi,p_x,p_y,p_z) = r\ \sin\theta\ \cos\phi\ p_x + r\ \sin\theta\ \sin\phi\ p_y + r\ \cos\theta\ p_z \quad .$$

One finds $p_r = \partial G_2/\partial r = \sin\theta\ \cos\phi\ p_x + \sin\theta\ \sin\phi\ p_y + \cos\theta\ p_z$,

$$p_\theta = \partial G_2/\partial\theta = r\ \cos\theta\ \cos\phi\ p_x + r\ \cos\theta\ \sin\phi\ p_y - r\ \sin\theta\ p_z,$$
$$p_\phi = \partial G_2/\partial\phi = -r\ \sin\theta\ \sin\phi\ p_x + r\ \sin\theta\ \cos\phi\ p_y,$$
$$x = \partial G_2/\partial p_x = r\ \sin\theta\ \cos\phi, \text{ etc.}$$

The following may be useful:

$$\begin{cases} p_x = \cos\phi\ (\sin\theta\ p_r + \cos\theta\ p_\theta/r) - (\sin\phi\ /r\sin\theta)p_\phi \quad , \\ p_y = \sin\phi\ (\sin\theta\ p_r + \cos\theta\ p_\theta/r) + (\cos\phi\ /r\sin\theta)p_\phi \quad , \\ p_z = \cos\theta\ p_r - \sin\theta\ p_\theta/r \quad . \end{cases}$$

S7.7 (i) One finds

$$p_1 = \partial G_2/\partial x_1 = \sqrt{\lambda}\ P_2 + \lambda x_2\ ,\ p_2 = \partial G_2/\partial x_2 = \sqrt{\lambda}\ P_1 + \lambda x_1\ ,\ p_3 = \partial G_2/\partial x_3 = P_3,$$

$$Q_1 = \partial G_2/\partial P_1 = P_2 + \sqrt{\lambda}x_2\ ,\ Q_2 = \partial G_2/\partial P_2 = P_1 + \sqrt{\lambda}x_1\ ,\ Q_3 = \partial G_2/\partial P_3 = x_3\ .$$

Note that $p_1 = \sqrt{\lambda}\ Q_1$.

(ii) The transformed Hamiltonian is

$$K = \frac{\lambda}{2m}(P_1^2 + Q_1^2) + \frac{1}{2m}P_3^2$$

(harmonic oscillator and free particle in one dimesion). P_2 and Q_2 do not appear in K.

Hamilton's equations yield

$$\dot{P}_1 = -\frac{\lambda}{m}\ Q_1\ ,\ \dot{Q}_1 = \frac{\lambda}{m}\ P_1\ ,\ P_2 = \text{constant}\ ,\ Q_2 = \text{constant} \quad ,$$

with the solution

$$P_1 + \mathrm{i}Q_1 = A\ \exp(\mathrm{i}(\lambda t/m + \alpha))$$

(A and α real). Then, since

$$P_1 + iQ_1 = Q_2 + iP_2 - \sqrt{\lambda}\,(x_1 - ix_2) \ ,$$

we have

$$x_1 - ix_2 = [Q_2 + iP_2 - A\ \exp(i(\lambda t/m\ + \alpha))]/\sqrt{\lambda}\ .$$

If $\lambda > 0$, this gives

$$\begin{cases} x_1 = [\ Q_2 - A\ \cos(\lambda t/m\ + \alpha)]/\sqrt{\lambda} \ , \\ x_2 = [-P_2 + A\ \sin(\lambda t/m\ + \alpha)]/\sqrt{\lambda} \ . \end{cases}$$

(iii) Comparing the Hamiltonian in (i) with

$$H = [p_1^2 + (p_2 - eBx_1/c)^2 + p_3^2]/2m$$

we see that the equations of motion for the positron are

$$\begin{cases} x_1 = [\ Q_2 - A\ \cos(\omega t + \alpha)]/\sqrt{m\omega} \ , \\ x_2 = [-P_2 + A\ \sin(\omega t + \alpha)]/\sqrt{m\omega} \ , \\ x_3 = x_3(0) + vt \ , \end{cases}$$

with $\omega = eB/mc$.

S7.9 (i) $\delta f = \eta[f, A_i]$,

$$\delta x_k = (\eta/km)[2x_i p_k - (\mathbf{r}\cdot\mathbf{p})\delta_{ik} - x_k p_i] \ ,$$
$$\delta p_k = (\eta/kmr^3)[kmr^2\delta_{ik} - kmx_i x_k - r^3|\mathbf{p}|^2\delta_{ik} + r^3 p_i p_k] \ ,$$

(ii)

$$\delta|\mathbf{p}|^2 = (2\eta/r^3)[r^2 p_i - (\mathbf{p}\cdot\mathbf{r})x_i] + O(\eta^2) \ ,$$
$$\delta(1/r) = (\eta/kmr^3)[r^2 p_i - (\mathbf{p}\cdot\mathbf{r})x_i] + O(\eta^2) \ ,$$
$$\delta H = \delta(|\mathbf{p}|^2/2m\ -\ k/r) = O(\eta^2)\ .$$

S7.10 (i)

$$p = \partial G_2/\partial q = P + b,\ P = p - b,\ Q = \partial G_2/\partial P = q - at$$

(ii) $H(p(P,Q), q(P,Q)) = (P+b)^2/2m$. If we write (wrongly) $\dot{Q} = \partial H/\partial P$, we obtain $\dot{Q} = (P+b)/m$. If in this we substitute $P = p - b$, $Q = q - at$, we get $\dot{q} - a = p/m$.

We should have remembered that the new Hamiltonian is

$$K = H + \partial G_2/\partial t = (P+b)^2/2m\ - a(P+b)$$

and so

$$\dot{Q} = \partial K/\partial P = (P+b)/m\ - a\ ,\ \dot{q} - a = p/m\ - a.$$

S7.11 (i)

$$H = [(p_x - eBy/c)^2 + p_y^2 + p_z^2]/2m \quad ,$$
$$K = [(p'_x - eBy'/c)^2 + p_y'^2 + p_z'^2]/2m \; - (eBu/c)y'$$

(ii)

$$\mathbf{p} = (m\dot{x} + eBy/c, m\dot{y}, m\dot{z}),\ \mathbf{p}' = (m\dot{x}' + eBy'/c, m\dot{y}', m\dot{z}') = \mathbf{p} - m\mathbf{u}$$
$$\dot{p}_x = 0 \ ,\ \dot{p}_y = eB(p_x - eBy/c)/mc \ ,\dot{p}_z = 0$$
$$\dot{x} = (p_x - eBy/c)/m \ ,\ \dot{y} = p_y/m \ ,\ \dot{z} = p_z/m$$
$$m\ddot{x} = -eB\dot{y}/c \ ,\ m\ddot{y} = eB\dot{x}/c \ ,\ m\ddot{z} = 0$$
$$\dot{p}'_x = 0 \ ,\ \dot{p}'_y = eB(p'_x - eBy'/c)/mc + eBu/c \ ,\ \dot{p}'_z = 0$$
$$\dot{x}' = (p'_x - eBy'/c)/m \ ,\ \dot{y}' = p'_y/m \ ,\ \dot{z}' = p'_z/m$$

Since $\dot{x}' = \dot{x} - u$, $\ddot{x}' = \ddot{x}$, $\dot{y}' = \dot{y}$, $\dot{z}' = \dot{z}$, the equations $m\ddot{x}' = -eB\dot{y}'/c$ and $m\ddot{z}' = 0$ are clearly equivalent to the corresponding unprimed equations. We have also

$$m\ddot{y} = m\ddot{y}' = eB\dot{x}'/c \; - eE'_y = eB(\dot{x} - u)/c \; + eBu/c = eB\dot{x}/c \ .$$

S7.12

$$p \to p + \delta p,\ q \to q + \delta q, \text{ with } \delta p = \epsilon[u, p],\ \delta q = \epsilon[u, q],$$

$$\delta p = \epsilon\frac{im\omega}{p + im\omega q} \ ,\ \delta q = -\epsilon\frac{1}{p + im\omega q} \quad .$$

S7.13 (i)

$$\delta p = \epsilon[I, p] = -\epsilon \ \partial U(x - ut)/\partial x \ ,\ \delta x = \epsilon[I, x] = \epsilon(p/m \; - u)$$

(ii)

$$\delta H = \frac{\partial H}{\partial p}\ \delta p + \frac{\partial H}{\partial x}\ \delta x = \left[-\frac{p}{m}\frac{\partial U(x - ut)}{\partial x} + \frac{\partial U(x - ut)}{\partial x}\left(\frac{p}{m} \; - u\right)\right]\epsilon$$

$$= -\epsilon u\frac{\partial U(x - ut)}{\partial x} = +\epsilon\frac{\partial U(x - ut)}{\partial t} = -\epsilon\frac{\partial I}{\partial t}$$

Elementary remark: In a system moving with velocity u, we would have $m\dot{x}'^2/2 \; + U(x') = E'$ =constant. But now $x' = x - ut$, $\dot{x}' = \dot{x} - u$, and so $E' = (m/2)(\dot{x}^2 + u^2 - 2u\dot{x}) + U(x - ut) = [m\dot{x}^2/2 \; + U(x - ut)] - um\dot{x} + mu^2/2$ $= H - up + mu^2/2$. Thus $I = up - H = mu^2/2 \; - E'$ =constant.

S7.14

$$\tilde{d}p' \wedge \tilde{d}q' = \tilde{d}(p - s\ \partial H/\partial q) \wedge \tilde{d}(q + s\ \partial H/\partial p)$$
$$= \left(1 + s\ \frac{\partial^2 H}{\partial q \partial p} - s\ \frac{\partial^2 H}{\partial p \partial q}\right) \tilde{d}p \wedge \tilde{d}q = \tilde{d}p \wedge \tilde{d}q \quad .$$

Also, since $\tilde{d}\tilde{\omega}^2 = 0$ for $\tilde{\omega}^2 = \tilde{d}p \wedge \tilde{d}q$, the formula in the following problem 7.15 gives

$$L_{\bar{U}}\tilde{\omega}^2 = \tilde{d}[\tilde{\omega}^2(\bar{U})] = \left(\frac{\partial U_p}{\partial p} + \frac{\partial U_q}{\partial q}\right) \tilde{d}p \wedge \tilde{d}q = 0 \quad .$$

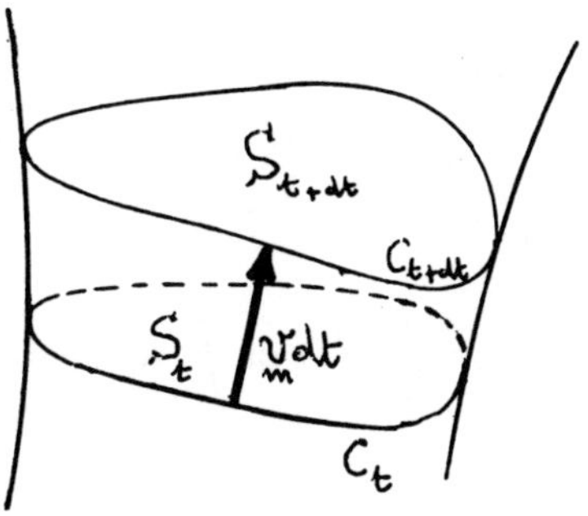

Figure 7.5: Problem 7.17

S7.15

$$\tilde{\omega} = \omega_j \tilde{d}x^j \ , \ \tilde{d}\tilde{\omega} = (\partial\omega_j/\partial x^i)\tilde{d}x^i \wedge \tilde{d}x^j \ , \ \tilde{d}[\tilde{\omega}(\bar{A})] = \frac{\partial(\omega_j A^j)}{\partial x^i}\,\tilde{d}x^i \quad ,$$

$$(\tilde{d}\tilde{\omega})(\bar{A}) = \frac{\partial\omega_j}{\partial x^i}\,(A^i\tilde{d}x^j - A^j\tilde{d}x^i) = \left(\frac{\partial\omega_i}{\partial x^j} - \frac{\partial\omega_j}{\partial x^i}\right) A^j\tilde{d}x^i \quad .$$

The sum of the last two expressions is

$$\left(\frac{\partial\omega_i}{\partial x^j}\,A^j + \omega_j\,\frac{\partial A^j}{\partial x^i}\right)\tilde{d}x^i = (L_{\bar{A}}\tilde{\omega})_i\tilde{d}x^i \quad .$$

S7.16 Acting with $\tilde{d}$ on the equation in the previous problem, and remembering that $\tilde{d}^2 = 0$, we have

$$\tilde{d}L_{\bar{A}}\tilde{\omega} = \tilde{d}[(\tilde{d}\tilde{\omega})(\bar{A})] = L_{\bar{A}}\tilde{d}\tilde{\omega} - (\tilde{d}\tilde{d}\tilde{\omega})(\bar{A}) = L_{\bar{A}}\tilde{d}\tilde{\omega}.$$

S7.17

$$L_{\bar{V}}\tilde{\omega}^1 = \tilde{d}[\tilde{\omega}^1(\bar{V})] + (\tilde{d}\tilde{\omega}^1)(\bar{V}),\ \oint \tilde{d}[...] = 0,$$

$$\mathrm{dt}\oint_{C_t} L_{\bar{V}}\tilde{\omega}^1 = \mathrm{dt}\oint_{C_t} \tilde{d}\tilde{\omega}^1(\bar{V})$$

$$= \oint_{C_t}\left(\frac{\partial A_j}{\partial x^i} - \frac{\partial A_i}{\partial x^j}\right)(v^i\ \mathrm{dt})\tilde{d}x^j \quad .$$

The above equation simply expresses the vanishing of the solenoidal $\nabla \times \mathbf{A}$ through the closed surface formed by the surfaces $S_{t+\mathrm{dt}}$, S_t, and the flow tube connecting them (see figure 7.5).

S7.18 For $\tilde{\omega}^1 = p\,\tilde{d}q$, it is easy to show that

$$L_{\bar{U}}\tilde{\omega}^1 = \tilde{d}[\tilde{\omega}^1(\bar{U}) - H] \quad .$$

Since $\oint \tilde{d}(...) = 0$, we have

$$\oint_{C_{t+\mathrm{dt}}} \tilde{\omega}^1 = \oint_{C_t} \tilde{\omega}^1 + \mathrm{dt} \oint_{C_t} L_{\bar{U}}\tilde{\omega}^1 = \oint_{C_t} \tilde{\omega}^1 \quad .$$

Chapter 8

ACTION-ANGLE VARIABLES

Periodic motions are mapped onto uniform circular motions by introducing the action-angle variables.
For a system executing periodic motion, a quantity can be defined which is invariant under slow changes of the system parameters. This is an "adiabatic invariant".

8.1 One dimension

In this section we study the mapping of a periodic one-dimensional motion onto a uniform circular motion.

This is realized by a canonical transformation $(p, q) \to (J, \phi/2\pi)$, where

$$J = \oint_{\mathrm{C}} p\,\mathrm{d}q = \left(\int \tilde{d}p \wedge \tilde{d}q\right)_{\mathrm{D}} = A(E) \tag{8.1}$$

is the action integral along the closed orbit "C" enclosing the region "D" of the (p, q)-plane.

Since $\tilde{d}p \wedge \tilde{d}q = \tilde{d}J \wedge \tilde{d}(\phi/2\pi)$ and J is a constant of motion, we have

$$J = \oint_{\mathrm{C}} p\mathrm{d}q = \oint_{\mathrm{C}} J\,\mathrm{d}(\phi/2\pi) = J\oint_{\mathrm{C}} \mathrm{d}(\phi/2\pi) \quad , \tag{8.2}$$

and so it is clear that ϕ is incremented by 2π while C is covered once.

From the Hamilton equation

$$\frac{\dot{\phi}}{2\pi} = \frac{\partial H}{\partial J} = \frac{\partial E}{\partial J} = \left(\frac{\mathrm{d}A(E)}{\mathrm{d}E}\right)^{-1} \tag{8.3}$$

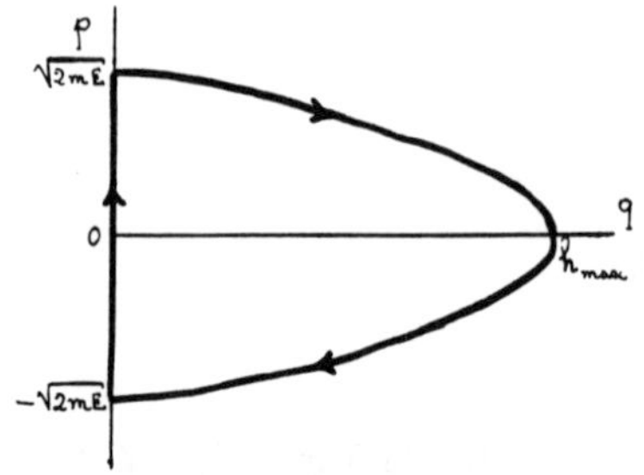

Figure 8.1: Ball bouncing up and down

we see that $\dot{\phi}$ is constant, and the period T is given by

$$T = \frac{2\pi}{\dot{\phi}} = \frac{\mathrm{d}A(E)}{\mathrm{d}E} \tag{8.4}$$

as we already know (see equation (1.10)).

Example 1: The mapping of a harmonic motion onto a uniform circular motion is the inverse of the elementary construction of the harmonic motion by projecting a uniform circular motion onto a diameter.

Substituting $p = \sqrt{m\omega J/\pi}\,\cos\phi$ and $q = \sqrt{J/m\omega\pi}\,\sin\phi$ in the Hamiltonian $H = (p^2 + m^2\omega^2 q^2)/2m$, we find $E = \omega J/2\pi$. The transformation is canonical because

$$\frac{\partial q}{\partial \phi}\frac{\partial p}{\partial J} - \frac{\partial q}{\partial J}\frac{\partial p}{\partial \phi} = 2\pi \quad .$$

Example 2: Ball bouncing elastically up and down on floor (see figure 8.1). The canonical transformation is

$$p = \left(\frac{3m^2 gJ}{2}\right)^{\frac{1}{3}}\left(1 - \frac{\phi}{\pi}\right) \quad , \quad q = \left(\frac{3J}{2m\sqrt{g}}\right)^{\frac{2}{3}}\frac{\phi}{\pi}\left(1 - \frac{\phi}{2\pi}\right) \quad .$$

Substituting in the Hamiltonian $H = p^2/2m + mgq$, one finds

$$E = \frac{1}{2m}\left(\frac{3m^2 gJ}{2}\right)^{\frac{2}{3}} \quad ,$$

$A(E) = J = 2(2mE)^{\frac{3}{2}}/3m^2 g$, $T = \mathrm{d}A(E)/\mathrm{d}E = 2v_0/g$, where $E = mv_0^2/2$.

It is natural to ask: "How did you get those canonical transformations?" The answer is simple. One must find the generating function $G_2(q, J)$. Then $p = \partial G_2/\partial q$ and $\phi/2\pi = \partial G_2/\partial J$.

The generating function can be obtained by solving the Hamilton-Jacobi equation

$$H\left(q, \frac{\partial G_2}{\partial q}\right) = E(J) \quad , \tag{8.5}$$

where we have stressed that the energy must be expressed in terms of J.

For the harmonic oscillator (Example 1), $G_2(q, J)$ is easily obtained from $S(q_2, q_1, E = ka^2/2)$ of equation (1.22) with the following changes: $E \to \omega J/2\pi$, $a \to \sqrt{J/\pi m\omega}$, $q_2 \to q$, $q_1 \to 0$. One obtains

$$G_2(q, J) = \frac{m\omega}{2}\left[q\sqrt{\frac{J}{\pi m\omega} - q^2} + \frac{J}{\pi m\omega}\sin^{-1}\left(q\sqrt{\frac{\pi m\omega}{J}}\right)\right] \quad . \tag{8.6}$$

Then

$$\frac{\phi}{2\pi} = \frac{\partial G_2}{\partial J} = \frac{1}{2\pi}\sin^{-1}\left(q\sqrt{\frac{\pi m\omega}{J}}\right) \ , \ q = \sqrt{\frac{J}{\pi m\omega}}\sin\phi \quad , \tag{8.7}$$

while

$$p = \frac{\partial G_2}{\partial q} = m\omega\sqrt{\frac{J}{\pi m\omega} - q^2} = \sqrt{\frac{Jm\omega}{\pi}}\cos\phi \quad . \tag{8.8}$$

For Example 2, we show the convenience of using $G_4(p, J)$, for which $q = -\partial G_4/\partial p$ and $\phi/2\pi = \partial G_4/\partial J$. Substituting the former in $p^2/2m + mgq = E(J)$, we have $(p^2/2m) - mg\partial G_4/\partial p = E(J)$, and, by an elementary integration,

$$G_4 = \frac{1}{mg}\left[\frac{1}{6m}(p^3 - p_0^3) - E(J)(p - p_0)\right] \quad . \tag{8.9}$$

Then $\phi/2\pi = \partial G_4/\partial J = -[(p - p_0)/mg]\mathrm{d}E(J)/\mathrm{d}J$ yields

$$p = p_0 - \left(\frac{3m^2 gJ}{2}\right)^{\frac{1}{3}}\frac{\phi}{\pi} \quad . \tag{8.10}$$

If we want $p = p_0 = \sqrt{2mE} = (3m^2gJ/2)^{\frac{1}{3}}$ for $\phi = 0$, we have

$$p = \left(\frac{3m^2 gJ}{2}\right)^{\frac{1}{3}}\left(1 - \frac{\phi}{\pi}\right) \quad . \tag{8.11}$$

Substituting this in $q = -\partial G_4/\partial p = -[p^2/2m - E(J)]/mg$, we find the equation for q.

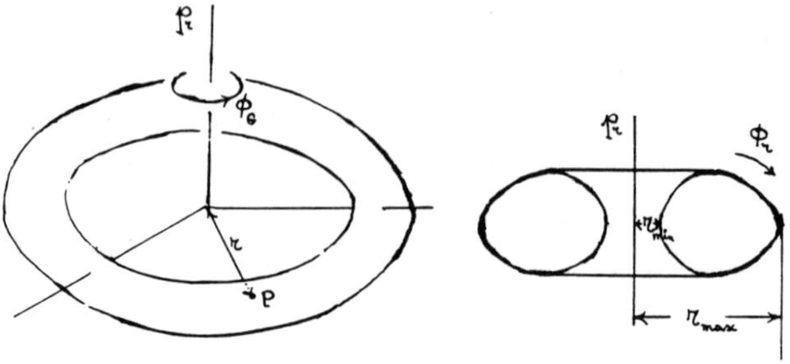

Figure 8.2: Torus for Kepler motion

8.2 Multiply periodic systems

A multiply periodic system with n degrees of freedom is one in which each of a set of n suitably chosen q_i $(i = 1, \ldots n)$ is simply periodic with period T_i.

The T_i's are in general different. If their ratios are irrational, the trajectory in $2n$-dimensional phase space is not closed. Remember the Lissajou figures from General Physics!

The motion in the $2n$-dimensional phase space can be mapped onto an n-torus, on which n angular coordinates $(\phi_1, \ldots \phi_n)$mod 2π can be defined. (For $n = 1$ the 1-torus is a circle.) Each coordinate changes uniformly with time,

$$\frac{\mathrm{d}\phi_i}{\mathrm{d}t} = \omega_i \text{ (constant)} \quad . \tag{8.12}$$

For each ϕ_i one can define a conjugate momentum J_i, so that

$$\sum_{i=1}^{n} \tilde{d}p_i \wedge \tilde{d}q_i = \sum_{i=1}^{n} \tilde{d}J_i \wedge \tilde{d}(\phi_i/2\pi) \quad . \tag{8.13}$$

Each J_i will be a constant of motion.

Example: Kepler problem in a plane with $U = -(k/r) - (h/r^2)$. For $E < 0$ the flow generated by the Hamiltonian

$$H = \frac{p_r^2}{2m} + \frac{p_\theta^2}{2mr^2} + U \tag{8.14}$$

is confined to the 2-torus shown in figure 8.2. Generalizing what we did in section 2.4, we first evaluate $J_r = \oint p_r \mathrm{d}r$ and $J_\theta = \oint p_\theta \mathrm{d}\theta$, finding (see problem 8.4)

$$J_r = 2\pi \left(\frac{mk}{\sqrt{2m|E|}} - l\alpha \right) \quad \text{with} \quad \alpha = \sqrt{1 - \frac{2mh}{l^2}} \tag{8.15}$$

and

$$J_\theta = 2\pi p_\theta = 2\pi l \quad . \tag{8.16}$$

Using these, we obtain

$$|E| = \frac{2\pi^2 mk^2}{\left(J_r + \sqrt{J_\theta^2 - 8\pi^2 mh}\,\right)^2} \quad . \tag{8.17}$$

We solve the Hamilton-Jacobi equation

$$\frac{1}{2m}\left(\frac{\partial G_2}{\partial r}\right)^2 + \frac{1}{2mr^2}\left(\frac{\partial G_2}{\partial \theta}\right)^2 - \frac{k}{r} - \frac{h}{r^2} = E(J_r, J_\theta) \tag{8.18}$$

finding

$$G_2(r,\theta,J_r,J_\theta) = \int_{r_0}^{r} \sqrt{2mE(J_r,J_\theta) + \frac{2mk}{r'} + \left(2mh - \frac{J_\theta^2}{4\pi^2}\right)\frac{1}{r'^2}}\; \mathrm{d}r'$$

$$+\frac{1}{2\pi}J_\theta(\theta - \theta_0) \quad . \tag{8.19}$$

Then

$$\frac{\phi_\theta}{2\pi} = \frac{\partial G_2}{\partial J_\theta} = \int_{r_0}^{r} \frac{1}{p_r}\left(m\frac{\partial E}{\partial J_\theta} - \frac{J_\theta}{4\pi^2 r'^2}\right)\mathrm{d}r' + \frac{\theta - \theta_0}{2\pi}$$

$$= m\frac{\partial E}{\partial J_\theta}\int_{r_0}^{r}\frac{\mathrm{d}r'}{p_r} - \frac{J_\theta}{4\pi^2}\int_{r_0}^{r}\frac{\mathrm{d}r'}{r'^2 p_r} + \frac{\theta - \theta_0}{2\pi} = m\frac{\partial E}{\partial J_\theta}\int_{r_0}^{r}\frac{\mathrm{d}r'}{p_r} \quad ,$$

since $J_\theta \int_{r_0}^{r} \mathrm{d}r'/r'^2 p_r = 2\pi(\theta - \theta_0)$.

On the other hand

$$\frac{\phi_r}{2\pi} = \frac{\partial G_2}{\partial J_r} = \frac{\partial E}{\partial G_2}\frac{\partial G_2}{\partial E} = \frac{\partial E}{\partial J_r}\int_{r_0}^{r}\frac{m\,\mathrm{d}r'}{p_r} \quad .$$

We note at once that neither ϕ_θ nor ϕ_r is equal to θ. We also note that the r-period is

$$T_r = \left(\frac{\partial E}{\partial J_r}\right)^{-1} = \frac{\left(J_r + \sqrt{J_\theta^2 - 8\pi^2 mh}\,\right)^3}{4\pi^2 mk^2} \quad , \tag{8.20}$$

while the θ-period is

$$T_\theta = \left(\frac{\partial E}{\partial J_\theta}\right)^{-1} = \frac{\sqrt{J_\theta^2 - 8\pi^2 mh}}{J_\theta}\,T_r \quad . \tag{8.21}$$

The $1/r^2$ force has removed the degeneracy. The periods T_r and T_θ are now different, and there is a precession with angular velocity

$$\omega_{\rm pr} = \frac{2\pi}{T_r} - \frac{2\pi}{T_\theta} = \frac{2\pi}{T_r}\left(1 - \sqrt{1 - \frac{8\pi^2 mh}{J_\theta^2}}\right) \simeq \omega_r \frac{mh}{l^2} \quad . \tag{8.22}$$

Therefore $\Delta\theta_{\rm pr} = 2\pi\omega_{\rm pr}/\omega_r \simeq 2\pi mh/l^2$, in agreement with (2.16).

But let us return to the pure Newtonian attraction. We are dying to know what $\phi_r = \phi_\theta$ is in this case. From $\phi_r/2\pi = (\partial E/\partial J_r)\int_{r_0}^{r} m\mathrm{d}r'/p_r$, we see that

$$\frac{\phi_r}{2\pi} = \frac{\partial E}{\partial J_r}\int_{r_0}^{r}\frac{\mathrm{d}r'}{\sqrt{2(E-U_{\mathrm{e}}(r'))/m}} = \frac{\partial E}{\partial J_r}(t-t_0) \quad ,$$

$$\frac{\phi_r}{2\pi} = \frac{4\pi^2 mk^2}{(J_r+J_\theta)^3}(t-t_0) = \frac{1}{2\pi}\sqrt{\frac{k}{ma^3}}\,(t-t_0) = \frac{n}{2\pi}(t-t_0) \quad , \tag{8.23}$$

with n as defined by equation (2.47). Hence, for $t_0 = 0$,

$$\phi_r = u - \epsilon\ \sin u \quad , \tag{8.24}$$

where u is the eccentric anomaly.

Remark: For the pure Kepler problem ("pure"= only Newtonian attraction) some authors evaluate

$$J_\phi = 2\pi l_3,\ J_\theta = \oint p_\theta \mathrm{d}\theta = \oint\sqrt{l^2 - l_3^2/\sin^2\theta}\ \mathrm{d}\theta = 2\pi(l-l_3),$$
$$J_r = \oint\sqrt{2m(E+k/r) - l^2/r^2}\ \mathrm{d}r = 2\pi[(mk/\sqrt{2m|E|}\,) - l].$$

Hence

$$J_r = 2\pi\left(\sqrt{\frac{m}{2|E|}}\,k - \frac{J_\theta + J_\phi}{2\pi}\right) \quad ,$$

from which

$$|E| = \frac{2\pi^2 mk^2}{(J_r + J_\theta + J_\phi)^2} \quad .$$

The calculation of J_r is presented as an interesting exercise in complex integration.

Goldstein attributes to Van Vleck the following method of evaluating J_θ. Notice first that comparing $H = \sum_i p_i\dot{q}_i - L = \sum_i p_i\dot{q}_i - K + U$ with $H = K + U$ one has $2K = \sum_i p_i\dot{q}_i$. Expressing the kinetic energy in terms of the original (r,θ,ϕ) variables and in terms of plane-motion variables (r,ψ), one finds

$$p_r\dot{r} + p_\theta\dot{\theta} + p_\phi\dot{\phi} = p_r\dot{r} + p_\psi\dot{\psi} \quad .$$

Note that ψ can be identified with the Euler angle defined in problem 8.3. Using the above equality one has

$$J_\theta = \oint p_\theta \mathrm{d}\theta = \oint p_\psi \mathrm{d}\psi - \oint p_\phi \mathrm{d}\phi = J_\psi - J_\phi = 2\pi(l - l_3) \quad .$$

Of course, if one likes to calculate integrals, J_θ can be obtained directly. Using problem 8.3, we have

$$J_\theta = \oint\sqrt{l^2 - l_3^2/\sin^2\theta}\ \mathrm{d}\theta = l\oint\sqrt{1-\cos^2\vartheta/\sin^2\theta}\ \mathrm{d}\theta \quad .$$

One can perform two changes of variable,
first $\cos\theta = \sin\vartheta\ \sin\psi$ ($\cos\theta = \hat{e}'_1 \cdot \hat{e}_3 = \sin\vartheta\ \sin\psi$), and then $v = \tan\psi$.
One gets

$$\begin{aligned} J_\theta = -l\sin^2\vartheta \oint \frac{\cos^2\psi\, d\psi}{1-\sin^2\vartheta\,\sin^2\psi} &= -l\sin^2\vartheta \oint \frac{dv}{(1+v^2)(1+v^2\cos^2\vartheta)} \\ &= -l\left[\oint \frac{dv}{1+v^2} - \cos^2\vartheta \oint \frac{dv}{1+v^2\cos^2\vartheta}\right] \\ &= -l\left(\oint d\psi - \oint d[\cos\vartheta\, \tan^{-1}(\cos\vartheta\ \tan\psi)]\right) \\ &= l(2\pi - 2\pi\ \cos\vartheta) = 2\pi(l - l_3) \quad . \end{aligned}$$

Why $\oint d\psi = -2\pi$ rather than $+2\pi$?

Summary of action-angle variables

We limit ourselves to the statement of the important Liouville's theorem on integrability [1]:

Let H be the time-independent Hamiltonian of a system with n degrees of freedom ($2n$-dimensional phase space). If we know n functions $P_i(p,q)$ $(i = 1, \ldots n)$ such that
(i) $[H, P_i] = 0$ (the P_i's are integrals of motion),
(ii) $[P_i, P_j] = 0$ (the P_i's are "in involution"),
(iii) the $\tilde{d}P_i$ are linearly independent,
then the system is "integrable", i.e. by algebraic manipulations the mathematical problem can be reduced to integrations.

For a system that satisfies the integrability conditions of Liouville's theorem <u>and</u> is "connected and compact", Arnold has shown [2] that each conserved P_i corresponds to an angular variable ϕ_i, so that the subspace $\{P_i = c_i, i = 1, \ldots n\}$ can be mapped onto (is diffeomorphic to) an n-torus $\{(\phi_1, \ldots \phi_n) \mathrm{mod}\ 2\pi\}$.

Although (P_i, ϕ_i) are not symplectic variables, action variables J_i can be constructed such that

$$\sum_i \tilde{d}p_i \wedge \tilde{d}q_i = \sum_i \tilde{d}J_i \wedge \tilde{d}(\phi_i/2\pi) \quad . \tag{8.25}$$

A J_i is defined as

$$J_i = \oint_{c_i} \sum_j p_j \tilde{d}q_j \tag{8.26}$$

[1] See section 3.3 in W. Thirring, *Classical Dynamical Systems* vol.1 of *A Course in Mathematical Physics* (Springer).
[2] See Thirring, *loc. cit.* p. 102.

along a curve in our subspace such that ϕ_i changes from zero to 2π, while each of the other ϕ_j's ($j \neq i$) returns to its initial value. Subject to this condition, c_i is arbitrary as can be proved by Stokes' theorem and the vanishing of $\tilde{\omega}^2$ in the subspace.

If the J_i's are known first, then the ϕ_i's are obtained by a canonical transformation with a generating function $G_2(q_1, \ldots q_n, J_1, \ldots J_n)$ solution of a Hamilton-Jacobi equation.

Since $\phi_i/2\pi = \partial G_2/\partial J_i$, the change of ϕ_i over a cycle in which ϕ_j changes by 2π is

$$\Delta\phi_i = \oint_{c_j} \sum_k \frac{\partial \phi_i}{\partial q_k} \tilde{d}q_k = 2\pi \oint_{c_j} \sum_k \frac{\partial G_2}{\partial q_k \partial J_i} \tilde{d}q_k$$

$$= 2\pi \frac{\partial}{\partial J_i} \oint_{c_j} \sum_k p_k \tilde{d}q_k = 2\pi \frac{\partial J_j}{\partial J_i} = 2\pi\delta_{ij} \quad . \tag{8.27}$$

8.3 Integrability, non-integrability, chaos

For a conservative system with $n = 2$, the "energy surface" $E =$ constant is a three-dimensional region embedded in the four-dimensional phase space. A Poincaré section is the intersection of this region with a plane, for instance the (p_1, q_1) plane.

The pattern of the points where a given phase space trajectory crosses the plane, obtained in most cases by numerical integration, provides useful information about the system.

For bounded trajectories, the crossing can be expected to occur at an infinite number of points, unless the motion is periodic.

For $n = 2$, a bounded trajectory for which an integral of motion other than the energy exists, lies on a two-dimensional torus. This intersects the Poincaré section on a closed curve. For instance, the left side of figure 8.2 shows a torus $E =$ constant, $l =$ constant in the four dimensional space $(p_r, r, p_\theta, \theta)$. The right side shows the intersection of the torus with the (p_r, r) plane.

There are two frequencies, ω_1 and ω_2 (ω_r and ω_θ for figure 8.2). If the motion is simply periodic ($\omega_1 = \omega_2$) there will be a single point on the intersection curve, through which the trajectory passes every time it crosses the Poincaré plane. If $\omega_1 \neq \omega_2$, but ω_1/ω_2 rational, there will be a finite number of points. If ω_1/ω_2 is irrational, the curve will be completely covered, telling us that the trajectory on the torus is "ergodic". In this case, given an arbitrary point on the torus and a number $\epsilon > 0$, the trajectory will pass at a distance less than ϵ from that point. The pure Kepler motion

($h = 0$) is simply periodic, while the last two cases are possible for $T_r \neq T_\theta$ ($h \neq 0$, equation (8.21)).

Another example: For the two-dimensional harmonic oscillator with Hamiltonian

$$H = [(p_1^2 + \omega_1^2 q_1^2) + (p_2^2 + \omega_2^2 q_2^2)]/2$$

consider a trajectory with energy E passing through a point P_0, for simplicity ($p_{10}, q_{10}, q_{20} = 0$). Then

$$p_{20} = \pm\sqrt{2E - p_{10}^2 - \omega_1^2 q_{10}^2} \quad .$$

Thus the trajectory is completely defined by P_0 up to a sign.
The trajectory will cross the (p_1, q_1) plane again and again. At which points? If as a second integral of motion besides the energy we take the energy of the "2" oscillator, all the intersections will lie on the curve

$$p_1^2 + \omega_1^2 q_1^2 = 2(E - E_2) \quad .$$

Using the solution

$$\begin{cases} q_1(t) = q_1(0)\cos(\omega_1 t) + (p_1(0)/\omega_1)\sin(\omega_1 t) \quad , \\ p_1(t) = p_1(0)\cos(\omega_1 t) - \omega_1 q_1(0)\sin(\omega_1 t) \quad , \end{cases}$$

we find for successive intersections the two-dimensional map

$$\begin{cases} q_{1i} = q_{1,i-1}\cos(2\pi\omega_1/\omega_2) + (p_{1,i-1}/\omega_1)\sin(2\pi\omega_1/\omega_2) \quad , \\ p_{1i} = p_{1,i-1}\cos(2\pi\omega_1/\omega_2) - q_{1,i-1}\omega_1\sin(2\pi\omega_1/\omega_2) \quad . \end{cases}$$

In the $(p_1, \omega_1 q_1)$ plane, this is a rotation through $2\pi\omega_1/\omega_2$. If $\omega_1/\omega_2 = m/n$ (m and n integers), the rotation, repeated n times, is the identity, and the system is multiply periodic. If the ratio of the frequencies is irrational, the intersection curve is covered ergodically.

Keeping E fixed and varying E_2 one has a family of nested tori, intersecting the (p_1, q_1) plane on a family of nested curves.

The Toda Hamiltonian $H_\mathrm{T} = (p_x^2 + p_y^2)/2 + U_\mathrm{T}$ (see equation (2.58)) is integrable because of the existence of the Hénon integral of motion I (2.59). For given E and varying I one has a family of nested tori whose intersections with the (p_x, x) plane are nested curves.

In contrast to this, the Hénon-Heiles Hamiltonian $H_\mathrm{HH} = (p_x^2 + p_y^2)/2 + U_\mathrm{HH}$ (see equation (2.60), problem 2.7, and figure 2.6) is non-integrable, the only integral of motion being the energy E. The nested tori structure of the Toda motion is progressively destroyed with increasing energy, as shown on the Poincaré section by the appearance of islands. A single trajectory jumps from island to island, in much the same way as in the logistic map for $r > 3$, where x_n, starting from a given x_0, jumps from one branch to another of figure 3.5 as n increases.

It will be enough for us to state that for $E = 1/6$ all but a few minute islands have disappeared. Apart from these, a single trajectory wanders chaotically in the three-dimensional region $E = 1/6$, and crosses the Poincaré section at random (see,for instance, M. Tabor, *Chaos and integrability in nonlinear dynamics* (Wiley,1989) p. 122).
We note that as E increases with consequent extension of the available (x, y) region, the approximation $U_{\text{HH}} \simeq U_{\text{T}}$ becomes increasingly bad. We might write $H_{\text{HH}} = H_{\text{T}} + \delta H$, and attribute the destruction of the tori to δH.

We must mention the KAM (Kolmogorov, Arnold, Moser) theorem. For $n = 2$, part of the theorem states that if, among other technical conditions, the Jacobian $|\partial\omega_i/\partial J_j|$ is not zero, then those tori whose frequency ratio ω_2/ω_1 is "sufficiently irrational" are stable under the perturbation δH for $|\delta H|$ very small (see, for instance, H.G. Schuster, *Deterministic chaos. An introduction* (VCH, New York,1988).

At this point we rest our brief account of chaos and suggest a visit to any moderately stocked library. The reader will find an abundant supply of books on the subject.

8.4 Adiabatic invariants

It is widely known that the shortening (lengthening) of the string of a pendulum causes the amplitude θ_{max} to increase (decrease). It may also be known that if the length of a Galilei pendulum (small amplitude oscillations) is changed slowly, the amplitude varies according to the law

$$\theta_{\text{max}}(t) = \left(\frac{\ell(0)}{\ell(t)}\right)^{\frac{3}{4}} \theta_{\text{max}}(0) \quad . \tag{8.28}$$

"Slowly" means that $|\mathrm{d}\ell/\mathrm{d}t|$ is small, and that $t \ll \ell/|\mathrm{d}\ell/\mathrm{d}t|$. If for t we take $t = T(0) = 2\pi\sqrt{\ell(0)/g}$, the initial period of the pendulum, we must have $|\mathrm{d}\ell/\mathrm{d}t|_0 T(0) \ll \ell(0)$.
From equation (8.28) we infer that

$$\frac{E(t)}{\omega(t)} = \frac{E(0)}{\omega(0)} \quad , \tag{8.29}$$

with

$$E(t) = mg\ell(t)\theta^2_{\text{max}}(t)/2 \ \text{ and } \ \omega(t) = \sqrt{g/\ell(t)} \quad , \tag{8.30}$$

namely

$$J(t) = J(0) \quad , \tag{8.31}$$

where $J(t) = 2\pi E(t)/\omega(t)$ is the action variable. Thus $J(t)$ is an "adiabatic invariant".

Let us attempt a little theory. Differentiating $J = 2\pi H/\omega$ with respect to t, we have

$$\dot{J} = -\frac{2\pi}{\omega}\left(\frac{\partial L}{\partial t} + \frac{\dot{\omega}}{\omega}H\right) \tag{8.32}$$

since $\mathrm{d}H/\mathrm{d}t = -\partial L/\partial t$. For the oscillator with $L = m[\dot{q}^2 - \omega(t)^2 q^2]/2$ this gives

$$\dot{J} = \frac{\pi\dot{\omega}}{m\omega^2}(m^2\omega^2 q^2 - p^2) \quad . \tag{8.33}$$

which already shows that $\dot{J}$ is small if $\dot{\omega}$ is so.

Taking the average of $\dot{J}$ over a period of the motion with $\omega(t)$ and $\dot{\omega}(t)$ treated as constants and replaced by their values at $t = 0$, we have

$$\langle\dot{J}\rangle = 2\pi\omega^{-2}\dot{\omega}(\langle m\omega^2 q^2/2\rangle - \langle p^2/2m\rangle) = 0 \quad , \tag{8.34}$$

since the average values of the kinetic and potential energies are equal. This indicates that $\langle\dot{J}\rangle$ is of the order of $\dot{\omega}^2$.

Some authors derive equation (8.29) in the case of the pendulum by considering the work done by the tension of the string when the length ℓ is changed by $\Delta\ell$. This work is $\Delta W \simeq -(mg + m\ell\dot{\theta}^2 - mg\theta^2/2)\Delta\ell$, and its average value is $\langle\Delta W\rangle \simeq -mg\Delta\ell - (E_\theta/2\ell)\Delta\ell$, where E_θ is the energy of the oscillations. The second term of $\langle\Delta W\rangle$ changes E_θ by $\Delta E_\theta = -(E_\theta/2\ell)\Delta\ell$, and so $\Delta(E_\theta\sqrt{\ell}) = 0$, $E_\theta\sqrt{\ell}$ =constant, E_θ/ω =constant.

A more sophisticated approach makes use of the generating function $G_1(q, \phi/2\pi) = (m\omega q^2/2)\mathrm{cotan}\phi$ to transform from the canonical variables (p, q) to $(J, \phi/2\pi)$. This is still possible although ω is now a function of t, but the transformed Hamiltonian is $K(J, \phi, t) = H + \partial G_1/\partial t$,

$$K = \frac{\omega J}{2\pi} + \frac{\dot{\omega}J}{2\pi\omega}\sin\phi\,\cos\phi \quad . \tag{8.35}$$

Hamilton's equations for K yield

$$\dot{J} = -J\frac{\dot{\omega}}{\omega}\cos(2\phi) \tag{8.36}$$

and

$$\dot{\phi} = \omega + \frac{\dot{\omega}}{2\omega}\sin(2\phi) \quad . \tag{8.37}$$

In chapter 8 on pertubation theory it will be shown that if $\dot{\omega}_0/\omega_0$ is small

$$J(t) \simeq J(0)\left(1 - \frac{\dot{\omega}_0}{2\omega_0^2}[\sin(2(\phi_0 + \omega_0 t)) - \sin(2\phi_0)]\right) \quad . \tag{8.38}$$

Thus the value of $J(t)$ oscillates about $J(0)$ with a small amplitude.

8.5 Outline of rigorous theory

We follow sections 51 and 52 of V.I. Arnold, *Mathematical Methods of Classical Mechanics* (Springer,1978), but confine ourselves to one-dimensional systems.

Definition: The quantity $J(p,q;\lambda)$ is an adiabatic invariant of a system with Hamiltonian $H(p,q;\lambda)$ ($\lambda = \epsilon t$, ϵ constant) if for every $\kappa > 0$ there is an $\epsilon_0 > 0$ such that, if $\epsilon < \epsilon_0$ and $0 < t < 1/\epsilon$, then

$$|J(p(t),q(t);\epsilon t) - J(p(0),q(0);0)| < \kappa \quad .$$

Averaging theorem: Consider the system of equations

$$\dot{\phi} = \omega(J) + \epsilon f(J,\phi) \quad , \quad \dot{J} = \epsilon g(J,\phi) \quad ,$$

where $f(J,\phi) = f(J,\phi+2\pi)$ and $g(J,\phi) = g(J,\phi+2\pi)$.

The gist of the theorem is the comparison of $J(t)$ with $J'(t)$ [3]satisfying the equation

$$\dot{J}' = \epsilon\bar{g}(J') \quad ,$$

where

$$\bar{g}(J') = (2\pi)^{-1}\int_0^{2\pi} g(J',\phi)\mathrm{d}\phi \quad .$$

Assuming that in a certain region of the (J,ϕ) plane $0 < c < \omega < c_1$ and $|f| < c_1$, $|g| < c_1$, then for $J(0) = J'(0)$ and $0 < t < 1/\epsilon$ one has

$$|J(t) - J'(t)| < c_0\epsilon \quad ,$$

where c_0 depends on c and c_1, but not on ϵ.

* * * * * * * * * *

Quick proof: Define $P = J + \epsilon k(J,\phi)$, where

$$k(J,\phi) = -\int_0^{\phi} \frac{g(J,\phi') - \bar{g}(J)}{\omega(J)}\,\mathrm{d}\phi' \quad ,$$

and assume that this relation can be inverted, $J = P + \epsilon h(P,\phi,\epsilon)$. Then

$$\dot{P} = \dot{J} + \epsilon\frac{\partial k}{\partial J}\dot{J} + \epsilon\frac{\partial k}{\partial \phi}\dot{\phi}$$

$$= \epsilon g(J,\phi) + \epsilon^2\frac{\partial k}{\partial J}g(J,\phi) - \frac{\epsilon}{\omega(J)}[g(J,\phi) - \bar{g}(J)]\dot{\phi}$$

$$= \epsilon g(J,\phi) + \epsilon^2\frac{\partial k}{\partial J}g(J,\phi) - \frac{\epsilon}{\omega(J)}[g(J,\phi) - \bar{g}(J)][\omega(J) + \epsilon f(J,\phi)]$$

[3]Our J and J' are Arnold's I and J.

$$= \epsilon\bar{g}(J) + O(\epsilon^2) = \epsilon\bar{g}(P) + O(\epsilon^2) \quad .$$

First compare $\dot{J}' = \epsilon\bar{g}(J')$ with $\dot{P} = \epsilon\bar{g}(P) + O(\epsilon^2)$.
Starting from $J'(0) = P(0)$, it is clear that $|P(t) - J'(t)| < \epsilon c_3$ for $0 < t < 1/\epsilon$.
On the other hand, $|J(t) - P(t)| = \epsilon|k| < \epsilon c_2$, since $|\bar{g}| < c_1$ and so $|k| < (2\pi c_1/c) + (2\pi c_1/c) = 4\pi c_1/c = c_2$. Then

$$|J(t) - J'(t)| \le |J(t) - P(t)| + |P(t) - J'(t)| < \epsilon c_2 + \epsilon c_3 = \epsilon c_0$$

for $0 < t < 1/\epsilon$.

* * * * * * * * * *

Let us apply this to the oscillator with $\omega(t) = \omega_0 + \lambda$, $\lambda = \epsilon t$, for which $f(J, \phi, \lambda) = \sin(2\phi)/2(\omega_0 + \lambda)$ and $g(J, \phi, \lambda) = -J\ \cos(2\phi)/(\omega_0 + \lambda)$.
Since in this case $\bar{g}(J') = 0$, we have $\dot{J}' = 0$, $J'(t) = J'(0)$.
Therefore $\dot{J}' = 0$, $J'(t) = J'(0)$. The theorem yields

$$c_0\epsilon \ge |J(t) - J'(t)| = |J(t) - J'(0)| = |J(t) - J(0)|$$

for $0 < t < 1/\epsilon$.

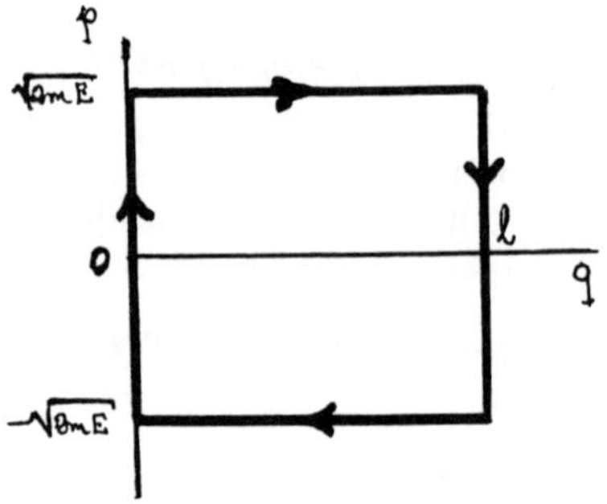

Figure 8.3: Particle in box

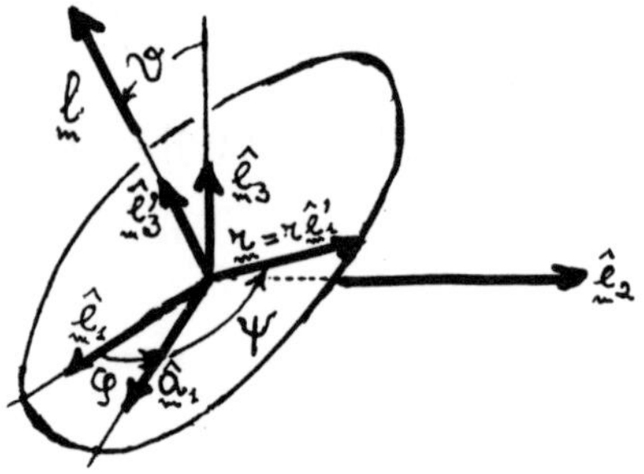

Figure 8.4: Kepler problem in three dimensions

8.6 Chapter 8 problems

8.1 Apply the material of section 8.1 to the simple pendulum,

$$H = \frac{p_\theta^2}{2m\ell^2} + mg\ell(1 - \cos\theta) = \frac{p_\theta^2}{2m\ell^2} + 2mg\ell \sin^2(\theta/2) \quad .$$

8.2 A particle of energy E bounces elastically to and fro in a one-dimensional box of size ℓ (see figure 8.3).
Find the expressions for the $(p, q) \to (J, \phi/2\pi)$ canonical transformation, and a generating function.
Which one?

8.3 Perhaps this should have been one of the chapter 7 problems. We place it here in preparation for problem 8.5.

To treat the Kepler problem in three dimensions choose a system of Euler angles $(\varphi, \vartheta, \psi)$ such that $\mathbf{l} = l\ \hat{\mathbf{e}}'_3$ and $\mathbf{r} = r\ \hat{\mathbf{e}}'_1$, see figure 8.4.
Denoting by l_i $(i = 1, 2, 3)$ the components of $\mathbf{l}$ with respect to the unprimed

axes, show that (i) $[\varphi, l_3] = 1$ (not to be confused with $[\phi, l_3] = 1$), and (ii) $[\psi, l] = 1$.

8.4 Perform the calculation leading to equation (8.15).

8.5 For the Kepler problem in three dimensions with $U = -(k/r) - (h/r^2)$, find the periods T_r, T_ψ, and T_φ.

8.6 For an electron in the field of a nucleus of charge $+Ze$, the relativistic Hamiltonian is $H = c\sqrt{\mathbf{p}^2 + m_0^2c^2} - Ze^2/r$. (i) Find the energy E_{rel} of closed orbits ($0 < E_{\text{rel}} < m_0c^2$) as a function of J_r, J_θ, and J_ϕ.
(ii) Check your result against the non-relativistic expression.

Solutions to ch. 8 problems

S8.1 For $0 < E < 2mg\ell$ we find

$$J_\theta = A(E) = 2\int_{-\theta_0}^{\theta_0} \ell\sqrt{2m}\sqrt{E - 2mg\ell\sin^2(\theta/2)}\,\mathrm{d}\theta$$

$$= 4\ell\sqrt{2m}\int_0^{\theta_0} \sqrt{E - 2mg\ell\sin^2(\theta/2)}\,\mathrm{d}\theta \quad ,$$

where

$$\theta_0 = 2\sin^{-1}\left(\sqrt{\frac{E}{2mg\ell}}\right) \quad .$$

The period is

$$T = \frac{\mathrm{d}A(E)}{\mathrm{d}E} = 2\sqrt{\frac{\ell}{g}}\int_0^{\theta_0} \frac{\mathrm{d}\theta}{\sqrt{\sin^2(\theta_0/2) - \sin^2(\theta/2)}} \quad .$$

The change of variables $\theta = 2\sin^{-1}(\sin(\theta_0/2)\sin\psi)$ casts this expression into the standard form

$$T = 4\sqrt{\frac{\ell}{g}}\int_0^{\frac{\pi}{2}} \frac{\mathrm{d}\psi}{\sqrt{1 - k^2\sin^2\psi}}$$

with

$$k^2 = \sin^2(\theta_0/2) = E/2mg\ell \quad .$$

Using the generating function

$$G_2(\theta, J_\theta) = \ell\sqrt{2m}\int_0^{\theta} \sqrt{E(J_\theta) - 2mg\ell\sin^2(\theta'/2)}\,\mathrm{d}\theta'$$

we find

$$\frac{\phi}{2\pi} = \frac{\partial G_2}{\partial J_\theta} = \frac{\mathrm{d}E}{\mathrm{d}J_\theta}\,\ell\sqrt{\frac{m}{2}}\int_0^{\theta} \frac{\mathrm{d}\theta'}{\sqrt{E - 2mg\ell\sin^2(\theta'/2)}} \quad ,$$

yielding

$$\frac{\phi}{2\pi} = \frac{1}{4}\,\frac{\int_0^{\psi}\mathrm{d}\psi'/\sqrt{1 - k^2\sin^2\psi'}}{\int_0^{\frac{\pi}{2}}\mathrm{d}\psi'/\sqrt{1 - k^2\sin^2\psi'}} \quad .$$

Note that the expression for ϕ might have been found more simply by noticing that $\dot\phi$ must be constant and $p_\theta = m\ell^2\dot\theta$. This requires that

$$\phi = C\int_0^{\theta} \frac{\mathrm{d}\theta'}{p_\theta(E, \theta')} \quad (C = \text{constant}) \quad .$$

Then $\dot\phi/2\pi = C/2\pi m\ell^2$ is equal to $1/T = \mathrm{d}E/\mathrm{d}J_\theta$ for $C = 2\pi m\ell^2\,\mathrm{d}E/\mathrm{d}J_\theta$. Thus

$$\phi = 2\pi m\ell^2\frac{\mathrm{d}E}{\mathrm{d}J_\theta}\int_0^{\theta} \frac{\mathrm{d}\theta'}{p_\theta(E, \theta')} \quad ,$$

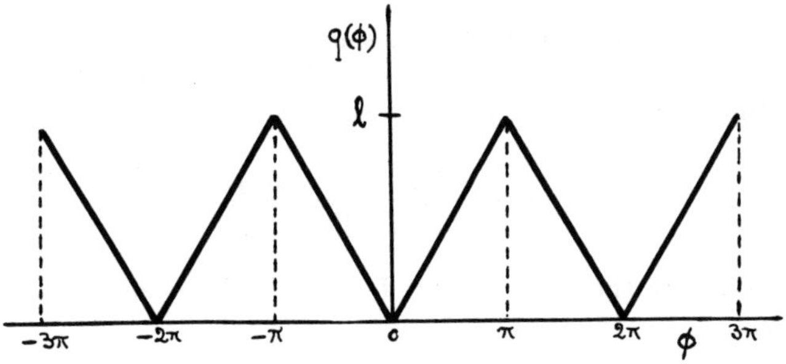

Figure 8.5: Function $q(\phi)$, problem 8.2

in agreement with the expression previously found.

The cases $E > 2mg\ell$ and $E = 2mg\ell$ are left to the reader.

S8.2 Clearly $J = A(E) = 2\ell\sqrt{2mE}$, $T = \mathrm{d}A(E)/\mathrm{d}E = 2\ell/v$, where v is the velocity. The canonical transformation is

$$\begin{cases} p = J/2\ell \;, \; q = \ell\phi/\pi \quad \text{for } 0 \le \phi \le \pi \;, \\ p = -J/2\ell \;, \; q = \ell(2\pi - \phi)/\pi \quad \text{for } \pi \le \phi \le 2\pi \;. \end{cases}$$

It is convenient to use the generating function $G_3 = -p\, q(\phi)$, where

$$q(\phi) = \frac{\ell}{2} - \frac{4\ell}{\pi^2}\sum_{n=0}^{\infty}\frac{\cos((2n+1)\phi)}{(2n+1)^2}$$

is the function shown in figure 8.5. Then

$$q = -\partial G_3/\partial p = q(\phi),$$

$$J = -\partial G_3/\partial(\phi/2\pi) = 2\pi p\, \mathrm{d}q(\phi)/\mathrm{d}\phi \;, \quad p = \pm J/2\ell \;,$$

$$H = p^2/2m = J^2/8m\ell^2 \;,$$

$$\dot{J} = -\partial H/\partial(\phi/2\pi) = 0 \;,$$

$$\dot{\phi}/2\pi = \partial H/\partial J = J/4m\ell^2 = \ell^{-1}\sqrt{E/2m} \;,$$

$$T = 2\pi/\dot{\phi} = 2\ell/v \;.$$

S8.3

$$l_1 = l\, \sin\varphi\, \sin\vartheta \;, \; l_2 = -l\, \cos\varphi\, \sin\vartheta \;, \; l_3 = l\, \cos\vartheta \;.$$

In figure 8.4, $l_1 > 0$, $l_2 < 0$, $\sin\varphi = l_1/\sqrt{l^2 - l_3^2} > 0$, $\cos\varphi = -l_2/\sqrt{l^2 - l_3^2} > 0$, $\tan\varphi = -l_1/l_2 > 0$.

(i) Since $\varphi = -\tan^{-1}(l_1/l_2)$, we have

$$[\varphi, l_3] = -\frac{1}{1 + (l_1/l_2)^2}\left(\frac{1}{l_2}[l_1, l_3] - \frac{l_1}{l_2^2}[l_2, l_3]\right) = 1 \quad .$$

(ii) It is not surprising that $[\psi, l] = 1$, since l is the generator of infinitesimal rotations around $\mathbf{l}$, and ψ is an angle around $\mathbf{l}$ in a plane normal to $\mathbf{l}$. If we want to prove it the hard way, we can use the expression

$$\psi = \cos^{-1}\left(\frac{l_1 x_2 - l_2 x_1}{r\sqrt{l^2 - l_3^2}}\right) \quad ,$$

where x_i and l_i are the components of $\mathbf{r}$ and $\mathbf{l}$ with respect to the unprimed system. Thus

$$[\psi, l] = -\frac{l_1[x_2, l] - l_2[x_1, l]}{r \sin\psi \sqrt{l_1^2 + l_2^2}} = -\frac{-x_3(l_1^2 + l_2^2) + l_3(l_1 x_1 + l_2 x_2)}{rl \sin\psi \sqrt{l_1^2 + l_2^2}} \quad ,$$

which leads to the desired result by expressing the coordinates and the angular momentum components in terms of the Euler angles.

S8.4 With the notation of section 2.4 we have

$$p_r = -l\, \mathrm{d}s/\mathrm{d}\theta = l\alpha(s_1 - s_2)\sin(\alpha\theta)/2 = l\alpha\sqrt{(s_1 - s)(s - s_2)} \quad ,$$

$$J_r = \oint p_r \mathrm{d}r = 2\int_{r_1}^{r_2} p_r \mathrm{d}r = 2l\alpha \int_{s_2}^{s_1} \frac{\sqrt{R}}{s^2}\, \mathrm{d}s \quad ,$$

where

$$R = a + bs + cs^2,\ a = -s_1 s_2 < 0,\ b = s_1 + s_2 > 0,$$
$$c = -1,\ \Delta = 4ac - b^2 = -(s_1 - s_2)^2 < 0\ .$$

Using eq. 2.267.2 in I.S. Gradshtein and I.M. Ryzhik, *Table of integrals, series, and products*(Academic Press,1980) pp. 81 to 84, one finds

$$\int_{s_2}^{s_1} \frac{\sqrt{R}}{s^2}\, \mathrm{d}s = \pi\left(\frac{s_1 + s_2}{2\sqrt{s_1 s_2}} - 1\right)$$

etc.

S8.5 J_r is given by equation (8.15), $J_\varphi = 2\pi\, l_3$, and $J_\psi = 2\pi\, l$. The Hamiltonian is

$$H = \frac{p_r^2}{2m} + \frac{l^2}{2mr^2} - \frac{k}{r} - \frac{h}{r^2} = -\frac{2\pi^2 m k^2}{\left(J_r + \sqrt{J_\psi^2 - 8\pi^2 m h}\,\right)^2} \quad .$$

The periods are

$$T_r = \left(\frac{\partial E}{\partial J_r}\right)^{-1} = \frac{\left(J_r + \sqrt{J_\psi^2 - 8\pi^2 m h}\,\right)^3}{4\pi^2 m k^2} \quad ,$$

$$T_\psi = \left(\frac{\partial E}{\partial J_\psi}\right)^{-1} = \frac{\sqrt{J_\psi^2 - 8\pi^2 m h}}{J_\psi} T_r \quad , \quad T_\varphi = \left(\frac{\partial E}{\partial J_\varphi}\right)^{-1} = \infty \quad .$$

We expect the last since φ is a constant determining the position of the orbital plane.

S8.6 (i) $c^2\mathbf{p}^2 = (E_{\rm rel} + Ze^2/r)^2 - (m_0c^2)^2$, and, with $\mathbf{p}^2 = p_r^2 + l^2/r^2$,

$$p_r = \sqrt{\frac{E_{\rm rel}^2 - (m_0c^2)^2}{c^2} + \frac{2E_{\rm rel}Ze^2}{c^2r} - \frac{l^2 - (Z^2e^4/c^2)}{r^2}} \quad ,$$

$$J_r = \oint p_r \, \mathrm{d}r = -2\pi\sqrt{l^2 - Z^2e^4/c^2} + 2\pi E_{\rm rel}Ze^2/c\sqrt{(m_0c^2)^2 - E_{\rm rel}^2} \quad ,$$

$$J_r + \sqrt{(J_\theta + J_\phi)^2 - 4\pi^2Z^2e^4/c^2} = 2\pi E_{\rm rel}Ze^2/c\sqrt{(m_0c^2)^2 - E_{\rm rel}^2} \quad ,$$

$$E_{\rm rel} = m_0c^2\left[1 + \frac{4\pi^2Z^2e^4/c^2}{\left(J_r + \sqrt{(J_\theta + J_\phi)^2 - 4\pi^2Z^2e^4/c^2}\,\right)^2}\right]^{-\frac{1}{2}} \quad .$$

(ii) Putting $E_{\rm rel} = m_0c^2 + E$, we have $E = \mathbf{p}^2/2m - k/r - h/r^2 + \ldots$ with $k = Ze^2$ and $h = Z^2e^4/2m_0c^2$. Comparing with equation (8.17), we have

$$E \simeq -\frac{2\pi^2m_0Z^2e^4}{\left(J_r + \sqrt{(J_\theta + J_\phi)^2 - 4\pi^2Z^2e^4/c^2}\,\right)^2} \quad ,$$

which agrees with the expansion of $E_{\rm rel}$ found in (i).

Chapter 9

PERTURBATION THEORY

Perturbation expansions are presented emphasizing their similarity to those of Quantum Mechanics.

9.1 The operator Ω

Let $H(p,q)$ be a Hamiltonian not explicitly dependent on t. Therefore $H(p(t),q(t)) = H(p_0,q_0)$, where $p_0 = p(0)$ and $q_0 = q(0)$. As we know, the evolution of a function $f(p,q)$ is governed by the equation

$$\frac{\mathrm{d}f}{\mathrm{d}t} = \bar{U}f = [f,H] \quad . \tag{9.1}$$

Unfortunately, the operator

$$\bar{U} = \sum_i \left(-\frac{\partial H}{\partial q_i}\frac{\partial}{\partial p_i} + \frac{\partial H}{\partial p_i}\frac{\partial}{\partial q_i} \right) \tag{9.2}$$

requires the knowledge of $p(t)$ and $q(t)$, i.e. the solution of the dynamical problem.

In this section we shall express $f(p,q)$ in the form

$$f(p(t),q(t)) = \mathrm{e}^{t\Omega} f(p_0,q_0) \quad , \tag{9.3}$$

where the operator Ω acts on the initial values (p_0,q_0).

This is reminiscent of the Schroedinger-picture formulation of Quantum Mechanics, in which the wavefunction at time t is expressed in the form

$\psi(q,t) = \exp(-itH(p,q))\psi(q,0)$, with p and q time-independent operators. (Of course, $f(p(t),q(t))$ has nothing to do with the wavefunction. If we feel the need of a quantum mechanical analogue, $f(p(t),q(t))$ can be said to correspond to a Heisenberg-picture operator.) Similarly, Ω is an operator acting on the initial values of the canonical variables and, therefore, time-independent. To save space we shall write x for $(p(t),q(t))$ and x_0 for (p_0,q_0). Then equation (9.1) can be written more precisely in the form

$$\frac{\mathrm{d}f(x)}{\mathrm{d}t} = [f(x), H(x)]_x \quad , \tag{9.4}$$

where the Poisson bracket $[\bullet,\bullet]_x$ is understood to be in terms of p and q,

$$[f(x),H]_x = [f(p,q),H(p,q)]_{p,q} = \sum_i \left(\frac{\partial f(x)}{\partial q_i}\frac{\partial H(x)}{\partial p_i} - \frac{\partial f(x)}{\partial p_i}\frac{\partial H(x)}{\partial q_i} \right) \quad . \tag{9.5}$$

We now use the property of the Poisson brackets $[f,g]_{p,q} = [f,g]_{P,Q}$, where (P,Q) are canonical variables like (p,q). Since (p_0,q_0) and $(p(t),q(t))$ are related by a canonical transformation (the Hamiltonian flow), we can write

$$\frac{\mathrm{d}f(x)}{\mathrm{d}t} = [f(x), H(x)]_{x_0} \quad . \tag{9.6}$$

Then, since the Hamiltonian is a constant of motion, we can also write

$$\frac{\mathrm{d}f(x)}{\mathrm{d}t} = [f(x(x_0,t)), H(x_0)]_{x_0} \quad , \tag{9.7}$$

where $x(x_0,t)$ stands for $(p(p_0,q_0,t), q(p_0,q_0,t))$, namely p and q as functions of the time and of their initial values. Taking $t=0$ in both sides of (9.7), we have

$$\left(\frac{\mathrm{d}f(x)}{\mathrm{d}t}\right)_{t=0} = [f(x_0, H(x_0)]_{x_0} \quad . \tag{9.8}$$

We define the operator Ω by the equation

$$\Omega f(x_0) = \left(\frac{\mathrm{d}f(x)}{\mathrm{d}t}\right)_{t=0} = [f(x_0, H(x_0)]_{x_0} \quad . \tag{9.9}$$

A trivial extension of this formula reads

$$\Omega^n f(x_0) = \left(\frac{\mathrm{d}^n f(x)}{\mathrm{d}t^n}\right)_{t=0} = [[[[f(x_0, H(x_0)], H(x_0)]\ldots]] \quad . \tag{9.10}$$

Hence

$$f(p(t),q(t)) = \sum_n \frac{t^n}{n!}\left(\frac{d^n f(p(t),q(t))}{\mathrm{d}t^n}\right)_{t=0}$$

$$= \sum_{n=0}^{\infty} \frac{t^n}{n!} \Omega^n f(p_0, q_0) = \mathrm{e}^{t\Omega} f(p_0, q_0) \quad . \tag{9.11}$$

Example: For $H = p$ (a dimensional constant factor omitted) we have

$$[f(p_0, q_0), H(p_0, q_0)] = \frac{\partial f}{\partial q_0}\frac{\partial p_0}{\partial p_0} - \frac{\partial f}{\partial p_0}\frac{\partial p_0}{\partial q_0} = \frac{\partial f}{\partial q_0} \quad , \tag{9.12}$$

$$[[f(p_0, q_0), p_0], p_0] = \frac{\partial^2 f}{\partial q_0^2} \quad , \ldots \tag{9.13}$$

Hence

$$\Omega = \frac{\partial}{\partial q_0} \quad , \quad \mathrm{e}^{t\Omega} = \mathrm{e}^{t\frac{\partial}{\partial q_0}} \quad . \tag{9.14}$$

Example: For $H = q$ we have $\Omega = -\partial/\partial p_0$.

These two examples are reminiscent of the quantum mechanical momentum and coordinate operators in the coordinate and in the momentum representation, respectively.

Example: For $H = (p^2 + m^2\omega^2 q^2)/2m$ one finds
$\Omega p_0 = [p_0, H(p_0, q_0)] = -m\omega^2 q_0$, $\Omega q_0 = p_0/m$,

$$\Omega \begin{pmatrix} p_0 \\ q_0 \end{pmatrix} = \omega \begin{pmatrix} 0 & -m\omega \\ 1/m\omega & 0 \end{pmatrix} \begin{pmatrix} p_0 \\ q_0 \end{pmatrix} \quad . \tag{9.15}$$

The powers of the matrix ($a = m\omega$)

$$\mathsf{A} = \begin{pmatrix} 0 & -a \\ 1/a & 0 \end{pmatrix}$$

are

$$\mathsf{A}^{2n} = (-)^n \mathsf{I} \quad \text{and} \quad \mathsf{A}^{2n+1} = (-)^n \mathsf{A} \quad .$$

Hence

$$\mathrm{e}^{t\Omega} = \mathsf{I}\ \cos(\omega t) + \mathsf{A}\ \sin(\omega t) \quad , \tag{9.16}$$

$$\begin{pmatrix} p(t) \\ q(t) \end{pmatrix} = \begin{pmatrix} p_0 \cos(\omega t) - m\omega q_0 \sin(\omega t) \\ q_0 \cos(\omega t) + (1/m\omega) p_0 \sin(\omega t) \end{pmatrix} \quad . \tag{9.17}$$

In terms of J and ϕ one has $H = \omega J/2\pi$, $\Omega J_0 = 0$, $\Omega\phi_0 = [\phi_0, \omega J_0/2\pi] = \omega$, $\Omega^n \phi_0 = 0$ $(n > 1)$, $J(t) = \exp(t\Omega) J_0 = J_0$, $\phi(t) = \exp(t\Omega)\phi_0 = \phi_0 + \omega t$.

9.2 Perturbation expansions

Let $H = H^{(0)} + H^{(1)}$. We assume that we can solve exactly the dynamical problem for $H^{(0)}$ alone, while $H^{(1)}$ is a perturbation ($|H^{(1)}| \ll |H^{(0)}|$).

The Hamiltonian flow for $H^{(0)}$ yields the functions $(p^{(0)}(t), q^{(0)}(t))$ from initial values $(p_0 = p^{(0)}(0), q_0 = q^{(0)}(0))$. Thus $\Omega^{(0)}$ such that

$$\mathrm{e}^{t\Omega^{(0)}} f(p_0, q_0) = f(p^{(0)}(t), q^{(0)}(t)) \tag{9.18}$$

is a known operator.

Let Ω be the corresponding operator for the total Hamiltonian $H = H^{(0)} + H^{(1)}$. Starting from the same initial values as in the unperturbed problem, we write

$$e^{t\Omega} f(p_0, q_0) = f(p(t), q(t)) \quad . \tag{9.19}$$

We can also write

$$f(p(t), q(t)) = e^{t\Omega} e^{-t\Omega^{(0)}} e^{t\Omega^{(0)}} f(p_0, q_0) = S(t) f(p^{(0)}(t), q^{(0)}(t)) \quad , \tag{9.20}$$

where

$$S(t) = e^{t\Omega} e^{-t\Omega^{(0)}} \quad . \tag{9.21}$$

Differentiating with respect to t we have

$$\frac{dS(t)}{dt} = e^{t\Omega}(\Omega - \Omega^{(0)}) e^{-t\Omega^{(0)}} = S(t)\Omega_I(t) \quad , \tag{9.22}$$

where

$$\Omega_I(t) = e^{t\Omega^{(0)}}(\Omega - \Omega^{(0)}) e^{-t\Omega^{(0)}} \quad . \tag{9.23}$$

The subscript "I" stands for "interaction" to remind the reader of the kinship of this method with the "interaction picture" of Quantum Mechanics.

Note that, in contrast to $\Omega^{(0)}$, Ω_I is a function of time. The differential equation for $S(t)$ and the initial condition $S(0) = 1$ can be embodied in the integral equation

$$S(t) = 1 + \int_0^t S(t')\Omega_I(t')dt' \quad . \tag{9.24}$$

This can be iterated, yielding

$$S(t) = 1 + \int_0^t dt_1 \Omega_I(t_1) + \int_0^t dt_1 \int_0^{t_1} dt_2\, \Omega_I(t_2)\Omega_I(t_1) + \ldots \tag{9.25}$$

Example: $H^{(0)} = (p^2 + m^2\omega^2 q^2)/2m, \qquad H^{(1)} = \gamma q$

$$\Omega^{(0)} = -m\omega^2 q_0 \frac{\partial}{\partial p_0} + \frac{1}{m} p_0 \frac{\partial}{\partial q_0} \tag{9.26}$$

$$(\Omega - \Omega^{(0)})p_0 = [p_0, \gamma q_0] = -\gamma \quad , \quad (\Omega - \Omega^{(0)})q_0 = 0 \quad . \tag{9.27}$$

Therefore

$$\Omega - \Omega^{(0)} = -\gamma \frac{\partial}{\partial p_0} \quad , \tag{9.28}$$

$$\left(\int_0^t e^{t'\Omega^{(0)}}(\Omega - \Omega^{(0)}) e^{-t'\Omega^{(0)}} dt'\right) p^{(0)}(t) = \left(\int_0^t dt'\, e^{t'\Omega^{(0)}} \left(-\gamma \frac{\partial}{\partial p_0}\right) e^{(t-t')\Omega^{(0)}} p_0\right)$$

$$= \int_0^t \mathrm{d}t' \, \mathrm{e}^{t'\Omega^{(0)}} \left(-\gamma \frac{\partial}{\partial p_0}\right) [p_0 \cos(\omega(t-t')) - m\omega q_0 \sin(\omega(t-t'))]$$

$$= -\gamma \int_0^t \mathrm{e}^{t'\Omega^{(0)}} \cos(\omega(t-t'))\mathrm{d}t' = -\gamma \int_0^t \cos(\omega(t-t'))\mathrm{d}t'$$

$$= -(\gamma/\omega)\sin(\omega t) \ , \tag{9.29}$$

$$\left(\int_0^t \mathrm{e}^{t'\Omega^{(0)}} (\Omega - \Omega^{(0)}) \mathrm{e}^{-t'\Omega^{(0)}} \mathrm{d}t'\right) q^{(0)}(t) = -\frac{\gamma}{m\omega^2}(1 - \cos(\omega t)) \ . \tag{9.30}$$

Higher order terms of the perturbation expansion give zero. Therefore

$$\begin{pmatrix} p(t) \\ q(t) \end{pmatrix} = \begin{pmatrix} p^{(0)}(t) - (\gamma/\omega)\sin(\omega t) \\ q^{(0)}(t) - (\gamma/m\omega^2)(1 - \cos(\omega t)) \end{pmatrix}$$

$$= \begin{pmatrix} p_0 \cos(\omega t) - (m\omega q_0 + \gamma/\omega)\sin(\omega t) \\ (q_0 + \gamma/m\omega^2)\cos(\omega t) + (p_0/m\omega)\sin(\omega t) - \gamma/m\omega^2 \end{pmatrix} \tag{9.31}$$

Note that the Hamiltonian $H = (p^2 + m^2\omega^2 q^2)/2m + \gamma q$ differs only by an additional constant from the Hamiltonian

$$H_{\mathrm{D}} = \frac{p^2}{2m} + \frac{m\omega^2}{2}\left(q + \frac{\gamma}{m\omega^2}\right)^2 \tag{9.32}$$

for a displaced harmonic oscillator, $H_{\mathrm{D}} = H + \gamma^2/2m\omega^2$. Hamilton's equations for H_{D},

$$\dot{p} = -\frac{\partial H_{\mathrm{D}}}{\partial q} = -m\omega^2 \left(q + \frac{\gamma}{m\omega^2}\right) \ , \ \dot{q} = \frac{\partial H_{\mathrm{D}}}{\partial p} = \frac{p}{m} \tag{9.33}$$

are satisfied by our perturbation solution.

9.3 Perturbed periodic systems

In section 7.10 and elsewhere, we glibly stated that all one had to do was to find a canonical transformation to new variables (P_i, Q_i) with the P_i's constants of motion. This is possible only for a small number of exactly integrable systems.

This section deals with cases where the Hamiltonian $H(p, q)$ differs by a small perturbation from a Hamiltonian H_0 for which exact periodic solutions of Hamilton's equations are known.

We consider a periodic system in one dimension with

$$H(J_0, \phi_0) = H_0(J_0) + \lambda H_1(J_0, \phi_0) \quad (|\lambda| < 1) \ . \tag{9.34}$$

The "unperturbed" Hamiltonian H_0 depends only on J_0. If the second term were missing ($\lambda = 0$), the frequency would be

$$\omega_0(J_0) = 2\pi \frac{\mathrm{d}H_0(J_0)}{\mathrm{d}J_0} \tag{9.35}$$

and the energy $E_0(J_0) = H_0(J_0)$.

For the harmonic oscillator

$$H_0 = \frac{p^2}{2m} + \frac{m\omega_0^2 q^2}{2} = \frac{\omega_0 J_0}{2\pi} \quad , \tag{9.36}$$

ω_0 does not depend on J_0, and p and q are the well-known functions of (J_0, ϕ_0)

$$q = \sqrt{\frac{J_0}{\pi m \omega_0}} \sin\phi_0 \quad \text{and} \quad p = \sqrt{\frac{m\omega_0 J_0}{\pi}} \cos\phi_0 \quad .$$

Our program is to find new variables (J, ϕ) such that the "perturbed" Hamiltonian

$$H(J_0(J,\phi), \phi_0(J,\phi)) = K(J,\phi) \tag{9.37}$$

is a function of J only with an error $\mathrm{O}(\lambda^n)$,

$$H(J_0, \phi_0) = K(J) + \mathrm{O}(\lambda^n) \quad . \tag{9.38}$$

The greater n the better is the approximation.

The perturbation Hamiltonian $H_1(J_0, \phi_0) = H_1(p(J_0,\phi_0), q(J_0,\phi_0))$ is by assumption periodic in ϕ_0 and can be expanded in Fourier series[1]

$$H_1(J_0, \phi_0) = \sum_k h_1(k, J_0) \mathrm{e}^{\mathrm{i}k\phi_0} \quad , \tag{9.39}$$

$$h_1(k, J_0) = \frac{1}{2\pi} \int_0^{2\pi} \mathrm{d}\phi_0 \; \mathrm{e}^{-\mathrm{i}k\phi_0} H_1(J_0, \phi_0) \quad . \tag{9.40}$$

Note that the average of H_1 over the unperturbed motion is

$$\langle H_1 \rangle = h_1(0, J_0) \quad . \tag{9.41}$$

We now perform a canonical transformation with a generating function of the G_2-type [2]

$$J_0 = J_1 + 2\pi\lambda \frac{\partial S_1(\phi_0, J_1)}{\partial \phi_0} \quad , \tag{9.42}$$

---

[1] For typographical convenience we do not write k as a subscript of h_1. Our notation $h_1(k, J_0)$ does not imply that k is a continuous variable.

[2] $J_0 = 2\pi \partial S(\phi_0, J_1)/\partial\phi_0$, $S(\phi_0, J_1) = (\phi_0/2\pi)J_1 + \lambda S_1(\phi_0, J_1)$, where $(\phi_0/2\pi)J_1$ is the generating function of the identity transformation.

$$\phi_1 = \phi_0 + 2\pi\lambda \frac{\partial S_1(\phi_0, J_1)}{\partial J_1} \quad . \tag{9.43}$$

Choose S_1 so that

$$\omega_0(J_1)\frac{\partial S_1(\phi_0, J_1)}{\partial \phi_0} + H_1(J_1, \phi_0) = h_1(0, J_1) \quad , \tag{9.44}$$

where

$$H_1(J_1, \phi_0) = [H_1(J_0, \phi_0)]_{J_0 \to J_1} \tag{9.45}$$

and

$$\omega_0(J_1) = 2\pi \left[\frac{\mathrm{d}H_0(J_0)}{\mathrm{d}J_0}\right]_{J_0 \to J_1} \quad . \tag{9.46}$$

Then

$$H(J_0, \phi_0) = H_0\left(J_1 + 2\pi\lambda\frac{\partial S_1(\phi_0, J_1)}{\partial \phi_0}\right) + \lambda H_1\left(J_1 + 2\pi\lambda\frac{\partial S_1(\phi_0, J_1)}{\partial \phi_0}, \phi_0\right)$$

$$\simeq H_0(J_1) + \frac{\mathrm{d}H_1(J_1)}{\mathrm{d}J_1} 2\pi\lambda \frac{\partial S_1(\phi_0, J_1)}{\partial \phi_0} + \lambda H_1(J_0, \phi_0) + \mathrm{O}(\lambda^2)$$

$$= H_0(J_1) + \lambda h_1(0, J_1) + \mathrm{O}(\lambda^2) \quad ,$$

$$H(J_0, \phi_0) \simeq H_0(J_1) + \lambda\langle H_1(J_0, \phi_0)\rangle_{J_0 \to J_1} + \mathrm{O}(\lambda^2) \quad . \tag{9.47}$$

If we want the approximation quadratic in λ we must go one step further with the expansion in powers of λ. Neglecting terms $O(\lambda^3)$, we have

$$H(J_0, \phi_0) - H_0(J_1) = \frac{\mathrm{d}H_0(J_1)}{\mathrm{d}J_1} 2\pi\lambda \frac{\partial S_1(\phi_0, J_1)}{\partial \phi_0}$$

$$+\frac{1}{2}\frac{\mathrm{d}^2 H_0(J_1)}{\mathrm{d}J_1^2}\left(2\pi\lambda\frac{\partial S_1(\phi_0, J_1)}{\partial \phi_0}\right)^2 + \lambda H_1(J_1, \phi_0)$$

$$+\lambda\frac{\partial H_1(J_1, \phi_0)}{\partial J_1} 2\pi\lambda \frac{\partial S_1(\phi_0, J_1)}{\partial \phi_0}$$

$$= \lambda h_1(0, J_1) + \lambda^2 \left[2\pi^2 \frac{\mathrm{d}^2 H_0(J_1)}{\mathrm{d}J_1^2}\left(\frac{\partial S_1(\phi_0, J_1)}{\partial \phi_0)}\right)^2 + 2\pi \frac{\partial H_1(J_1, \phi_0)}{\partial J_1}\frac{\partial S_1(\phi_0, J_1)}{\partial \phi_0}\right]$$

$$= \lambda h_1(0, J_1) + A_1 \left[\frac{1}{2}\frac{\mathrm{d}\omega_0(J_1)}{\mathrm{d}J_1}\frac{1}{\omega_0(J_1)}(h_1(0, J_1) - H_1(J_1, \phi_1))^2 \right.$$

$$\left. + \frac{\partial H_1(J_1, \phi_1)}{\partial J_1}(h_1(0, J_1) - H_1(J_1, \phi_1))\right] ,$$

where

$$A_1 = \lambda^2 \frac{2\pi}{\omega_0(J_1)} \ .$$

We now perform a second transformation $(J_1, \phi_1) \to (J_2, \phi_2)$ etc. If we are content with the approximation quadratic in λ, we just rename $(J_1, \phi_1) \to (J, \phi)$ and go no further. We then have

$$H(J_0, \phi_0) = H_0(J) + \lambda \langle H_1(J, \phi) \rangle$$

$$+A \left[\frac{1}{2} \frac{\mathrm{d}\omega_0(J)}{\mathrm{d}J} \frac{1}{\omega_0(J)} (\langle H_1(J,\phi)^2 \rangle - \langle H_1(J,\phi) \rangle^2) \right]$$

$$+A \left[\langle H_1(J,\phi) \rangle \left\langle \frac{\partial H_1(J,\phi)}{\partial J} \right\rangle - \left\langle \frac{\partial H_1(J,\phi)}{\partial J} H_1(J,\phi) \right\rangle \right] \ .$$

where

$$A = \lambda^2 \frac{2\pi}{\omega_0(J)} \ .$$

Suppose the Hamiltonian depends on several action-angle variables (J_{0i}, ϕ_{0i}) $(i = 1, \ldots n)$,

$$H(\mathbf{J}_0, \phi_0) = H_0(\mathbf{J}_0) + \lambda H_1(\mathbf{J}_0, \phi_0) \ , \tag{9.48}$$

where $\mathbf{J}_0 = (J_{01}, \ldots J_{0n})$, $\phi_0 = (\phi_{01}, \ldots \phi_{0n})$, and

$$H_1(\mathbf{J}_0, \phi_0) = \sum_{\mathbf{k}} h_1(\mathbf{k}, \mathbf{J}_0) \, \exp(\mathrm{i}\mathbf{k} \cdot \phi_0) \ . \tag{9.49}$$

Requiring

$$H_0(\mathbf{J}_0) + \lambda H_1(\mathbf{J}_0, \phi_0) = K(\mathbf{J}) + O(\lambda^2) \tag{9.50}$$

one finds

$$K(\mathbf{J}) = H_0(\mathbf{J}) + \lambda \langle H_1(\mathbf{J}_0, \phi_0) \rangle \tag{9.51}$$

and that the generating function is

$$S(\phi_0, \mathbf{J}) = \frac{1}{2\pi} \phi_0 \cdot \mathbf{J} + \mathrm{i} \sum_{\mathbf{k} \neq 0} \frac{h_1(\mathbf{k}, \mathbf{J}) \, \exp(\mathrm{i}\mathbf{k} \cdot \phi_0)}{\mathbf{k} \cdot \omega_0(\mathbf{J})} \ , \tag{9.52}$$

where

$$\omega_0(\mathbf{J}) = \left(\frac{\partial H_0(\mathbf{J})}{\partial J_1}, \ldots \frac{\partial H_0(\mathbf{J})}{\partial J_n} \right) \ . \tag{9.53}$$

The generating function diverges if the zero-order frequencies $\omega_{0i} = \partial H_0 / \partial J_i$ are commensurable, i.e. there exists a relation $m_i \omega_{0i} = 0$ with m_i integers. In this case, perturbation theory is unsuitable for the integration.

9.4 Chapter 9 problems

9.1 Do the calculation in section 9.2 (Example) using action-angle variables.

9.2 Oscillator with time-dependent frequency using action-angle variables.

9.3 In section 9.3 we only needed to assume the existence of $S_1(\phi_0, J_1)$. Find an explicit expression.

9.4 Let us apply the method of section 9.3 to the harmonic oscillator with the perturbation

$$\lambda H_1 = \lambda q^3 = \lambda \left(\frac{J_0}{\pi m \omega_0} \right)^{\frac{3}{2}} \sin^3 \phi_0 \quad .$$

9.5 Express the Hamiltonian $H = p_\theta^2/2 + \omega^2(1 - \cos\theta)$ for the plane pendulum ($\ddot{\theta} + \omega^2 \sin\theta = 0$, $\omega = \sqrt{g/l}$) in terms of the harmonic oscillator variables

$$p_\theta = \sqrt{J_0 \omega/\pi} \, \cos\phi_0 \quad , \quad \theta = \sqrt{J_0/\pi\omega} \, \sin\phi_0 \quad ,$$

and find the θ_0^2 correction to the period.

9.6 Expressing the coupled-oscillators Hamiltonian

$$H = (p_1^2 + \omega_1^2 q_1^2)/2 \; + \; (p_2^2 + \omega_2^2 q_2^2)/2 \; + \; \lambda q_1 q_2$$

in terms of harmonic-oscillator variables, we have

$$H = \frac{\omega_1 J_{01} + \omega_2 J_{02}}{2\pi} + \frac{\lambda}{\pi} \sqrt{\frac{J_{01} J_{02}}{\omega_1 \omega_2}} \, \sin\phi_{01} \sin\phi_{02} \quad .$$

Find $K(J_1, J_2)$ to the second order in λ.

Solutions to ch. 9 problems

S9.1

$$H^{(0)} = \omega J/2\pi \quad , \quad H^{(1)} = \gamma\sqrt{J/\pi m\omega}\ \sin\phi$$
$$J^{(0)}(t) = J_0 \quad , \quad \phi^{(0)}(t) = \phi_0 + \omega t$$

$$(\Omega - \Omega^{(0)})J_0 = [J_0, H_0^{(1)}] = -2\gamma\sqrt{\pi J_0/m\omega}\ \cos\phi_0$$
$$(\Omega - \Omega^{(0)})\phi_0 = [\phi_0, H_0^{(1)}] = \gamma\sqrt{\pi/m\omega J_0}\ \sin\phi_0$$

$$\left(\int_0^t e^{t'\Omega^{(0)}}(\Omega - \Omega^{(0)})e^{-t'\Omega^{(0)}}\ dt'\right) J^{(0)}(t) = \int_0^t e^{t'\Omega^{(0)}}(\Omega - \Omega^{(0)})J_0\ dt'$$

$$= -2\gamma\sqrt{\pi J_0/m\omega}\int_0^t e^{t'\Omega^{(0)}}\cos\phi_0\ dt' = -2\gamma\sqrt{\pi J_0/m\omega}\int_0^t \cos(\phi_0 + \omega t')\ dt'$$

$$= -2\gamma\sqrt{\pi J_0/m\omega^3}\ [\sin(\phi_0 + \omega t) - \sin\phi_0]$$

$$J(t) = J_0 - 2\gamma\sqrt{\pi J_0/m\omega^3}\ [\sin(\phi_0 + \omega t) - \sin\phi_0] + \ldots$$

$$\left(\int_0^t e^{t'\Omega^{(0)}}\ (\Omega - \Omega^{(0)})e^{-t'\Omega^{(0)}}\ dt'\right) \phi^{(0)}(t) = (\ldots)(\phi_0 + \omega t)$$

$$= \int_0^t e^{t'\Omega^{(0)}}[\phi_0 + \omega(t - t')]\ dt' = \gamma\sqrt{\pi/m\omega J_0}\int_0^t e^{t'\Omega^{(0)}}\ \sin\phi_0\ dt'$$

$$= \gamma\sqrt{\pi/m\omega^3 J_0}\int_0^t \sin(\phi_0 + \omega t')\ dt' = \gamma\sqrt{\pi/m\omega J_0}[\cos\phi_0 - \cos(\phi_0 + \omega t)]$$

$$\phi(t) = \phi_0 + \gamma\sqrt{\pi/mJ_0\omega^3}\ [\cos\phi_0 - \cos(\phi_0 + \omega t)] + \ldots$$

Check that Hamilton's equations for J and ϕ,
$$\dot{J} = -\partial H/\partial(\phi/2\pi) = -2\gamma\sqrt{\pi J/m\omega}\ \cos\phi$$
and
$$\dot{\phi}/2\pi = \partial H/\partial J = \omega/2\pi + [\gamma/2\sqrt{\pi m\omega J}\ ]\ \sin\phi,$$
are satisfied in the present approximation.

S9.2 The Hamiltonian is

$$H = \frac{\omega J}{2\pi} + \frac{\dot{\omega} J}{2\pi\omega} \sin\phi \, \cos\phi \quad .$$

With $\omega \simeq \omega_0 + \dot{\omega}_0 t$, we have $H = H^{(0)} + H^{(1)}$,

$$H^{(0)} = \frac{\omega_0 J}{2\pi} \quad \text{and} \quad H^{(1)} = \frac{\dot{\omega}_0 J}{2\pi} t + \frac{\dot{\omega}_0 J}{4\pi\omega_0} \sin(2\phi) + \dots$$

We have

$$(\Omega - \Omega^{(0)}) J_0 = [J_0, H_0^{(1)}] = \frac{\dot{\omega}_0 J_0 \cos(2\phi_0)}{\omega_0} \left[J_0, \frac{\phi_0}{2\pi}\right] = -\frac{\dot{\omega}_0 J_0 \cos(2\phi_0)}{\omega_0} \quad ,$$

$$(\Omega - \Omega^{(0)}) \phi_0 = [\phi_0, H_0^{(1)}] = \dot{\omega}_0 \left(t + \frac{\sin(2\phi_0)}{2\omega_0}\right) \left[\frac{\phi_0}{2\pi}, J_0\right] = \dot{\omega}_0 \left(t + \frac{\sin(2\phi_0)}{2\omega_0}\right) \quad ,$$

$$\left(\int_0^t e^{t'\Omega^{(0)}} (\Omega - \Omega^{(0)}) e^{-t'\Omega^{(0)}} \, dt'\right) J^{(0)}(t) = \int_0^t e^{t'\Omega^{(0)}} (\Omega - \Omega^{(0)}) J_0 \, dt'$$

$$= \int_0^t e^{t'\Omega^{(0)}} \left(-\frac{\dot{\omega}_0}{\omega_0} J_0 \cos(2\phi_0)\right) dt' = -\frac{\dot{\omega}_0}{\omega_0} J_0 \int_0^t \cos(2(\phi_0 + \omega_0 t')) \, dt' \quad ,$$

$$J(t) = J_0 \left(1 - \frac{\dot{\omega}_0}{\omega_0} \int_0^t \cos(2(\phi_0 + \omega_0 t')) \, dt'\right) + \dots \quad ,$$

$$J(t) = J_0 \left(1 - \frac{\dot{\omega}_0}{2\omega_0^2} [\sin(2(\phi_0 + \omega_0 t)) - \sin(2\phi_0)]\right) + \dots$$

Similarly we find

$$\phi(t) = \phi_0 + \omega_0 t + \frac{\dot{\omega}_0 t^2}{2} + \frac{\dot{\omega}_0}{4\omega_0^2} [\cos(2\phi_0) - \cos(2(\phi_0 + \omega_0 t))] + \dots \quad ,$$

in agreement with eq. (3), solution of problem 13.9 in G.L. Kotkin and V.G. Serbo, *Collection of Problems in Classical Mechanics* (Pergamon Press,1971).

S9.3

$$S_1(\phi_0, J_1) = -\sum_{k \neq 0} \frac{h_1(k, J_1)}{i\omega_0(J_1) k} \exp(ik\phi_0)$$

In fact,

$$\omega_0(J_1) \frac{\partial S_1(\phi_0, J_1)}{\partial \phi_0} + H_1(J_1, \phi_0)$$

$$= -\sum_{k \neq 0} h_1(k, J_1) \exp(ik\phi_0) + \sum_k h_1(k, J_1) \exp(ik\phi_0) = h_1(0, J_1) \quad .$$

S9.4 We find at once that $\langle H_1 \rangle = 0$, and so there is no correction to the Hamiltonian linear in λ. Furthermore, since ω_0 is a constant, $\mathrm{d}\omega_0(J)/\mathrm{d}J = 0$, and so we have only

$$K = H_0(J) - \lambda^2 \frac{2\pi}{\omega_0} \left\langle \frac{\partial H_1(J,\phi)}{\partial J} H_1(J,\phi) \right\rangle + O(\lambda^3) \quad .$$

Since

$$\left\langle \frac{\partial H_1(J,\phi)}{\partial J} H_1(J,\phi) \right\rangle = \frac{3J^2}{2(\pi m\omega_0)^3} \langle \sin^6\phi \rangle$$

and

$$\frac{1}{2\pi} \int_0^{2\pi} \sin^6\phi = \frac{5}{16} \quad ,$$

we have

$$K(J) = \frac{\omega_0 J}{2\pi} - \lambda^2 \frac{15J^2}{16\pi^2 m^3 \omega_0^4} + O(\lambda^3) \quad .$$

Let us go back to p and q. From the unperturbed solution we have $p = p(J_0, \phi_0)$ and $q = q(J_0, \phi_0)$. In these formulae we must express J_0 and ϕ_0 in terms of J and ϕ.

For the harmonic oscillator with the perturbation $\lambda H_1 = \lambda q^3$, in the approximation linear in λ we have

$$q = \sqrt{\frac{J_0}{\pi m\omega_0}} \sin\phi_0 = \frac{1}{\sqrt{\pi m\omega_0}} \sqrt{J + 2\pi\lambda \frac{\partial S_1}{\partial \phi_0}} \sin\left(\phi - 2\pi\lambda \frac{\partial S_1}{\partial J}\right)$$

$$= \sqrt{\frac{J}{\pi m\omega_0}} \sin\phi + \frac{2\pi\lambda}{\sqrt{\pi m\omega_0}} \left[\frac{1}{2\sqrt{J}} \frac{\partial S_1}{\partial \phi_0} \sin\phi - \sqrt{J} \frac{\partial S_1}{\partial J} \cos\phi \right] + O(\lambda^2) \quad .$$

Now

$$\frac{\partial S_1}{\partial \phi_0} = -\frac{1}{\omega_0} H_1(J, \phi_0) = -\frac{1}{\omega_0} \left(\frac{J_0}{\pi m\omega_0} \right)^{\frac{3}{2}} \sin^3\phi_0 \quad ,$$

$$S_1 = \frac{1}{\omega_0} \left(\frac{J_0}{\pi m\omega_0} \right)^{\frac{3}{2}} \left(\cos\phi_0 - \frac{1}{3}\cos^3\phi_0 \right) \quad ,$$

$$\frac{\partial S_1}{\partial J_0} = \frac{3\sqrt{J_0}}{2\omega_0 (\pi m\omega_0)^{\frac{3}{2}}} \left(\cos\phi_0 - \frac{1}{3}\cos^3\phi_0 \right) \quad ,$$

and so

$$q = \sqrt{\frac{J}{\pi m\omega_0}} \sin\phi + \lambda \frac{J}{\pi m^2 \omega_0^3} [-\sin^4\phi - 3\cos^2\phi + \cos^4\phi] + O(\lambda^2) \quad ,$$

$$= \sqrt{\frac{J}{\pi m\omega_0}} \sin\phi - \frac{\lambda J}{2\pi m^2 \omega_0^3} (3 + \cos(2\phi)) + O(\lambda^2) \quad .$$

Similarly one finds

$$p = \sqrt{\frac{m\omega_0 J}{\pi}} \cos\phi + \frac{\lambda J}{\pi m\omega_0^2} \sin(2\phi) + O(\lambda^2) \quad .$$

Check that $(p^2 + m^2\omega^2 q^2)/2m + \lambda q^3 = \omega_0 J/2\pi + O(\lambda^2)$, and, with $[\phi, J] = 2\pi$, that $[q,p] = 1 + O(\lambda^2)$.

S9.5 We have

$$H = \frac{J_0\omega}{2\pi}\cos^2\phi_0 + \omega^2\left[1 - \cos\left(\sqrt{\frac{J_0}{\pi\omega}}\sin\phi_0\right)\right] = \frac{J_0\omega}{2\pi} - \frac{J_0^2}{4!\pi^2}\sin^4\phi_0 + \dots$$

From action-angle perturbation theory

$$K(J) = \frac{J\omega}{2\pi} - \frac{J^2}{64\pi^2} + \dots$$

Substituting $J = \pi\omega\theta_0^2$, where θ_0 is the amplitude, we have

$$\dot{\phi} = 2\pi\frac{\partial H}{\partial J} = \sqrt{\frac{g}{\ell}}\left(1 - \frac{\theta_0^2}{16} + \dots\right) \quad ,$$

$$T = \frac{2\pi}{\dot{\phi}} = 2\pi\sqrt{\frac{\ell}{g}}\left(1 + \frac{\theta_0^2}{16} + \dots\right) \quad .$$

S9.6 We have

$$K_0(\mathbf{J}) = H_0(\mathbf{J}) = \frac{\omega_1 J_1 + \omega_2 J_2}{2\pi} \quad ,$$

$$S_1(\phi_0, \mathbf{J}) = \frac{\lambda}{2\pi}\sqrt{\frac{J_1 J_2}{\omega_1\omega_2}}\left(\frac{\sin(\phi_{01} + \phi_{02})}{\omega_1 + \omega_2} - \frac{\sin(\phi_{01} - \phi_{02})}{\omega_1 - \omega_2}\right) \quad ,$$

$$K_1(\mathbf{J}) = \langle H_1\rangle + 2\pi\left\langle \frac{\partial H_0}{\partial J_1}\frac{\partial S_1}{\partial\phi_{01}} + \frac{\partial H_0}{\partial J_2}\frac{\partial S_1}{\partial\phi_{02}}\right\rangle = 0 + 0 \quad ,$$

$$K_2(\mathbf{J}) = \langle H_2\rangle + 2\pi\left\langle \frac{\partial H_1}{\partial J_1}\frac{\partial S_1}{\partial\phi_{01}} + \frac{\partial H_1}{\partial J_2}\frac{\partial S_1}{\partial\phi_{02}}\right\rangle$$

$$+2\pi^2\left[\frac{\partial^2 H_0}{\partial J_1^2}\left\langle\left(\frac{\partial S_1}{\partial\phi_{01}}\right)^2\right\rangle + 2\frac{\partial^2 H_0}{\partial J_1\partial J_2}\left\langle\frac{\partial S_1}{\partial\phi_{01}}\frac{\partial S_1}{\partial\phi_{02}}\right\rangle + \frac{\partial^2 H_0}{\partial J_2^2}\left\langle\left(\frac{\partial S_1}{\partial\phi_{02}}\right)^2\right\rangle\right]$$

$$+2\pi\left\langle\frac{\partial H_0}{\partial J_1}\frac{\partial S_2}{\partial\phi_{01}} + \frac{\partial H_0}{\partial J_2}\frac{\partial S_2}{\partial\phi_{02}}\right\rangle \quad .$$

This reduces to

$$K_2(\mathbf{J}) = 2\pi\left\langle \frac{\partial H_1}{\partial J_1}\frac{\partial S_1}{\partial\phi_{01}} + \frac{\partial H_1}{\partial J_2}\frac{\partial S_1}{\partial\phi_{02}}\right\rangle$$

$$= \frac{\lambda^2}{\pi\omega_1\omega_2}\,\frac{\omega_2 J_1 - \omega_1 J_2}{\omega_1^2 - \omega_2^2}\langle\sin^2\phi_{01}\sin^2\phi_{02}\rangle = \frac{\lambda^2}{4\pi(\omega_1^2 - \omega_2^2)}\left(\frac{J_1}{\omega_1} - \frac{J_2}{\omega_2}\right) \quad .$$

Finally

$$\dot{\phi}_1 = 2\pi\frac{\partial K}{\partial J_1} = \omega_1 + \frac{\lambda^2}{2\omega_1(\omega_1^2 - \omega_2^2)} \quad ,$$

in agreement with problem 7.5,

$$\Omega_1 = \sqrt{\frac{\omega_1^2 + \omega_2^2 + \sqrt{(\omega_1^2 - \omega_2^2)^2 + 4\lambda^2}}{2}}$$

$$\simeq \sqrt{\omega_1^2 + \frac{\lambda^2}{\omega_1^2 - \omega_2^2}} \simeq \omega_1 + \frac{\lambda^2}{2\omega_1(\omega_1^2 - \omega_2^2)} \quad .$$

Chapter 10

RELATIVISTIC DYNAMICS

Relativistic kinematics and dynamics of a particle, Lorentz transformations and their connection with the $SL(2)$ group, are presented first. The Thomas effect is discussed
The equations of motion of a charged particle with magnetic moment in an electromagnetic field are established.
The chapter ends with the Lagrangian and Hamiltonian equations of a charged particle.

10.1 Lorentz transformations

Consider an inertial system S characterized by a system of axes XYZ and by a clock C. Let $x^1 = x$, $x^2 = y$, $x^3 = z$ be the coordinates of a point particle with respect to XYZ at the time t shown by C.

We regard $x^0 = ct$, x^1, x^2, and x^3 as the components x^μ $(\mu = 0, 1, 2, 3)$[1] of a 4-vector

$$\mathsf{x} = \begin{pmatrix} x^0 \\ x^1 \\ x^2 \\ x^3 \end{pmatrix} . \tag{10.1}$$

Let x'^μ $(\mu = 0, 1, 2, 3)$ be the space-time coordinates of the particle with respect to another inertial system S'.

[1] Here c is the speed of light. In the following, Greek indices take the values from 0 to 3, Roman indices from 1 to 3. The sum convention will continually be used for both kinds of indices.

It is well known that x'^μ and x^μ are related by a Lorentz transformation

$$x^\mu = \Lambda^\mu{}_\nu x'^\nu \tag{10.2}$$

such that the expression $(x^0)^2 - (x^1)^2 - (x^2)^2 - (x^3)^2$ is invariant,

$$(x'^0)^2 - (x'^1)^2 - (x'^2)^2 - (x'^3)^2 = (x^0)^2 - (x^1)^2 - (x^2)^2 - (x^3)^3 \quad .$$

A transformation in which $S' = S$, $t' = t$, while (x^1, x^2, x^3) and (x'^1, x'^2, x'^3) are spatial coordinates with respect to XYZ and $X'Y'Z'$, the latter obtained from XYZ by a rotation, will be regarded as a Lorentz transformation. The group of space rotations is a subgroup of the Lorentz group.

The invariant expression can be written more concisely as

$$(x^0)^2 - (x^1)^2 - (x^2)^2 - (x^3)^2 = g_{\mu\nu} x^\mu x^\nu = x^\mu x_\mu \quad . \tag{10.3}$$

Here $g_{\mu\nu}$ are the components of the metric tensor

$$g_{00} = 1 \ , \ g_{11} = g_{22} = g_{33} = -1 \ , g_{\mu\nu} = 0 \text{ if } \mu \neq \nu \tag{10.4}$$

and

$$x_\mu = g_{\mu\nu} x^\nu \ , \ \text{i.e. } x_0 = x^0 \ , \ x_i = -x^i \quad (i = 1, 2, 3) \quad . \tag{10.5}$$

x_μ and x^μ are called the "covariant" and "contravariant" components of x.

The contravariant components $g^{\mu\nu}$ of the metric tensor can be defined as solutions of the equations

$$g^{\mu\rho} g_{\rho\nu} = g^\mu{}_\nu = (1 \text{ if } \mu = \nu \ , \ 0 \text{ if } \mu \neq \nu) \tag{10.6}$$

These conditions make it possible to invert equation (10.5) to give

$$x^\mu = g^{\mu\nu} x_\nu \quad . \tag{10.7}$$

Of course, one sees at once that $g^{\mu\nu} = g_{\mu\nu}$, while $g^\mu{}_\nu$ is defined in equation (10.6).

In general, a 4-vector

$$\mathsf{a} = \begin{pmatrix} a^0 \\ a^1 \\ a^2 \\ a^3 \end{pmatrix} \tag{10.8}$$

is an object satisfying all the definitions and relations written above for the space-time coordinates,

$$a'^\mu = \Lambda^\mu{}_\nu a^\nu \ , \ a_\mu = g_{\mu\nu} a^\nu \ , \ a'_\mu a'^\mu = a_\mu a^\mu \ldots \tag{10.9}$$

A tensor T has components $T^{\mu\nu}$ transforming according to $T^{\mu\nu} = \Lambda^\mu{}_\rho \Lambda^\nu{}_\sigma T'^{\rho\sigma}$. Other components are $T^\mu_\nu = T^{\mu\rho} g_{\rho\nu}$, $T_\mu{}^\nu = g_{\mu\rho} T^{\rho\nu}$.

10.2 Dynamics of a particle

The 4-velocity of a point particle is a 4-vector v with components

$$v^\mu = \frac{\mathrm{d}x^\mu}{\mathrm{d}s} \quad . \tag{10.10}$$

Here

$$\mathrm{d}s = c^{-1}\sqrt{(\mathrm{d}x^0)^2 - (\mathrm{d}x^1)^2 - (\mathrm{d}x^2)^2 - (\mathrm{d}x^3)^2} = c^{-1}\sqrt{\mathrm{d}x_\mu \mathrm{d}x^\mu} \tag{10.11}$$

is an invariant representing an infinitesimal time change as measured by an observer moving with the particle. In fact, if $\mathrm{d}x^i = 0$ $(i = 1, 2, 3)$ and $\mathrm{d}x^0 > 0$, $\mathrm{d}s = c^{-1}\mathrm{d}x^0 = \mathrm{d}t$.

If $\mathbf{v} = (\mathrm{d}x/\mathrm{d}t, \mathrm{d}y/\mathrm{d}t, \mathrm{d}z/\mathrm{d}t)$ is the familiar velocity vector, we have

$$v^0 = \frac{\mathrm{d}x^0}{\mathrm{d}s} = \gamma c \ , \ v^i = \frac{\mathrm{d}x^i}{\mathrm{d}s} = \gamma \frac{\mathrm{d}x^i}{\mathrm{d}t} \quad (v^1 = \gamma v_x, \ldots) \ , \tag{10.12}$$

where

$$\gamma = \frac{1}{\sqrt{1-\beta^2}} \ , \ \beta = \frac{v}{c} \ , \ v = |\mathbf{v}| \quad . \tag{10.13}$$

In a frame in which the particle is at rest, one has

$$v^0 = c \quad , \quad v^i = 0 \ \ (i = 1, 2, 3) \quad . \tag{10.14}$$

The norm of v is $v_\mu v^\mu = c^2$ in any inertial frame.

The acceleration 4-vector is defined as

$$\mathsf{a} = \frac{\mathrm{d}\mathsf{v}}{\mathrm{d}s} \quad . \tag{10.15}$$

Differentiating $v_\mu v^\mu = c^2$, we find

$$a_\mu v^\mu = 0 \quad . \tag{10.16}$$

We now want to establish the relativistic version of Newton's second law. We start from the experimentally established formula

$$\frac{\mathrm{d}\mathbf{p}}{\mathrm{d}t} = \frac{\mathrm{d}}{\mathrm{d}t}(m\gamma\mathbf{v}) = \mathbf{f} \quad , \tag{10.17}$$

which tells us that the relativistic momentum is given by the product *mass* $\times$ *velocity*, but with a velocity-dependent mass

$$m(v) = m\gamma = \frac{m}{\sqrt{1-\beta^2}} \quad (m = \text{rest mass, constant}) \quad . \tag{10.18}$$

The rest mass was previously denoted by m_0.
Using $\mathrm{d}t/\mathrm{d}s = \gamma$, we can write

$$\frac{\mathrm{d}}{\mathrm{d}s}(m\gamma\mathbf{v}) = \gamma\mathbf{f} \quad \text{or} \quad ma^i = f^i \ \ (f^1 = \gamma f_x, \ldots) \quad . \tag{10.19}$$

From $a^\mu v_\mu = 0$ we get

$$ma^0 c\gamma = -ma^i v_i = -f^i v_i = \gamma^2(f_x v_x + \ldots) = \gamma^2 \mathbf{f}\cdot\mathbf{v} \quad . \tag{10.20}$$

Hence

$$ma^0 = \frac{\gamma}{c}\mathbf{f}\cdot\mathbf{v} \quad , \tag{10.21}$$

and so

$$m\mathsf{a} = \mathsf{f} \quad , \tag{10.22}$$

where

$$\mathsf{f} = \begin{pmatrix} c^{-1}\gamma\mathbf{f}\cdot\mathbf{v} \\ \gamma f_x \\ \gamma f_y \\ \gamma f_z \end{pmatrix} \tag{10.23}$$

is the 4-force. In terms of a 4-momentum with components $p^0 = mc\gamma$ and $p^i = mv^i$ $(p^1 = m\gamma v_x, \ldots)$, we have

$$\frac{\mathrm{d}p^\mu}{\mathrm{d}s} = f^\mu \quad .$$

The relativistic energy is
$E = cp^0 = mc^2\gamma$ ($E = mc^2 + mv^2/2 + \ldots$ for $v \ll c$).
As we would expect, its rate of change equals the power,

$$\frac{\mathrm{d}E}{\mathrm{d}t} = \frac{c}{\gamma}a^0 = \mathbf{f}\cdot\mathbf{v} \quad . \tag{10.24}$$

Longitudinal and transversal masses:
Suppose $\mathbf{f}$ is parallel to the velocity $\mathbf{v}$, $\mathbf{f} = \mathbf{f}_\parallel$. Then $\mathrm{d}(m\gamma\mathbf{v})/\mathrm{d}t = \mathbf{f}_\parallel$ gives

$$m\frac{\mathrm{d}\gamma}{\mathrm{d}t}\mathbf{v} + m\gamma\frac{\mathrm{d}\mathbf{v}}{\mathrm{d}t} = \mathbf{f}_\parallel \quad , \tag{10.25}$$

which tells us that $\mathrm{d}\mathbf{v}/\mathrm{d}t$ is parallel to $\mathbf{v}$, and so $\mathrm{d}\mathbf{v}/\mathrm{d}t = (\mathbf{v}/v)\mathrm{d}v/\mathrm{d}t$.
Since $1 - \beta^2 = 1/\gamma^2$, $-2\beta\mathrm{d}\beta/\mathrm{d}t = -(2/\gamma^3)\mathrm{d}\gamma/\mathrm{d}t$, $\mathrm{d}\gamma/\mathrm{d}t = (v\gamma^3/c^2)\mathrm{d}v/\mathrm{d}t$,
we have

$$m\frac{\mathbf{v}}{v}\frac{\mathrm{d}v}{\mathrm{d}t}\left(\frac{v^2}{c^2}\gamma^3 + \gamma\right) = \mathbf{f}_\parallel \quad , \quad (m\gamma^3)\frac{\mathbf{v}}{v}\frac{\mathrm{d}v}{\mathrm{d}t} = \mathbf{f}_\parallel \quad ,$$

$$m_{\|}\frac{\mathbf{v}}{v}\frac{\mathrm{d}v}{\mathrm{d}t} = \mathbf{f}_{\|} \quad , \tag{10.26}$$

where $m_{\|}$ is the "longitudinal mass" $m_{\|} = m\gamma^3$.

If $\mathbf{f}$ is normal to $\mathbf{v}$, then

$$m\frac{\mathrm{d}\gamma}{\mathrm{d}t}\mathbf{v} + m\gamma\frac{\mathrm{d}\mathbf{v}}{\mathrm{d}t} = \mathbf{f}_{\perp} \quad . \tag{10.27}$$

From $\mathrm{d}E/\mathrm{d}t = \mathbf{f}\cdot\mathbf{v} = \mathbf{f}_{\perp}\cdot\mathbf{v} = 0$, we have $\mathrm{d}\gamma/\mathrm{d}t = 0$. Then

$$m_{\perp}\frac{\mathrm{d}\mathbf{v}}{\mathrm{d}t} = \mathbf{f}_{\perp} \tag{10.28}$$

with $m_{\perp} = m\gamma$ ("transversal mass").

Orbital angular momentum

Since $p^\mu = m\mathrm{d}x^\mu/\mathrm{d}s$, we see that the tensor

$$L^{\mu\nu} = x^\mu p^\nu - x^\nu p^\mu \tag{10.29}$$

satisfies the equation

$$\frac{\mathrm{d}L^{\mu\nu}}{\mathrm{d}s} = x^\mu f^\nu - x^\nu f^\mu \quad . \tag{10.30}$$

With $\gamma = \mathrm{d}t/\mathrm{d}s$ and $f^1 = \gamma f_x$ etc. , we have

$$\frac{\mathrm{d}L^{12}}{\mathrm{d}t} = xf_y - yf_x \quad , \tag{10.31}$$

where

$$L^{12} = x^1p^2 - x^2p^1 = m\gamma(xv_y - yv_x) \quad . \tag{10.32}$$

On the other hand

$$\frac{\mathrm{d}L^{01}}{\mathrm{d}s} = x^0 f^1 - x^1 f^0 \quad . \tag{10.33}$$

For a free particle this gives

$$ctp^1 - xE/c = \text{const} \quad , \quad x = (c^2p^1/E)t + \text{const} = v_xt + \text{const} \quad . \tag{10.34}$$

10.3 Formulary of Lorentz transformations

Let a and b be two 4-vectors, and L a Lorentz transformation matrix,

$$\mathsf{L} = \begin{pmatrix} \Lambda^0{}_0 & \Lambda^0{}_1 & \Lambda^0{}_2 & \Lambda^0{}_3 \\ \Lambda^1{}_0 & \Lambda^1{}_1 & \Lambda^1{}_2 & \Lambda^1{}_3 \\ \Lambda^2{}_0 & \Lambda^2{}_1 & \Lambda^2{}_2 & \Lambda^2{}_3 \\ \Lambda^3{}_0 & \Lambda^3{}_1 & \Lambda^3{}_2 & \Lambda^3{}_3 \end{pmatrix} \quad . \tag{10.35}$$

The product

$$a_\mu b^\mu = g_{\mu\nu} a^\mu b^\nu = \mathsf{a}^{\mathrm{T}} \mathsf{g} \mathsf{b} \tag{10.36}$$

with

$$\mathsf{g} = \begin{pmatrix} 1 & 0 & 0 & 0 \\ 0 & -1 & 0 & 0 \\ 0 & 0 & -1 & 0 \\ 0 & 0 & 0 & -1 \end{pmatrix}$$

must be invariant under Lorentz transformations,

$$\mathsf{a}'^{\mathrm{T}} \mathsf{g} \mathsf{b}' = \mathsf{a}^{\mathrm{T}} \mathsf{g} \mathsf{b} \quad . \tag{10.37}$$

Substituting $\mathsf{b} = \mathsf{L}\mathsf{b}'$ and $\mathsf{a}^{\mathrm{T}} = \mathsf{a}'^{\mathrm{T}} \mathsf{L}^{\mathrm{T}}$, we find

$$\mathsf{L}^{\mathrm{T}} \mathsf{g} \mathsf{L} = \mathsf{g} \quad . \tag{10.38}$$

Taking the (0,0) element of both sides of $\mathsf{L}^{\mathrm{T}}\mathsf{g}\mathsf{L} = \mathsf{g}$, one finds

$$(\Lambda^0{}_0)^2 - (\Lambda^1{}_0)^2 - (\Lambda^2{}_0)^2 - (\Lambda^3{}_0)^2 = 1 \quad . \tag{10.39}$$

From equation (10.38) and $\mathsf{g}^2 = \mathsf{I}$, we have $\mathsf{g}\mathsf{L}^{\mathrm{T}}\mathsf{g}\mathsf{L} = \mathsf{I}$,

$$\mathsf{L}^{-1} = \mathsf{g}\mathsf{L}^{\mathrm{T}}\mathsf{g} \quad , \tag{10.40}$$

$$\mathsf{L}^{-1} = \begin{pmatrix} \Lambda^0{}_0 & -\Lambda^1{}_0 & -\Lambda^2{}_0 & -\Lambda^3{}_0 \\ -\Lambda^0{}_1 & \Lambda^1{}_1 & \Lambda^2{}_1 & \Lambda^3{}_1 \\ -\Lambda^0{}_2 & \Lambda^1{}_2 & \Lambda^2{}_2 & \Lambda^3{}_2 \\ -\Lambda^0{}_3 & \Lambda^1{}_3 & \Lambda^2{}_3 & \Lambda^3{}_3 \end{pmatrix} \quad . \tag{10.41}$$

If the elements of L^{-1} are denoted by $(\Lambda^{-1})^\mu{}_\nu$, we have

$$(\Lambda^{-1})^0{}_0 = \Lambda^0{}_0 \, , \; (\Lambda^{-1})^0{}_i = -\Lambda^i{}_0 \, , \; (\Lambda^{-1})^i{}_0 = -\Lambda^0{}_i \, , \; (\Lambda^{-1})^i{}_j = \Lambda^j{}_i \quad .$$

Equation (10.39) for L^{-1} gives

$$(\Lambda^0{}_0)^2 - (\Lambda^0{}_1)^2 - (\Lambda^0{}_2)^2 - (\Lambda^0{}_3)^2 = 1 \quad . \tag{10.42}$$

How many conditions does $\mathsf{L}^T\mathbf{g}\mathsf{L} = \mathbf{g}$ impose? Since both sides are symmetric, just $4+6=10$ conditions imposed on the 16 elements of L. Hence a Lorentz transformation depends on 6 parameters.

Space rotations

For these ${\Lambda^0}_0 = 1$, ${\Lambda^0}_i = {\Lambda^i}_0 = 0$. The remaining elements ${\Lambda^i}_j$ can be expressed in terms of three parameters. We can use the formulae of section 4.1, replacing r by x, n_i by n^i [2], and extending the J_i's to the 4×4 matrices

$$\mathsf{J}_1 = \left(\begin{array}{c|ccc} 0 & 0 & 0 & 0 \\ \hline 0 & 0 & 0 & 0 \\ 0 & 0 & 0 & 1 \\ 0 & 0 & -1 & 0 \end{array}\right), \quad \mathsf{J}_2 = \left(\begin{array}{c|ccc} 0 & 0 & 0 & 0 \\ \hline 0 & 0 & 0 & -1 \\ 0 & 0 & 0 & 0 \\ 0 & 1 & 0 & 0 \end{array}\right), \quad \mathsf{J}_3 = \left(\begin{array}{c|ccc} 0 & 0 & 0 & 0 \\ \hline 0 & 0 & 1 & 0 \\ 0 & -1 & 0 & 0 \\ 0 & 0 & 0 & 0 \end{array}\right).$$

These matrices obey the same commutation relations as their 3×3 counterparts, $[\mathsf{J}_1, \mathsf{J}_2] = -\mathsf{J}_3$ and cyclic permutations. For finite rotations we have also $\mathsf{x} = \exp(-\mathsf{J}(\hat{\mathbf{n}})\phi)\mathsf{x}'$ with $\mathsf{J}(\hat{\mathbf{n}}) = n^i \mathsf{J}_i$.

Pure Lorentz transformations

Pure Lorentz transformations are those for which L is symmetric $({\Lambda^\mu}_\nu = {\Lambda^\nu}_\mu)$, ${\Lambda^0}_i = {\Lambda^i}_0$ $(i = 1,2,3)$ are not all zero, and

$${\Lambda^0}_0 = \sqrt{1 + ({\Lambda^1}_0)^2 + ({\Lambda^2}_0)^2 + ({\Lambda^3}_0)^2}.$$

It follows that L depends only on the three parameters ${\Lambda^1}_0$, ${\Lambda^2}_0$, and ${\Lambda^3}_0$, and can be expressed in the form [3]

$$\mathsf{L} = \left(\begin{array}{c|c} {\Lambda^0}_0 & {\Lambda^1}_0 \; {\Lambda^2}_0 \; {\Lambda^3}_0 \\ \hline {\Lambda^1}_0 & \\ {\Lambda^2}_0 & \delta_{ij} + \frac{{\Lambda^i}_0 {\Lambda^j}_0}{{\Lambda^0}_0 + 1} \\ {\Lambda^3}_0 & \end{array}\right), \tag{10.43}$$

$$\mathsf{L}^{-1} = \left(\begin{array}{c|c} {\Lambda^0}_0 & -{\Lambda^1}_0 \; -{\Lambda^2}_0 \; -{\Lambda^3}_0 \\ \hline -{\Lambda^1}_0 & \\ -{\Lambda^2}_0 & \delta_{ij} + \frac{{\Lambda^i}_0 {\Lambda^j}_0}{{\Lambda^0}_0 + 1} \\ -{\Lambda^3}_0 & \end{array}\right). \tag{10.44}$$

From the elements ${\Lambda^i}_0$ one can construct the unit vector

$$n^i = -n_i = \frac{{\Lambda^i}_0}{\sqrt{({\Lambda^0}_0)^2 - 1}} \tag{10.45}$$

[2] In chapter 4 we did not need to distinguish between covariant and contravariant indices. Here $n^1 = n_x, n^2 = n_y, n^3 = n_z$.

[3] δ_{ij} is the Kronecker symbol ($\delta_{11} = \delta_{22} = \delta_{33} = 1$, $\delta_{ij} = 0$ if $i \neq j$).

and the relative velocity

$$u = c\,\frac{\sqrt{({\Lambda^0}_0)^2 - 1}}{{\Lambda^0}_0} \;. \tag{10.46}$$

With $\beta = u/c$ and $\gamma = 1/\sqrt{1-\beta^2}$, we have

$${\Lambda^0}_0 = \gamma \quad , \quad {\Lambda^i}_0 = \beta\gamma n^i \quad , \quad {\Lambda^i}_j = \delta_{ij} + (\gamma - 1)n^i n^j \quad . \tag{10.47}$$

If you are not satisfied by i and j not being in the same position in the two members, write

$${\Lambda^i}_j = {g^i}_j - (\gamma - 1)n^i n_j \tag{10.48}$$

with ${g^i}_j$ defined by equation (10.6).

More explicitly the matrix for a pure Lorentz transformation with velocity $\mathbf{u}=(un_x,un_y,un_z)$, $n_x = n^1$ etc. , is

$$\mathsf{L}(\mathbf{u}) = \left(\begin{array}{c|c|c|c} \gamma & \beta\gamma n_x & \beta\gamma n_y & \beta\gamma n_z \\ \hline \beta\gamma n_x & 1+(\gamma-1)n_x^2 & (\gamma-1)n_x n_y & (\gamma-1)n_x n_z \\ \hline \beta\gamma n_y & (\gamma-1)n_y n_x & 1+(\gamma-1)n_y^2 & (\gamma-1)n_y n_z \\ \hline \beta\gamma n_z & (\gamma-1)n_z n_x & (\gamma-1)n_z n_y & 1+(\gamma-1)n_z^2 \end{array}\right) \tag{10.49}$$

Special case: The $X'Y'Z'$ axes are parallel to the XYZ axes. The origin O' of the $X'Y'Z'$ axes moves with velocity u with respect to XYZ in the positive x^1 direction ($u > 0$, $n_x = 1$, $n_y = n_z = 0$):

$$\mathsf{L}(u,0,0) = \begin{pmatrix} \gamma & \beta\gamma & 0 & 0 \\ \beta\gamma & \gamma & 0 & 0 \\ 0 & 0 & 1 & 0 \\ 0 & 0 & 0 & 1 \end{pmatrix} \;, \tag{10.50}$$

$\mathsf{x} = \mathsf{L}(u,0,0)\mathsf{x}'$ gives

$$x^0 = \gamma(x'^0 + \beta x'^1)\;,\; x^1 = \gamma(x'^1 + \beta x'^0)\;,\; x^2 = x'^2\;,\; x^3 = x'^3\;,$$

$$x'^0 = \gamma(x^0 - \beta x^1)\;,\; x'^1 = \gamma(x^1 - \beta x^0)\;,\; x'^2 = x^2\;,\; x'^3 = x^3\;.$$

For $u \ll c$, $\beta \ll 1$, $x'^0 = x^0$, $x'^1 = x^1 - ut$, $x'^2 = x^2$, $x'^3 = x^3$ (Galilean transformation).

It is useful to express β and γ as $\beta = \tanh\alpha$, $\gamma = \cosh\alpha$, $\beta\gamma = \sinh\alpha$. Then

$${\Lambda^0}_0 = \cosh\alpha\;,\; {\Lambda^i}_0 = n^i\sinh\alpha\;,\; {\Lambda^i}_j = \delta_{ij} + (\cosh\alpha - 1)n^i n^j \quad . \tag{10.51}$$

The components a^μ of a 4-vector with respect to S are related to the components a'^μ with respect to S' by the formulae

$$\begin{cases} a'^0 = \cosh\alpha\, a^0 + \sinh\alpha\,(n_j a^j) \;, \\ a'^i = a^i - n^i[\sinh\alpha\, a^0 + (\cosh\alpha - 1)(n_j a^j)] \;. \end{cases} \tag{10.52}$$

Remember that $n_j a^j = -(n_x a^1 + \ldots)$.

Perform two Lorentz transformations in the same direction, say

$$\mathsf{L}(u_1,0,0) = \begin{pmatrix} \gamma_1 & \beta_1\gamma_1 & 0 & 0 \\ \beta_1\gamma_1 & \gamma_1 & 0 & 0 \\ 0 & 0 & 1 & 0 \\ 0 & 0 & 0 & 1 \end{pmatrix} = \begin{pmatrix} \cosh\alpha_1 & \sinh\alpha_1 & 0 & 0 \\ \sinh\alpha_1 & \cosh\alpha_1 & 0 & 0 \\ 0 & 0 & 1 & 0 \\ 0 & 0 & 0 & 1 \end{pmatrix}$$

and $\mathsf{L}(u_2,0,0)$. We find

$$\mathsf{L}(u_1,0,0)\,\mathsf{L}(u_2,0,0) = \begin{pmatrix} \cosh(\alpha_1+\alpha_2) & \sinh(\alpha_1+\alpha_2) & 0 & 0 \\ \sinh(\alpha_1+\alpha_2) & \cosh(\alpha_1+\alpha_2) & 0 & 0 \\ 0 & 0 & 1 & 0 \\ 0 & 0 & 0 & 1 \end{pmatrix}$$

$$= \mathsf{L}(u,0,0) = \begin{pmatrix} \gamma & \beta\gamma & 0 & 0 \\ \beta\gamma & \gamma & 0 & 0 \\ 0 & 0 & 1 & 0 \\ 0 & 0 & 0 & 1 \end{pmatrix} \tag{10.53}$$

with

$$\gamma = \cosh(\alpha_1+\alpha_2) = \gamma_1\gamma_2(1+\beta_1\beta_2) \quad ,$$

$$\beta\gamma = \sinh(\alpha_1\alpha_2) \doteq \gamma_1\gamma_2(\beta_1+\beta_2) \quad ,$$

$$\beta = \frac{\beta_1+\beta_2}{1+\beta_1\beta_2} \quad .$$

It is easy to remember that when performing two pure Lorentz transformations in the same direction one must simply sum the α's.

A pure infinitesimal Lorentz transformation with $\mathbf{u} = \hat{\mathbf{n}}\,\delta\alpha$ can be expressed in the form

$$\mathsf{L}(\hat{\mathbf{n}}\delta\alpha) = \mathsf{I} + n^i\mathsf{K}_i\,\delta\alpha = \mathsf{I} + (n_x\mathsf{K}_1 + \ldots)\delta\alpha \quad , \tag{10.54}$$

where the K_i's are the symmetric matrices

$$\mathsf{K}_1 = \begin{pmatrix} 0 & 1 & 0 & 0 \\ 1 & 0 & 0 & 0 \\ 0 & 0 & 0 & 0 \\ 0 & 0 & 0 & 0 \end{pmatrix} , \; \mathsf{K}_2 = \begin{pmatrix} 0 & 0 & 1 & 0 \\ 0 & 0 & 0 & 0 \\ 1 & 0 & 0 & 0 \\ 0 & 0 & 0 & 0 \end{pmatrix} , \; \mathsf{K}_3 = \begin{pmatrix} 0 & 0 & 0 & 1 \\ 0 & 0 & 0 & 0 \\ 0 & 0 & 0 & 0 \\ 1 & 0 & 0 & 0 \end{pmatrix} .$$

Hence for a finite transformation we have

$$\mathsf{L}(\hat{\mathbf{n}}\alpha) = \lim_{N\to\infty}\left(\mathsf{I} + n^i\mathsf{K}_i\frac{\alpha}{N}\right)^N = e^{\mathsf{K}(\hat{\mathbf{n}})\alpha} \tag{10.55}$$

with $\mathsf{K}(\hat{\mathbf{n}}) = n^i\mathsf{K}_i$

While two pure Lorentz transformations in the same direction give the same result irrespective of the order in which they are performed,

$$\mathsf{L}(\mathbf{u}_1)\mathsf{L}(\mathbf{u}_2) \neq \mathsf{L}(\mathbf{u}_2)\mathsf{L}(\mathbf{u}_1)$$

if $\mathbf{u}_1$ and $\mathbf{u}_2$ are not parallel.

This is similar to the fact that two successive rotations around different axes yield a result that depends on the order in which they are performed. However, while the product of two space rotations is a space rotation, the product of two pure Lorentz transformations in different directions is not a pure Lorentz transformation.

This is connected with the fact that the K_i's do not form a closed algebra. One has

$$[\mathsf{K}_1, \mathsf{K}_2] = \mathsf{J}_3 \ , \ [\mathsf{K}_i, \mathsf{K}_j] = \epsilon_{ijk}\mathsf{K}_k \ , \tag{10.56}$$

$$[\mathsf{J}_1, \mathsf{K}_2] = -\mathsf{K}_3 \ , \ [\mathsf{J}_2, \mathsf{K}_1] = \mathsf{K}_3 \ , \ [\mathsf{J}_i, \mathsf{K}_j] = -\epsilon_{ijk}\mathsf{K}_k \ , \tag{10.57}$$

and the already known

$$[\mathsf{J}_1, \mathsf{J}_2] = -\mathsf{J}_3 \quad \text{and cyclic permutations.} \tag{10.58}$$

It can be shown that

$$\mathsf{L}(\mathbf{u}_1)\mathsf{L}(\mathbf{u}_2) = \mathsf{L}(\mathbf{u})\mathsf{R} \quad , \tag{10.59}$$

where $\mathsf{L}(\mathbf{u})$ is a pure Lorentz transformation and R is a space rotation.

In section 10.4, using the $SL(2)$ representation of the Lorentz group, we shall prove the approximate formula with an error of the order of $|\delta\mathbf{u}|^2$

$$\mathsf{L}(\mathbf{u} + \delta\mathbf{u})\ \mathsf{L}(-\mathbf{u}) \simeq \mathsf{L}(\delta\mathbf{w})\ \mathsf{R}(\hat{\mathbf{n}}\delta\phi) \tag{10.60}$$

with $\delta\mathbf{w} = \gamma^2\delta\mathbf{u}_{\|} + \gamma\delta\mathbf{u}_{\perp}$, where $\delta\mathbf{u}_{\|}$ and $\delta\mathbf{u}_{\perp}$ are the components of $\delta\mathbf{u}$ parallel and normal to $\mathbf{u}$, and

$$\mathsf{R}(\hat{\mathbf{n}}\delta\phi) = \mathsf{I} + \mathsf{J}\cdot\left(\frac{\gamma^2}{c^2(\gamma+1)}\,\mathbf{u}\times\delta\mathbf{u}\right) \quad . \tag{10.61}$$

Here is a quick check for $u \ll c$ ($\gamma \simeq 1$) by using the Hausdorff identity

$$e^{\mathsf{A}}e^{\mathsf{B}} = e^{\mathsf{A}+\mathsf{B}+\frac{1}{2}[\mathsf{A},\mathsf{B}]+\dots} \quad . \tag{10.62}$$

We have

$$\exp\left(\frac{1}{c}(u^i+\delta u^i)\mathsf{K}_i\right)\exp\left(-\frac{1}{c}u^i\mathsf{K}_i\right)\simeq\exp\left(\frac{1}{c}\delta u^i\mathsf{K}_i-\frac{1}{2c^2}[(u^i+\delta u^i)\mathsf{K}_i,u^j\mathsf{K}_j]\right)$$

$$\simeq\mathsf{L}(\delta\mathbf{u})\exp\left(-\frac{1}{2c^2}\delta u^i\,u^j\epsilon_{ijl}\mathsf{J}_l\right)=\mathsf{L}(\delta\mathbf{u})\exp\left(\frac{1}{2c^2}(\mathbf{u}\times\delta\mathbf{u})^l\mathsf{J}_l\right)\quad . \tag{10.63}$$

Other useful relations are

$$\mathsf{L}(-\mathbf{u})\;\mathsf{L}(\mathbf{u}+\delta\mathbf{u})=\mathsf{L}(\delta\mathbf{w})\;\mathsf{R}(-\hat{\mathbf{n}}\delta\phi) \tag{10.64}$$

and

$$\mathsf{L}(\mathbf{u}+\delta\mathbf{u})\;\mathsf{R}(\hat{\mathbf{n}}\delta\phi)\;\mathsf{L}(-\mathbf{u})=\mathsf{I}+\gamma^2c^{-1}\mathsf{K}\cdot\delta\mathbf{u}+\gamma^2c^{-2}\mathsf{J}\cdot(\mathbf{u}\times\delta\mathbf{u}) \tag{10.65}$$

with $\mathsf{K}=(\mathsf{K}_1,\mathsf{K}_2,\mathsf{K}_3)$.

Boosts and active space rotations

The reader may have wondered why we wrote $\mathsf{x}=\mathsf{L}\mathsf{x}'$ rather than having the "prime" appear on the left side. With our way of writing a pure Lorentz transformation $\mathsf{L}(\mathbf{u})$ is a "boost". If $p'^0=mc$, $p'^i=0$ $(i=1,2,3)$ (particle at rest in S'), then, taking $\mathbf{u}$ in the x^1 direction for simplicity's sake,

$$p^0=\gamma mc \quad , \quad p^1=\gamma mu \quad , \quad p^2=p^3=0 \tag{10.66}$$

is the momentum with respect to S.

The formula $\mathsf{a}=\exp(-\mathsf{J}(\hat{\mathbf{n}})\phi)\mathsf{a}'$ relating the components of a fixed vector with respect to XYZ and $X'Y'Z'$ (obtained from XYZ by a counterclockwise rotation through ϕ around $\hat{\mathbf{n}}$) can also be given an "active" meaning. If (a^1,a^2,a^3) are the components of a vector $\mathbf{a}$ with respect to XYZ, then (a'^1,a'^2,a'^3) are the components, also with respect to XYZ, of a vector obtained by rotating $\mathbf{a}$ around $\hat{\mathbf{n}}$ through ϕ clockwise.

10.4 The spinor connection

Consider the Hermitian matrix

$$\mathsf{X}' = \begin{pmatrix} x'^0 + x'^3 & x'^1 - \mathrm{i}x'^2 \\ x'^1 + \mathrm{i}x'^2 & x'^0 - x'^3 \end{pmatrix} = x'^\mu \sigma_\mu \quad , \tag{10.67}$$

where σ_0 is the 2×2 unit matrix I, and

$$\sigma_1 = \begin{pmatrix} 0 & 1 \\ 1 & 0 \end{pmatrix} \; , \; \sigma_2 = \begin{pmatrix} 0 & -\mathrm{i} \\ \mathrm{i} & 0 \end{pmatrix} \; , \; \sigma_3 = \begin{pmatrix} 1 & 0 \\ 0 & -1 \end{pmatrix} \tag{10.68}$$

are the Pauli matrices. These obey the relations

$$[\sigma_i, \sigma_j]_+ = 2\delta_{ij}\mathsf{I} \quad , \quad [\sigma_i, \sigma_j]_- = 2\mathrm{i}\epsilon_{ijk}\sigma_k \quad , \tag{10.69}$$

$$\sigma_i \sigma_j = \mathsf{I}\;\delta_{ij} + \mathrm{i}\epsilon_{ijk}\;\sigma_k \quad . \tag{10.70}$$

The Pauli matrices are all traceless. Using this property, we find

$$\mathrm{trace}(\sigma_\mu \sigma_\nu) = \begin{pmatrix} 2 \text{ if } \mu = \nu \\ 0 \text{ if } \mu \neq \nu \end{pmatrix} \tag{10.71}$$

and

$$x'^\mu = (1/2)\mathrm{trace}(\sigma_\mu \mathsf{X}') \quad . \tag{10.72}$$

Note that

$$\det(\mathsf{X}') = (x'^0)^2 - (x'^1)^2 - (x'^2)^2 - (x'^3)^2 = x'_\mu x'^\mu \tag{10.73}$$

Let the matrix

$$\mathsf{M} = \begin{pmatrix} m_{11} & m_{12} \\ m_{21} & m_{22} \end{pmatrix} \tag{10.74}$$

with complex elements have an inverse ($\det(\mathsf{M}) \neq 0$). Then M is an element of $L(2)$, the group of linear transformations in two dimensions.

If we restrict M by requiring that $\det \mathsf{M} = 1$, $m_{11}m_{22} - m_{12}m_{21} = 1$, then M is an element of $SL(2)$, the group of the special linear transformations in two dimensions.

Consider the matrix

$$\mathsf{X} = \mathsf{M}\mathsf{X}'\mathsf{M}^\dagger \quad , \tag{10.75}$$

where M is an element of $SL(2)$ and $\mathsf{M}^\dagger$ is the Hermitian conjugate of M,

$$\mathsf{M}^\dagger = \begin{pmatrix} \bar{m}_{11} & \bar{m}_{21} \\ \bar{m}_{12} & \bar{m}_{22} \end{pmatrix} \quad . \tag{10.76}$$

The bar denotes complex conjugation.

X is Hermitian like X'. In fact $\mathsf{X}^\dagger = \mathsf{M}^{\dagger\dagger}\mathsf{X}'^\dagger\mathsf{M}^\dagger = \mathsf{M}\mathsf{X}'\mathsf{M}^\dagger = \mathsf{X}$. The quantities

$$x^\mu = (1/2)\mathrm{trace}(\sigma_\mu \mathsf{X}) \tag{10.77}$$

are real. In fact,

$$\bar{x}^\mu = (1/2)\mathrm{trace}(\mathsf{X}^\dagger\sigma_\mu^\dagger) = (1/2)\mathrm{trace}(\mathsf{X}\sigma_\mu) = (1/2)\mathrm{trace}(\sigma_\mu\mathsf{X}) = x^\mu \quad .$$

Therefore X can be expressed in the form

$$\mathsf{X} = \begin{pmatrix} x^0 + x^3 & x^1 - \mathrm{i}x^2 \\ x^1 + \mathrm{i}x^2 & x^0 - x^3 \end{pmatrix} \quad . \tag{10.78}$$

But now

$$\det(\mathsf{X}) = \det(\mathsf{M})\cdot\det(\mathsf{X}')\cdot\det\mathsf{M}^\dagger = |\det(\mathsf{M})|^2\cdot\det(\mathsf{X}') = \det(\mathsf{X}') \quad ,$$

and so

$$(x'^0)^2 - (x'^1)^2 - (x'^2)^2 - (x'^3)^2 = (x^0)^2 - (x^1)^2 - (x^2)^2 - (x^3)^2 \quad .$$

This shows that x'^μ and x^μ are related by a Lorentz transformation

$$x^\mu = \Lambda(\mathsf{M})^\mu{}_\nu x'^\nu \quad . \tag{10.79}$$

From

$$x^\mu = (1/2)\mathrm{trace}(\sigma_\mu\mathsf{X}) = (1/2)\mathrm{trace}(\sigma_\mu\mathsf{M}\mathsf{X}'\mathsf{M}^\dagger) = (1/2)\mathrm{trace}(\sigma_\mu\mathsf{M}\sigma_\nu\mathsf{M}^\dagger)x'^\nu$$

we see that

$$\Lambda(\mathsf{M})^\mu{}_\nu = (1/2)\mathrm{trace}(\sigma_\mu\mathsf{M}\sigma_\nu\mathsf{M}^\dagger) \quad . \tag{10.80}$$

The following formulae for the traces of products of Pauli matrices will be used:

$$\mathrm{trace}(\sigma_a) = 0 \ , \ \mathrm{trace}(\sigma_a\sigma_b) = 2\delta_{ab} \ , \ \mathrm{trace}(\sigma_a\sigma_b\sigma_c) = 2\mathrm{i}\epsilon_{abc} \ ,$$

$$\mathrm{trace}(\sigma_a\sigma_b\sigma_c\sigma_d) = 2(\delta_{ab}\delta_{cd} - \delta_{ac}\delta_{bd} + \delta_{ad}\delta_{bc}) \quad .$$

It is easy to verify the correspondence

$$\mathsf{L}(\mathbf{u}) = \mathrm{e}^{\mathsf{K}(\hat{\mathbf{n}})\alpha} \to \mathsf{M}(\mathbf{u}) = \cosh(\alpha/2) + n^i\sigma_i\sinh(\alpha/2) \tag{10.81}$$

between the pure Lorentz transformation matrix $\mathsf{L}(\mathbf{u})$ and the Hermitian matrix $\mathsf{M}(\mathbf{u})$. In problem 10.2 it will be shown that

$$\Lambda(\mathsf{M})^0{}_0 = \cosh\alpha \ , \ \Lambda(\mathsf{M})^0{}_i = \Lambda(\mathsf{M})^i{}_0 = n^i\sinh\alpha \ , \tag{10.82}$$

$$\Lambda(\mathsf{M})^i{}_j = \delta_{ij} + n^i n^j (\cosh\alpha - 1) \quad . \tag{10.83}$$

For a space rotation must be

$$x^0 = x'^0,\ \Lambda(\mathsf{M})^0{}_0 = 1\ ,\ \Lambda(\mathsf{M})^0{}_i = \Lambda(\mathsf{M})^i{}_0 = 0.$$

Since

$$\Lambda(\mathsf{M})^0{}_0 = (1/2)\mathrm{trace}(\mathsf{M}\mathsf{M}^\dagger) \quad ,$$

we see that space rotations correspond to $SL(2)$ matrices satisfying the additional condition $\mathsf{M}^\dagger = \mathsf{M}^{-1}$, namely to $SU(2)$ matrices. The special unitary group $SU(2)$ is a subgroup of $SL(2)$, as the group of space rotations is a subgroup of the Lorentz group.

We have

$$\mathsf{R}(\hat{\mathbf{n}},\phi) \to \mathsf{M}(\hat{\mathbf{n}},\phi) = \cos(\phi/2) - \mathrm{i}\,\hat{\mathbf{n}}\cdot\sigma\sin(\phi/2) \quad . \tag{10.84}$$

In section 10.3 the product $\mathsf{L}(\mathbf{u}+\delta\mathbf{u})\mathsf{L}(-\mathbf{u})$ with $|\delta\mathbf{u}| \ll |\mathbf{u}|$ was expressed in the form of equation (10.60). Here we prove that result by using the $SL(2)$ correspondence:

$$\mathsf{M}(\mathbf{u}) = \cosh(\alpha/2)\mathsf{I} + n^i\sigma_i\sinh(\alpha/2) = \sqrt{\frac{\gamma+1}{2}}\,\mathsf{I} + \frac{\gamma}{c\sqrt{2(\gamma+1)}}\,u^i\sigma_i \quad ,$$

$$\mathsf{M}(\mathbf{u}+\delta\mathbf{u}) = \mathsf{M}(\mathbf{u}) + \frac{\delta\gamma}{2\sqrt{2(\gamma+1)}}\,\mathsf{I} + \frac{\gamma}{c\sqrt{2(\gamma+1)}}\,\sigma_i\delta u^i + \frac{\gamma+2}{c[2(\gamma+1)]^{\frac{3}{2}}}\delta\gamma\, u^i\sigma_i \quad ,$$

$$\mathsf{M}(-\mathbf{u}) = \sqrt{\frac{\gamma+1}{2}}\,\mathsf{I} - \frac{\gamma}{c\sqrt{2(\gamma+1)}}\,u^i\sigma_i = (\mathsf{M}(\mathbf{u}))^{-1} \quad .$$

Since ("T" for Thomas)

$$\mathsf{L}(\mathbf{u}+\delta\mathbf{u})\mathsf{L}(-\mathbf{u}) \to \mathsf{M}(\mathbf{u}+\delta\mathbf{u})\mathsf{M}(-\mathbf{u}) = \mathsf{M}_{\mathrm{T}} \quad ,$$

a simple calculation yields

$$\mathsf{M}_{\mathrm{T}} \simeq \left(\mathsf{I} + \mathrm{i}\frac{\gamma^2}{2c^2(\gamma+1)}\sigma_i\epsilon_{ijk}u^j\delta u^k\right)\left(\mathsf{I} + \frac{\mathrm{d}\gamma}{2c(\gamma+1)}\sigma_i u^i + \frac{\gamma}{2c}\sigma_i\delta u^i\right) \quad .$$

Comparing the first factor with

$$\mathsf{I} - \mathrm{i}\,n^i\sigma_i\delta\phi/2 \to \mathsf{I} - n^i\mathsf{J}_i\delta\phi \quad ,$$

we recognize a rotation around the unit vector in the direction of $\mathbf{u}\times\delta\mathbf{u}$ with angle $\delta\phi = -[\gamma^2/(\gamma+1)c^2]|\mathbf{u}\times\delta\mathbf{u}|$.
On the other hand, the second factor can be written as

$$\mathsf{I} + \frac{\gamma^2}{2c}\sigma_i\delta u^i_{\parallel} + \frac{\gamma}{2c}\sigma_i\delta u^i_{\perp} \quad ,$$

where $\delta\mathbf{u}_{\|}$ and $\delta\mathbf{u}_{\perp}$ are the components of $\delta\mathbf{u}$ parallel and normal to $\mathbf{u}$. Hence

$$\mathsf{I} + \sigma_i \left(\frac{\gamma^2}{2c} \delta u^i_{\|} + \frac{\gamma}{2c} \delta u^i_{\perp} \right) \to \mathsf{I} + \frac{1}{c} \mathsf{K}_i \delta w^i$$

with

$$\delta\mathbf{w} = \gamma^2 \delta\mathbf{u}_{\|} + \gamma \delta\mathbf{u}_{\perp} \quad .$$

10.5 The spin

A point particle with an intrinsic angular momentum (spin) is in non-uniform motion with respect to the (inertial) laboratory system K[4]. At the time t (lab clock) the particle has velocity $\mathbf{u}$ with respect to K. At that time there will be an inertial system K' in which the particle is instantaneously at rest.

If $C(XYZ)$ is a system of Cartesian axes in K, let $C'(X'Y'Z')$ be the system in K' obtained from C by a boost with velocity $\mathbf{u}$. The axes of C' are parallel to those of C.

At the time $t+\delta t$, the particle has velocity $\mathbf{u}+\delta\mathbf{u}$ with respect to K, and will be instantaneously at rest in a system K''. What system of Cartesian axes are we going to use in K''?

We can either use the system $C''(X''Y''Z'')$ obtained from C of K by a boost with velocity $\mathbf{u} + \delta\mathbf{u}$, or a system $C'''(X'''Y'''Z''')$ obtained from C' of K' by a boost with velocity $\delta\mathbf{w} = \gamma^2\delta\mathbf{u}_{\|} + \gamma\delta\mathbf{u}_{\perp}$, where $\delta\mathbf{u}_{\|}$ and $\delta\mathbf{u}_{\perp}$ are the components of $\delta\mathbf{u}$ parallel and normal to $\mathbf{u}$, respectively, and $\gamma = 1/\sqrt{1-u^2/c^2}$. We propose as an exercise to use the formula for addition of velocities to verify that K'' (moving with velocity $\delta\mathbf{w}$ with respect to K'), moves with velocity $\mathbf{u} + \delta\mathbf{u}$ with respect to K.

Let a be a 4-vector. Its components with respect to $K(C)$, $K'(C')$, $K''(C'')$, and $K''(C''')$ are related as follows:

$$\mathsf{a} = \mathsf{L}(\mathbf{u})\mathsf{a}' \ , \ \mathsf{a} = \mathsf{L}(\mathbf{u}+\delta\mathbf{u})\mathsf{a}'' \ , \ \mathsf{a}' = \mathsf{L}(\delta\mathbf{w})\mathsf{a}''' \quad . \tag{10.85}$$

By equation (10.64), $\mathsf{L}(-\mathsf{u})\mathsf{L}(\mathbf{u}+\delta\mathbf{u}) = \mathsf{L}(\delta\mathbf{w})\mathsf{R}(-\hat{\mathbf{n}}\delta\phi)$, we have also

$$\mathsf{a}' = \mathsf{L}(-\mathbf{u})\mathsf{L}(\mathbf{u}+\delta\mathbf{u})\mathsf{a}'' = \mathsf{L}(\delta\mathbf{w})\mathsf{R}(-\hat{\mathbf{n}}\delta\phi)\mathsf{a}'' \tag{10.86}$$

and

$$\mathsf{a} = \mathsf{L}(\mathbf{u}+\delta\mathbf{u})\mathsf{R}(\hat{\mathbf{n}}\delta\phi)\mathsf{a}''' \quad . \tag{10.87}$$

[4] We are obliged to denote inertial systems by K rather than the more usual S. In this section there are too many S's meaning "spin", and there might be confusion.

Using this last equation, we find

$$\mathsf{a}'' = \mathsf{L}(-\mathbf{u}-\delta\mathbf{u})\mathsf{a} = \mathsf{L}(-\mathbf{u}-\delta\mathbf{u})\mathsf{L}(\mathbf{u}+\delta\mathbf{u})\mathsf{R}(\hat{\mathbf{n}}\delta\phi)\mathsf{a}''' = \mathsf{R}(\hat{\mathbf{n}}\delta\phi)\mathsf{a}''' \quad . \tag{10.88}$$

Postulated properties of the spin:

(i) The spin S is a 4-vector such that its time component in the rest frame of the particle is zero (see problem 10.8).

(ii) Consider a particle subject to forces and, therefore, in non-uniform motion, carrying a free spin, namely a spin subject to no torques. Construct a sequence of inertial systems, such that at any time the particle is at rest in one of them. Assume that each of these inertial systems, say that at time $t+\delta t$, is obtained from that at time t by a pure Lorentz transformation (no space rotation).

Then the components of the free spin with respect to the Cartesian system at the time $t + \delta t$ are equal to those with respect to the Cartesian system at the time t.

Let S^μ ($S_x = S^1 = -S_1$ etc.) be the spin components in a generic system. Property (i) can be cast into the invariant form

$$u^\mu S_\mu = 0 \quad , \tag{10.89}$$

which reduces to $S_0 = 0$ if the system is the rest system. Note that with respect to $K'(C')$, $K''(C'')$, and $K''(C''')$

$$\mathsf{S}'(t) = \begin{pmatrix} 0 \\ S'_x(t) \\ S'_y(t) \\ S'_z(t) \end{pmatrix} \quad , \tag{10.90}$$

$$\mathsf{S}''(t+\delta t) = \begin{pmatrix} 0 \\ S''_x(t+\delta t) \\ S''_y(t+\delta t) \\ S''_z(t+\delta t) \end{pmatrix} \quad , \tag{10.91}$$

$$\mathsf{S}'''(t+\delta t) = \begin{pmatrix} 0 \\ S'''_x(t+\delta t) \\ S'''_y(t+\delta t) \\ S'''_z(t+\delta t) \end{pmatrix} \tag{10.92}$$

The last two are consistent, since $K''(C'')$ and $K''(C''')$ differ by a space rotation.

In the lab system, $u^\mu S_\mu = 0$ reduces to

$$u^0 S_0 - u^1 S_x - u^2 S_y - u^3 S_z = 0, \; \gamma(cS_0 - \mathbf{u}\cdot\mathbf{S}) = 0,$$

$$S_0 = \mathbf{u}\cdot\mathbf{S}/c \quad . \tag{10.93}$$

Property (ii) implies that[5]

$$\mathsf{S}'''(t+\delta t) = \mathsf{S}'(t) \quad .$$

Thus $S'''^{\mu}(t+\delta t) = S'^{\mu}(t)$.

Rate of change of free spin components in lab system:

$$\mathsf{S}(t+\delta t) = \mathsf{L}(\mathbf{u}+\delta\mathbf{u})\mathsf{R}(\hat{\mathbf{n}}\delta\phi)\mathsf{S}'''(t+\delta t) = \mathsf{L}(\mathbf{u}+\delta\mathbf{u})\mathsf{R}(\hat{\mathbf{n}}\delta\phi)\mathsf{S}'(t)$$

$$= \mathsf{L}(\mathbf{u}+\delta\mathbf{u})\mathsf{R}(\hat{\mathbf{n}}\delta\phi)\mathsf{L}(-\mathbf{u})\mathsf{S}(t) = \left(\mathsf{I} + \frac{\gamma^2}{c}\mathsf{K}\cdot\delta\mathbf{u} + \frac{\gamma^2}{c^2}\mathsf{J}\cdot(\mathbf{u}\times\delta\mathbf{u})\right)\mathsf{S}(t) \ ,$$

having used equation (10.65). This gives

$$S_0(t+\delta t) = S_0(t) + (\gamma^2/c)\ \mathbf{S}(t)\cdot\delta\mathbf{u} \ ,$$

$$\frac{\mathrm{d}S_0(t)}{\mathrm{d}t} = \frac{\gamma^2}{c}\left(\mathbf{S}(t)\cdot\frac{\mathrm{d}\mathbf{u}}{\mathrm{d}t}\right) \quad , \tag{10.94}$$

and

$$\mathbf{S}(t+\delta t) = \mathbf{S}(t) + \frac{\gamma^2}{c}S_0\delta\mathbf{u} + \frac{\gamma^2}{c^2}\mathbf{S}(t)\times(\mathbf{u}\times\delta\mathbf{u})$$

$$= \mathbf{S}(t) + \frac{\gamma^2}{c^2}(\mathbf{u}\cdot\mathbf{S}(t))\delta\mathbf{u} + \frac{\gamma^2}{c^2}[(\mathbf{S}(t)\cdot\delta\mathbf{u})\mathbf{u} - (\mathbf{u}\cdot\mathbf{S}(t))\delta\mathbf{u}] \ ,$$

$$\frac{\mathrm{d}\mathbf{S}(t)}{\mathrm{d}t} = \frac{\gamma^2}{c^2}\left(\mathbf{S}(t)\cdot\frac{\mathrm{d}\mathbf{u}}{\mathrm{d}t}\right)\mathbf{u} \quad . \tag{10.95}$$

Equations (10.94,95) can be combined in the covariant equation for the free spin (see problem 10.9)

$$\frac{\mathrm{d}S^{\mu}}{\mathrm{d}s} = -\frac{1}{c^2}u^{\mu}\left(S_{\nu}\frac{\mathrm{d}u^{\nu}}{\mathrm{d}s}\right) \quad . \tag{10.96}$$

Multiplying this equation by u^{μ}, since $u_{\mu}u^{\mu} = c^2$ we have

$$\frac{\mathrm{d}}{\mathrm{d}s}(u_{\mu}S^{\mu}) = 0 \quad , \tag{10.97}$$

the 4-spin remains normal to the 4-velocity during the motion.

Note also that

$$\frac{\mathrm{d}}{\mathrm{d}s}(S_{\mu}S^{\mu}) = 0 \quad ,$$

as is easily shown multiplying equation (10.96) by S_{μ} and remembering that $S_{\mu}u^{\mu} = 0$. The magnitude of $\mathbf{S}$ is not constant during the motion.

[5] Not in contradiction with $\mathsf{a}''' = \mathsf{L}(-\delta\mathbf{w})\mathsf{a}'$, since S''' and S' are at different times in the two members of the equation.

10.6 Thomas precession

If each of the sequence of inertial systems in which the particle is instantaneously at rest uses Cartesian axes resulting from those of the lab by a pure Lorentz transformation, then the spin appears to precess,

$$\frac{d\mathbf{S}'}{dt} = \frac{\gamma^2}{c^2(\gamma+1)}\mathbf{S}' \times (\mathbf{u} \times \mathbf{a}) \quad , \tag{10.98}$$

where $\mathbf{a} = d\mathbf{u}/dt$ is the particle acceleration with respect to the lab system. In contrast to $\mathbf{S}$, $\mathbf{S}'$ has constant magnitude during the accelerated motion.

Define $\delta \mathsf{S}' = \mathsf{S}''(t+\delta t) - \mathsf{S}'(t)$. Then from

$$\mathsf{S}''(t+\delta t) = \mathsf{R}(\hat{\mathbf{n}}\delta\phi)\mathsf{S}'''(t+\delta t) = \mathsf{R}(\hat{\mathbf{n}}\delta\phi)\mathsf{S}'(t)$$

and (10.61) we have $\delta\mathsf{S}' = -\mathsf{J}\cdot\hat{\mathbf{n}}\,\delta\phi\,\mathsf{S}'$,

$$\delta\mathsf{S}' = -\mathsf{S}' \times \hat{\mathbf{n}}\,\delta\phi = \frac{\gamma^2}{c^2(\gamma+1)}\mathsf{S}' \times (\mathbf{u} \times \delta\mathbf{u}) \quad .$$

Let us consider the spin precession for a particle moving on a circular orbit, $\mathbf{u} = \omega \times \mathbf{r}$, $\mathbf{a} = -\omega^2\mathbf{r}$, $\mathbf{u} \times (d\mathbf{u}/dt) = u^2\omega = [(\gamma^2-1)c^2/\gamma^2]\omega$,

$$\frac{d\mathbf{S}'}{dt} = -(\gamma-1)\,\omega \times \mathbf{S}' \quad . \tag{10.99}$$

By definition, the motion of the particle around ω is counterclockwise. That of the spin is also counterclockwise. In one period of the orbital motion, the spin precesses by an angle

$$\Delta\varphi' = -2\pi(\gamma-1) = -2\pi\left(\frac{1}{\sqrt{1-r^2\omega^2/c^2}} - 1\right) \quad . \tag{10.100}$$

If $r\omega \ll c$,

$$\Delta\varphi' \simeq -\frac{\pi r^2\omega^2}{c^2} = -\frac{4\pi^3 r^2}{c^2T^2} \quad . \tag{10.101}$$

For simplicity's sake we assume that $\mathbf{S}'$ lies in the plane of the orbit (xy plane). Then equation (10.99) gives

$$\frac{dS'_x}{dt} = (\gamma-1)\omega S'_y \quad , \quad \frac{dS'_y}{dt} = -(\gamma-1)\omega S'_x \quad . \tag{10.102}$$

It is easy to verify that $S' = \sqrt{S'^2_x + S'^2_y}$ is a constant of motion.

The solution

$$S'_x = S' \cos((\gamma-1)\omega t) \quad , \quad S'_y = -S' \sin((\gamma-1)\omega t) \tag{10.103}$$

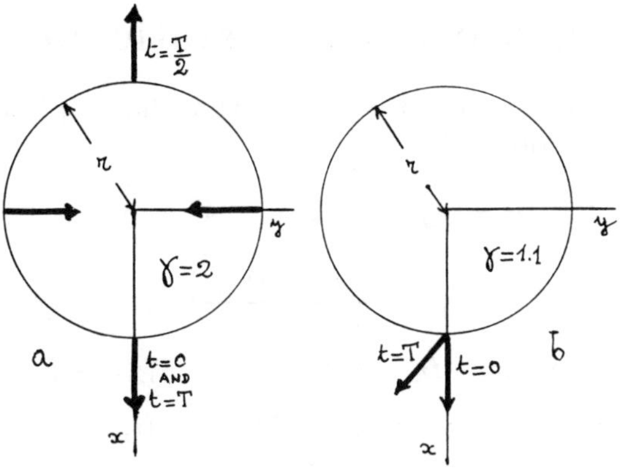

Figure 10.1: Spin precession along circular orbit

in conjunction with $x = r\cos(\omega t)$, $y = r\sin(\omega t)$, corresponds to the initial condition at $t = 0$ with the particle on the x axis and the spin along the x axis pointing in the positive direction.

For $t = T$, $S'_x = S'\cos(2\pi(\gamma - 1))$, $S'_y = -S'\sin(2\pi(\gamma - 1))$. In agreement with (10.100)

$$tan(\Delta\varphi') = S'_y/S'_x = -\tan(2\pi(\gamma - 1)).$$

What is $\mathbf{S}$ doing while $\mathbf{S}'$ precesses? Notice first that using (10.52) and $S'^0 = 0$ one has

$$\mathbf{S} = \mathbf{S}' + \frac{\gamma^2}{c^2(\gamma + 1)}(\mathbf{S}' \cdot \mathbf{u})\mathbf{u} \quad . \tag{10.104}$$

Then from the above solution for $\mathbf{S}'$ one finds

$$\begin{cases} S_x = S'[\cos(\omega t)\cos(\gamma\omega t) + \gamma\sin(\omega t)\sin(\gamma\omega t)] & , \\ S_y = S'[\sin(\omega t)\cos(\gamma\omega t) - \gamma\cos(\omega t)\sin(\gamma\omega t)] & , \end{cases} \tag{10.105}$$

$$|\mathbf{S}|^2 = S'^2[\cos^2(\gamma\omega t) + \gamma^2\sin^2(\gamma\omega t)] \quad .$$

For $t = T, S_x = S'\cos(2\pi(\gamma - 1))$, $(S_y = -\gamma S'\sin(2\pi(\gamma - 1))$.
With respect to the lab system, the spin has rotated through an angle $\Delta\varphi$, $\tan(\Delta\varphi) = S_y/S_x = -\gamma\tan(2\pi(\gamma - 1))$, and its magnitude has also changed.

Figure 10.1a shows what happens for $\gamma = 2$, $u = \sqrt{3}c/2$. Figure 10.1b is for $\gamma = 1.1$, $u = 0.42c$.

Note that $K'(C')$ is in motion with velocity $\mathbf{u} = (0, u)$ with respect to $K(C)$ (lab) both at $t = 0$ and $t = T$. One has $S_x = S'_x$ because $\mathbf{u}$ is normal to the x axis. Since $S'_0 = 0$, $S_y = \gamma S'_y$.

10.7 Charged particle in static em field

We study the motion of a point particle with a charge q in a static electromagnetic field of 4-potential

$$A_0 = V \ , \ A^1 = A_x \ , \ A^2 = A_y \ , \ A^3 = A_z \quad , \tag{10.106}$$

$$F_{\mu\nu} = \frac{\partial A_\nu}{\partial x^\mu} - \frac{\partial A_\mu}{\partial x^\nu} \quad ,$$

$$F_{01} = E_x, \ldots, F_{23} = -B_x \ , \ F_{31} = -B_y \ , \ F_{12} = -B_z \quad , \tag{10.107}$$

$$F^{0i} = -F_{0i} \ (i = 1, 2, 3) \ , \ F^{ij} = F_{ij} \ (i, j = 1, 2, 3)$$

$$F^{\mu\nu} = \begin{pmatrix} 0 & -E_x & -E_y & -E_z \\ E_x & 0 & -B_z & B_y \\ E_y & B_z & 0 & -B_x \\ E_z & -B_y & B_x & 0 \end{pmatrix} \ . \tag{10.108}$$

The covariant equations

$$m\frac{\mathrm{d}^2 x^\mu}{\mathrm{d}s^2} = \frac{q}{c} \ F^\mu{}_\nu \ \frac{\mathrm{d}x^\nu}{\mathrm{d}s} = \frac{q}{c} \ g^{\mu\lambda} \ F_{\lambda\nu} \ \frac{\mathrm{d}x^\nu}{\mathrm{d}s} \tag{10.109}$$

are easily seen to be equivalent to

$$m\frac{\mathrm{d}}{\mathrm{d}t}(\gamma\mathbf{u}) = q\left(\mathbf{E} + \frac{1}{c}\mathbf{u} \times \mathbf{B}\right) \tag{10.110}$$

and

$$mc\frac{\mathrm{d}\gamma}{\mathrm{d}t} = \frac{q}{c}\mathbf{E} \cdot \mathbf{u} \quad . \tag{10.111}$$

The former gives

$$m\gamma\frac{\mathrm{d}\mathbf{u}}{\mathrm{d}t} = q\left(\mathbf{E} - \frac{1}{c^2}(\mathbf{E} \cdot \mathbf{u})\mathbf{u} + \frac{1}{c}\mathbf{u} \times \mathbf{B}\right) \quad , \tag{10.112}$$

the latter $\mathrm{d}E/\mathrm{d}t = q\mathbf{E} \cdot \mathbf{u}$.

10.8 Magnetic moment in static em field

The magnetic moment is expressed as

$$\mu = \frac{gq}{2mc}\,\mathbf{S}' \quad . \tag{10.113}$$

A first form of its covariant equations of motion is

$$\frac{\mathrm{d}S^\mu}{\mathrm{d}s} = \frac{gq}{2mc}\left[F^{\mu\nu}S_\nu + \frac{1}{c^2}u^\mu(S_\rho F^{\rho\sigma}u_\sigma)\right] - \frac{1}{c^2}u^\mu\left(S_\nu\frac{\mathrm{d}u^\nu}{\mathrm{d}s}\right) \quad . \tag{10.114}$$

Multiplying by u_μ and using $u_\mu u^\mu = c^2$, we have

$$\frac{\mathrm{d}}{\mathrm{d}s}(u_\mu S^\mu) = \frac{gq}{2mc}\left[u_\mu F^{\mu\nu}S_\nu + S_\mu F^{\mu\nu}u_\nu\right]$$

$$= \frac{gq}{2mc}u_\mu S_\nu(F^{\mu\nu} + F^{\nu\mu}) = 0 \quad .$$

Using $\mathrm{d}u^\mu/\mathrm{d}s = (q/mc)F^\mu_{\ \nu}u^\nu$, equation (10.114) can be cast in a second form

$$\frac{\mathrm{d}S^\mu}{\mathrm{d}s} = \frac{gq}{2mc}F^{\mu\nu}S_\nu + \frac{(g-2)q}{2mc^3}(S_\rho F^{\rho\sigma}u_\sigma)u^\mu \quad . \tag{10.115}$$

For $g = 2$ the last equation gives in the lab system

$$\mathrm{d}\mathbf{S}/\mathrm{d}t = (q/mc\gamma)[\mathbf{S}\times\mathbf{B} + (1/c)(\mathbf{S}\cdot\mathbf{u})\mathbf{E}] \quad , \quad \mathrm{d}S_0/\mathrm{d}t = (q/mc\gamma)\,(\mathbf{S}\cdot\mathbf{E}) \quad . \tag{10.116}$$

Here are some calculations described by one author as "somewhat lengthy" and by another as "some tedious algebra".

By the equation $\mathbf{S}' = \mathbf{S} - [\gamma/c(\gamma+1)]S_0\mathbf{u}$ we find for $g = 2$

$$\frac{\mathrm{d}\mathbf{S}'}{\mathrm{d}t} = \frac{q}{mc\gamma}\left[\mathbf{S}\times\mathbf{B} + \frac{1}{c}(\mathbf{S}\cdot\mathbf{u})\mathbf{E}\right]$$

$$-\frac{q}{mc^2(\gamma+1)}(\mathbf{S}\cdot\mathbf{E})\mathbf{u} - \frac{1}{c^2}(\mathbf{S}\cdot\mathbf{u})\frac{\mathrm{d}}{\mathrm{d}t}\left(\frac{\gamma\mathbf{u}}{\gamma+1}\right) \quad .$$

Using $\mathbf{S} = \mathbf{S}' + [\gamma^2/c^2(\gamma+1)](\mathbf{S}'\cdot\mathbf{u})\mathbf{u}$, we find

$$\frac{\mathrm{d}\mathbf{S}'}{\mathrm{d}t} = \frac{q}{mc\gamma}\mathbf{S}'\times\mathbf{B} + \frac{q\gamma}{mc^3(\gamma+1)}(\mathbf{S}'\cdot\mathbf{u})(\mathbf{u}\times\mathbf{B})$$

$$+\frac{q}{mc^2}(\mathbf{S}'\cdot\mathbf{u})\mathbf{E} - \frac{q}{mc^2(\gamma+1)}(\mathbf{S}'\cdot\mathbf{E})\mathbf{u}$$

$$-\frac{q\gamma^2}{mc^4(\gamma+1)^2}(\mathbf{S}'\cdot\mathbf{u})(\mathbf{E}\cdot\mathbf{u})\mathbf{u} - \frac{\gamma}{c^2}(\mathbf{S}'\cdot\mathbf{u})\frac{\mathrm{d}}{\mathrm{d}t}\left(\frac{\gamma\mathbf{u}}{\gamma+1}\right) \quad .$$

Since $\mathrm{d}(\gamma\mathbf{u})/\mathrm{d}t = (q/m)(\mathbf{E} + (\mathbf{u}\times\mathbf{B})/c)$ and

$$\frac{\mathrm{d}}{\mathrm{d}t}\left(\frac{1}{\gamma+1}\right) = -\frac{q}{mc^2(\gamma+1)^2}(\mathbf{E}\cdot\mathbf{u}) \quad ,$$

we find

$$\frac{\mathrm{d}\mathbf{S}'}{\mathrm{d}t} = \frac{q}{mc\gamma}\mathbf{S}'\times\mathbf{B} - \frac{q}{mc^2(\gamma+1)}(\mathbf{E}\cdot\mathbf{S}')\mathbf{u}$$
$$-\frac{q\gamma}{mc^2(\gamma+1)}(\mathbf{S}'\cdot\mathbf{u})\mathbf{E} + \frac{q}{mc^2}(\mathbf{S}'\cdot\mathbf{u})\mathbf{E} \quad ,$$

$$\frac{\mathrm{d}\mathbf{S}'}{\mathrm{d}t} = \frac{q}{mc\gamma}\mathbf{S}'\times\left(\mathbf{B} - \frac{\gamma}{c(\gamma+1)}\mathbf{u}\times\mathbf{E}\right) \quad , \tag{10.117}$$

a formula valid for $g = 2$.

For $g = 0$ (only for computational convenience, since we do not know of a charged particle with spin and no magnetic moment) equation (10.115) gives

$$\frac{\mathrm{d}\mathbf{S}'}{\mathrm{d}t} = \frac{q\gamma}{mc^2(\gamma+1)}\mathbf{S}'\times(\mathbf{u}\times(\mathbf{E}+c^{-1}\mathbf{u}\times\mathbf{B})) \quad . \tag{10.118}$$

Now for any g

$$\frac{\mathrm{d}\mathbf{S}'}{\mathrm{d}t} = \frac{g}{2}\left(\frac{\mathrm{d}\mathbf{S}'}{\mathrm{d}t}\right)_{g=2} - \frac{g-2}{2}\left(\frac{\mathrm{d}\mathbf{S}'}{\mathrm{d}t}\right)_{g=0}$$

$$= \frac{gq}{2mc\gamma}\mathbf{S}'\times\left[\mathbf{B} - \frac{\gamma}{c(\gamma+1)}\mathbf{u}\times\mathbf{E}\right] - \frac{(g-2)q\gamma}{2mc^2(\gamma+1)}\mathbf{S}'\times(\mathbf{u}\times(\mathbf{E}+c^{-1}\mathbf{u}\times\mathbf{B})) \quad . \tag{10.119}$$

10.9 Lagrangian and Hamiltonian

The equation for a charged particle in an em field

$$m\frac{\mathrm{d}^2x^\mu}{\mathrm{d}s^2} = \frac{q}{c}g^{\mu\lambda}F_{\lambda\nu}\frac{\mathrm{d}x^\nu}{\mathrm{d}s} \tag{10.120}$$

can be derived from the Lagrangian

$$L = \frac{m}{2}\frac{\mathrm{d}x^\mu}{\mathrm{d}\tau}\frac{\mathrm{d}x_\mu}{\mathrm{d}\tau} + \frac{q}{c}\frac{\mathrm{d}x^\nu}{\mathrm{d}\tau}A_\nu(x) \quad , \tag{10.121}$$

where the parameter τ is not necessarily the proper time.
In fact

$$\frac{\mathrm{d}}{\mathrm{d}\tau}\left(\frac{\partial L}{\partial(\mathrm{d}x^\mu/\mathrm{d}\tau)}\right) - \frac{\partial L}{\partial x^\mu} = 0$$

gives

$$\frac{\mathrm{d}}{\mathrm{d}\tau}\left(m\frac{\mathrm{d}x_\mu}{\mathrm{d}\tau}+\frac{q}{c}A_\mu(x)\right)-\frac{q}{c}\frac{\partial A_\nu(x)}{\partial x^\mu}\frac{\mathrm{d}x^\nu}{\mathrm{d}\tau}=0 \quad ,$$

$$m\frac{\mathrm{d}^2x_\mu}{\mathrm{d}\tau^2}=\frac{q}{c}\left(\frac{\partial A_\nu}{\partial x^\mu}-\frac{\partial A_\mu}{\partial x^\nu}\right)\frac{\mathrm{d}x^\nu}{\mathrm{d}\tau}=\frac{q}{c}F_{\mu\nu}\frac{\mathrm{d}x^\nu}{\mathrm{d}\tau} \quad .$$

Multiplying this equation by $\mathrm{d}x^\mu/\mathrm{d}\tau$, we find

$$\frac{\mathrm{d}}{\mathrm{d}\tau}\left(\frac{\mathrm{d}x_\mu}{\mathrm{d}\tau}\frac{\mathrm{d}x^\mu}{\mathrm{d}\tau}\right)=0 \quad , \quad \frac{\mathrm{d}x_\mu}{\mathrm{d}\tau}\frac{\mathrm{d}x^\mu}{\mathrm{d}\tau}=\text{constant} \; . \tag{10.122}$$

We normalize τ so that constant $= c^2$. Thus $\tau = s$, the proper time. In the following, differentiation with respect to s will be denoted by a dot. Therefore $\dot{x}_\mu\dot{x}^\mu = c^2$.

Some people are disturbed by the fact that the first term in the Lagrangian turns out to be a constant. A cure for this is to use the method of the Lagrangian with constraints presented in problems 6.10 and 6.11.

For the Lagrangian

$$L = mc^2 + (q/c)\dot{x}^\mu A_\mu \text{ with } \dot{x}_\mu\dot{x}^\mu = c^2 \quad , \tag{10.123}$$

we obtain

$$\frac{\mathrm{d}}{\mathrm{d}s}\left(\frac{q}{c}A_\mu\right)-\frac{q}{c}\,\dot{x}^\nu\frac{\partial A_\nu}{\partial x^\mu}+\frac{\mathrm{d}}{\mathrm{d}s}\left[\frac{1}{c^2}\left(mc^2+\frac{q}{c}\,\dot{x}^\nu A_\nu-\frac{q}{c}\,\dot{x}^\nu A_\nu\right)\dot{x}_\mu\right]=0 \quad ,$$

$$\frac{q}{c}\left(\frac{\partial A_\mu}{\partial x^\nu}\dot{x}^\nu-\dot{x}^\nu\frac{\partial A_\nu}{\partial x^\mu}\right)+m\ddot{x}_\mu=0 \quad ,$$

$$m\ddot{x}_\mu=(q/c)F_{\mu\nu}\dot{x}^\nu \quad .$$

Returning to $L = (m/2)\dot{x}_\mu\dot{x}^\mu + (q/c)\dot{x}^\mu A_\mu$ we have

$$p_\mu=\frac{\partial L}{\partial\dot{x}^\mu}=m\dot{x}_\mu+\frac{q}{c}A_\mu \quad , \tag{10.124}$$

$$H=p_\mu\dot{x}^\mu-L=\frac{1}{2m}\left(p_\mu-\frac{q}{c}A_\mu\right)\left(p^\mu-\frac{q}{c}A^\mu\right) \quad , \tag{10.125}$$

$$H=\frac{m}{2}\,\dot{x}_\mu\dot{x}^\mu=\frac{mc^2}{2} \quad . \tag{10.126}$$

Hamilton's equations read

$$\dot{x}_\mu=\frac{\partial H}{\partial p^\mu}=\frac{1}{m}\left(p_\mu-\frac{q}{c}A_\mu\right) \quad , \tag{10.127}$$

$$\dot{p}_\mu=-\frac{\partial H}{\partial x^\mu}=\frac{q}{mc}\left(p_\nu-\frac{q}{c}A_\nu\right)\frac{\partial A^\nu}{\partial x^\mu}=\frac{q}{c}\dot{x}_\nu\frac{\partial A^\nu}{\partial x^\mu} \quad . \tag{10.128}$$

Hence

$$m\ddot{x}_\mu = \frac{q}{c}\,\dot{x}^\nu\left(\frac{\partial A_\nu}{\partial x^\mu} - \frac{\partial A_\mu}{\partial x^\nu}\right) \quad .$$

Poisson brackets:

$$[x^\mu, x_\nu] = 0 \quad , \quad [p^\mu, p_\nu] = 0 \quad , \quad [x^\mu, p_\nu] = g^\mu{}_\nu \quad . \tag{10.129}$$

Hence

$$[x^\mu, \dot{x}_\nu] = \frac{1}{m}\,g^\mu{}_\nu \quad , \tag{10.130}$$

but

$$[\dot{x}_\mu, \dot{x}_\nu] = \frac{1}{m^2}\left[p_\mu - \frac{q}{c}A_\mu, p_\nu - \frac{q}{c}A_\nu\right] = \frac{q}{m^2c}\left(-[p_\mu, A_\nu] - [A_\mu, p_\nu]\right)$$

$$= \frac{q}{m^2c}\left(\frac{\partial A_\nu}{\partial x^\lambda}[x^\lambda, p_\mu] - \frac{\partial A_\mu}{\partial x^\lambda}[x^\lambda, p_\nu]\right) = \frac{q}{m^2c}\left(\frac{\partial A_\nu}{\partial x^\mu} - \frac{\partial A_\mu}{\partial x^\nu}\right) = \frac{q}{m^2c}\,F_{\mu\nu} \quad .$$

The above formulation is based on the canonical form

$$\tilde{\omega}^2 = \tilde{d}p_\mu \wedge \tilde{d}x^\mu \quad , \tag{10.131}$$

in which p_μ is not a gauge invariant quantity.

The Hamiltonian is invariant under the combined transformations

$$A_\mu \to A_\mu + \frac{\partial \Lambda}{\partial x^\mu} \quad , \quad p_\mu \to p_\mu + \frac{q}{c}\frac{\partial \Lambda}{\partial x^\mu} \quad . \tag{10.132}$$

In fact

$$p_\mu + \frac{q}{c}\frac{\partial \Lambda}{\partial x^\mu} - \frac{q}{c}\left(A_\mu + \frac{\partial \Lambda}{\partial x^\mu}\right) = p_\mu - \frac{q}{c}A_\mu \quad .$$

The transformation $p_\mu \to p_\mu + (q/c)\,\partial\Lambda/\partial x^\mu$ is canonical. In fact

$$\left[p_\mu + \frac{q}{c}\frac{\partial \Lambda}{\partial x^\mu}\,,\, p_\nu + \frac{q}{c}\frac{\partial \Lambda}{\partial x^\nu}\right] = \frac{q}{c}\left(\left[p_\mu\,,\,\frac{\partial \Lambda}{\partial x^\nu}\right] + \left[\frac{\partial \Lambda}{\partial x^\mu}\,,\,p_\nu\right]\right)$$

$$= \frac{q}{c}\left(\frac{\partial^2 \Lambda}{\partial x^\nu \partial x^\lambda}\,[p_\mu, x^\lambda] + \frac{\partial^2 \Lambda}{\partial x^\mu \partial x^\lambda}\,[x^\lambda, p_\nu]\right)$$

$$= \frac{q}{c}\left(\frac{\partial^2 \Lambda}{\partial x^\nu \partial x^\mu} - \frac{\partial^2 \Lambda}{\partial x^\mu \partial x^\nu}\right) = 0 \quad ,$$

while the invariance of the other Poisson brackets is trivial.

Intermezzo:
F.J. Dyson, (*Am. J. Phys.*, **58**(1990)209) reminisced about the discovery by Feynman that the homogeneous Maxwell equations can be "derived" from Mechanics.

It amounts to the following:

$$[x^\mu, \dot{x}_\nu] = \frac{1}{m}\, g^\mu{}_\nu \to \left[x^\mu, \frac{\partial H}{\partial p^\nu}\right] = \frac{1}{m}\, g^\mu{}_\nu$$

giving

$$\frac{\partial^2 H}{\partial p^\nu \partial p_\mu} = \frac{1}{m} g^\mu{}_\nu \quad .$$

This requires the Hamiltonian to be of the form

$$H = (2m)^{-1}(p^\mu + a^\mu)(p_\mu + a_\mu) + b(x) \; ,$$

where the scalar field $b(x)$ has been added to Feynman's formula.
Therefore

$$\dot{x}_\nu = \frac{\partial H}{\partial p^\nu} = \frac{1}{m}(p_\nu + a_\nu) \quad , \quad \dot{p}_\mu = -\frac{\partial H}{\partial x^\mu} = -\dot{x}_\nu \frac{\partial a^\nu}{\partial x^\mu} - \frac{\partial b}{\partial x^\mu} \quad ,$$

$$m\ddot{x}_\mu = \dot{p}_\mu + \dot{a}_\mu = -\dot{x}_\nu \frac{\partial a^\nu}{\partial x^\mu} + \frac{\partial a_\mu}{\partial x^\nu}\dot{x}^\nu - \frac{\partial b}{\partial x^\mu} = \left(\frac{\partial a_\mu}{\partial x^\nu} - \frac{\partial a_\nu}{\partial x^\mu}\right)\dot{x}^\nu - \frac{\partial b}{\partial x^\mu} \quad .$$

For a function $f(p, q)$ we have

$$\frac{\mathrm{d}f}{\mathrm{d}s} = \dot{p}^\mu \frac{\partial f}{\partial p^\mu} + \dot{x}^\mu \frac{\partial f}{\partial x^\mu} = -\frac{\partial H}{\partial x_\mu}\frac{\partial f}{\partial p^\mu} + \frac{\partial H}{\partial p_\mu}\frac{\partial f}{\partial x^\mu}$$

$$= -\frac{1}{m}\left(p_\nu - \frac{q}{c}A_\nu\right)\left(-\frac{q}{c}\frac{\partial A^\nu}{\partial x_\mu}\right)\frac{\partial f}{\partial p^\mu} + \frac{1}{m}\left(p^\mu - \frac{q}{c}A^\mu\right)\frac{\partial f}{\partial x^\mu} = \bar{U}_H f$$

with

$$\bar{U}_H = \frac{q}{mc}\left(p_\nu - \frac{q}{c}A_\nu\right)\frac{\partial A^\nu}{\partial x^\mu}\frac{\partial}{\partial p_\mu} + \frac{1}{m}\left(p^\mu - \frac{q}{c}A^\mu\right)\frac{\partial}{\partial x^\mu} \quad . \tag{10.133}$$

One may prefer a formulation using the gauge-invariant kinetic momentum

$$p'_\mu = m\dot{x}_\mu = p_\mu - \frac{q}{c}A_\mu \; . \tag{10.134}$$

For this purpose we start with the Hamiltonian

$$H' = \frac{1}{2m}\, p'_\mu p'^\mu \tag{10.135}$$

and the canonical form

$$\tilde{\omega}'^2 = \tilde{d}p'_\mu \wedge \tilde{d}x^\mu - \frac{q}{2c}F_{\mu\nu}\, \tilde{d}x^\mu \wedge \tilde{d}x^\nu \quad . \tag{10.136}$$

This latter changes the definition of Poisson brackets.

Note that the condition $\tilde{d}\tilde{\omega}'^2 = 0$ gives

$$F_{\mu\nu,\lambda}\tilde{dx}^\lambda \wedge \tilde{dx}^\mu \wedge \tilde{dx}^\nu = 0 \quad ,$$

$$(1/3)(F_{\mu\nu,\lambda} + F_{\nu\lambda,\mu} + F_{\lambda\mu,\nu})\, \tilde{dx}^\lambda \wedge \tilde{dx}^\mu \wedge \tilde{dx}^\nu = 0 \quad ,$$

$$F_{\mu\nu,\lambda} + F_{\nu\lambda,\mu} + F_{\lambda\mu,\nu} = 0 \quad .$$

Compare this with Feynman's derivation of the homogeneous Maxwell equations.

If

$$\bar{U}_{H'} = \frac{\partial H'}{\partial p'_\mu}\frac{\partial}{\partial x^\mu} = \frac{p'^\mu}{m}\frac{\partial}{\partial x^\mu} \tag{10.137}$$

and

$$\bar{U}_f = \frac{\partial f}{\partial p'_\mu}\frac{\partial}{\partial x^\mu} - \frac{\partial f}{\partial x^\mu}\frac{\partial}{\partial p'_\mu} \tag{10.138}$$

we have

$$\frac{\mathrm{d}f}{\mathrm{d}s} = \tilde{\omega}'^2(\bar{U}_{H'}, \bar{U}_f) = \frac{\partial f}{\partial x^\mu}\frac{p'^\mu}{m} - \frac{q}{2c}F_{\mu\nu}\left(\frac{p'^\mu}{m}\frac{\partial f}{\partial p'_\nu} - \frac{p'^\nu}{m}\frac{\partial f}{\partial p'_\mu}\right)$$

$$= \frac{p'^\mu}{m}\frac{\partial f}{\partial x^\mu} - \frac{q}{mc}F_{\mu\nu}p'^\mu\frac{\partial f}{\partial p'_\nu} = \dot{x}^\mu\frac{\partial f}{\partial x^\mu} + \ddot{x}^\nu\frac{\partial f}{\partial \dot{x}^\nu} \quad . \tag{10.139}$$

10.10 Chapter 10 problems

10.1 A photon of 4-momentum $p = (\nu, \nu\hat{\mathbf{n}})$ ($c = 1, \hbar = 1$) is scattered by a particle at rest, which acquires a kinetic energy K in the collision. Calculate the angle θ formed by the momentum of the struck particle with the momentum of the incident photon.

10.2 (i) Show that equations (10.82,83) follow from (10.81) and (10.80). (ii) From (10.84) derive the matrix elements for a space rotation.

10.3 (i) Using the $SU(2)$ representation of space rotations, find the rotation which is the product of two space rotations. (ii) Check your result against the product $\mathsf{R}_1(\pi/2)\mathsf{R}_2(\pi/2)$ calculated by using equations (4.5,6).

10.4 Using the $SL(2)$ representation of the Lorentz group, verify that $\mathsf{L}(\mathbf{u}_1)\,\mathsf{L}(\mathbf{u}_2) = L(\mathbf{u})\mathsf{R}_3(\phi)$, where $\mathbf{u}_1 = u_0\hat{\mathbf{e}}_1$, $\mathbf{u}_2 = u_0\hat{\mathbf{e}}_2$, $\mathbf{u} = \mathbf{u}_1 + \mathbf{u}_2/\gamma_0$ ($\gamma_0 = 1/\sqrt{1-u_0^2}$, $c = 1$), and $\tan\phi = -u_0^2\gamma_0/2$.

10.5 (i) Find the generators δG of infinitesimal space-time translations, space rotations, and Lorentz transformations (Poincaré group).
(ii) Find the Poisson brackets of the above (algebra of the Poincaré group).

10.6 (i) The 4-potential for a constant electromagnetic field $F_{\mu\nu}$ is $A_\rho = (1/2)x^\sigma F_{\sigma\rho}$ up to a gauge transformation. For a particle of rest-mass m and charge q in this constant em field, verify that the quantities

$$C_\mu = p_\mu + (q/c)A_\mu = p_\mu - (q/2c)F_{\mu\nu}x^\nu$$

are constants of motion, $\mathrm{d}C_\mu/\mathrm{d}s = 0$.
Caution: The expressions $p_\mu - (q/c)A_\mu$, occurring in the Hamiltonian (10.125), differ by a sign from the C_μ's.
(ii) What is the physical meaning of the C_μ's if the particle is in the electrostatic field $\mathbf{E} = (0, 0, E)$?
(iii) Using the C_μ's in the expression (10.126) for the Hamiltonian, by analogy with problem 7.1 show that

$$C_\mu C^\mu + (2q/c)C^\mu F_{\mu\nu}x^\nu(s) + (q^2/c^2)x^\rho(s)x^\sigma(s)F_{\rho\mu}F_\sigma{}^\mu = m^2c^2 \quad .$$

10.7 (i) In the inertial frame (x', y', z', t') a particle of rest-mass m and charge $q > 0$ moves in a uniform magnetic field $\mathbf{B}' = (0, 0, B')$.
Show that if $x' = y' = z' = 0$, $\mathrm{d}x'/\mathrm{d}t' = -v'$ $(v' > 0)$, $\mathrm{d}y'/\mathrm{d}t' = \mathrm{d}z'/\mathrm{d}t' = 0$ at $t' = 0$, the particle is in uniform circular motion, clockwise, with

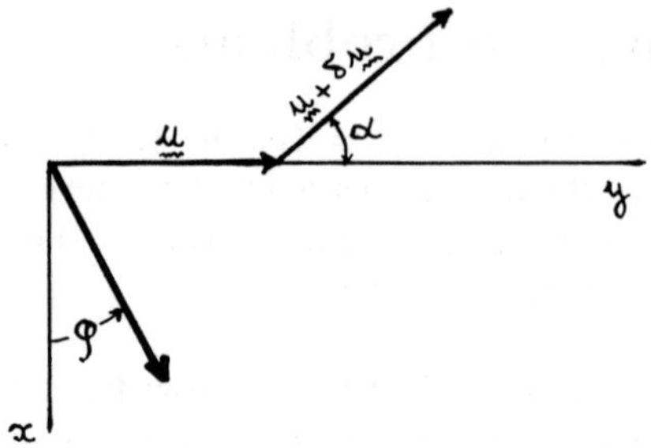

Figure 10.2: Spin under sudden acceleration

angular frequency $\omega' = qB'/mc\gamma'$ ($\gamma' = 1/\sqrt{1 - v'^2/c^2}$) and that the orbit equation is

$$x'^2 + (y' - r)^2 = r^2 \quad , \quad z' = 0 \quad ,$$

with $r = mc\gamma' v'/qB'$.
(ii) Perform the Lorentz transformation $x' = (x - v't)\gamma'$, $y' = y$, $z' = z$, $t' = (t - v'x/c^2)\gamma'$.
Show that the orbit equation in the (x, y, z, t) frame is

$$(x - v't)^2 + (y - r)^2/\gamma'^2 = r^2/\gamma'^2 \quad , \quad z = 0 \quad ,$$

where

$$r = \frac{mc^2 E}{qB^2(1 - E^2/B^2)}$$

and $E = \gamma' v' B'/c$, $B = B'\gamma'$ are the magnitudes of the electric field $\mathbf{E} = (0, E, 0)$ and the magnetic field $\mathbf{B} = (0, 0, B)$ in the (x, y, z, t) frame.

10.8 Show that

$$S_\mu = c^{-1} S' \epsilon_{\mu\nu\rho\sigma} u^\nu r_1^\rho r_2^\sigma$$

has the properties of a spin. Here $\epsilon_{\mu\nu\rho\sigma}$ is skewsymmetric with $\epsilon_{0123} = 1$, u^ν are the components of the 4-velocity, $\mathfrak{r}_1$ and $\mathfrak{r}_2$ are 4-vectors with components $(0, 1, 0, 0)$ and $(0, 0, 1, 0)$ in the rest system. In particular show that this S_μ satisfies equations (10.89,90), (10.93), and (10.104).

10.9 Show that equation (10.96) reduces to equations (10.94,95) in the lab frame.

10.10 Assume that the spin and the velocity are as shown in figure 10.2 (lab system). Show that due to the sudden change of direction of the velocity, the angle φ formed by the spin with the x-axis changes by $\delta\varphi = -\gamma^2\beta^2 \sin\alpha \cos^2\varphi$ with $\beta = |\mathbf{u}|/c$.

10.11 Derive equation (10.98) from equations (10.95) and (10.104).

10.12 What is $\mathrm{d}\mathbf{S}'/\mathrm{d}t$ in the following cases?
(i) $\mathbf{B} = 0$, $\mathbf{E} \parallel \mathbf{u}$, (ii) $\mathbf{E} = 0$, $\mathbf{B} \parallel \mathbf{u}$, (iii) $\mathbf{E} = 0$, $\mathbf{B} \perp \mathbf{u}$ ($\mathbf{B}$ uniform)

10.13 For an electron ($q = -e$) in an atom ($\mathbf{E} = -\nabla V(r) = -(\mathrm{d}V/\mathrm{d}r)(\mathbf{r}/r)$), equation (10.119) gives

$$\frac{\mathrm{d}\mathbf{S}'}{\mathrm{d}t} = \frac{e}{2m^2c^2r}\frac{\mathrm{d}V}{\mathrm{d}r}\left(\frac{g}{\gamma} - \frac{2}{1+\gamma}\right)\mathbf{S}' \times \mathbf{l} \quad (\mathbf{l} = m\gamma\mathbf{r} \times \mathbf{u}) \quad .$$

Show that for $\gamma \simeq 1$ this corresponds to the spin-orbit coupling interaction energy

$$-\frac{e}{2m^2c^2r}\frac{\mathrm{d}V}{\mathrm{d}r}(g-1)\mathbf{l} \cdot \mathbf{S}' \quad .$$

Solutions to ch. 10 problems

S10.1 Let $p' = (\nu', \nu'\hat{\mathbf{n}}')$, $q = (m, 0)$, $q' = (E' = K + m, \mathbf{q}')$ denote the 4-momenta of the photon after the collision and of the particle before and after the collision.

We have $p' + q' = p + q$, $p' - q = p - q'$. Squaring the last equation, and using $p'^2 = p^2 = 0$, $q'^2 = q^2 = m^2$, we have $m\nu' = E'\nu - \nu|\mathbf{q}'|\cos\theta$,

$$\cos\theta = (E'\nu - m\nu')/\nu|\mathbf{q}'| = (E'\nu - m(m + \nu - E'))/\nu\sqrt{E'^2 - m^2}$$
$$= (\nu + m)(E' - m)/\nu\sqrt{E'^2 - m^2},$$

$$\cos\theta = \frac{\nu + m}{\nu}\sqrt{\frac{E' - m}{E' + m}} = \frac{\nu + m}{\nu}\sqrt{\frac{K}{K + 2m}} \quad .$$

S10.2 (i) With $\mathsf{M} = \mathsf{M}(\mathbf{u})$, we have

$$2\,\Lambda(\mathsf{M})^0{}_0 = \mathrm{tr}(\mathsf{M}\mathsf{M}^\dagger) = \mathrm{tr}(\mathsf{M}^2) = \mathrm{tr}([\cosh^2(\alpha/2) + \sinh^2(\alpha/2)]\mathsf{I}) = \mathrm{tr}(\cosh\alpha\mathsf{I}) = 2\cosh\alpha \ ,$$

$$2\,\Lambda(\mathsf{M})^0{}_i = \mathrm{tr}(\mathsf{M}\sigma_i\mathsf{M}^\dagger) = \mathrm{tr}(\sigma_i\mathsf{M}^2) = \mathrm{tr}(\sigma_i[2n^j\sigma_j\sinh(\alpha/2)\cosh(\alpha/2)])$$
$$= \sinh\alpha\, n^j\,\mathrm{tr}(\sigma_i\sigma_j) = 2n^i\sinh\alpha \ ,$$
$$2\,\Lambda(\mathsf{M})^i{}_0 = \mathrm{tr}(\sigma_i\mathsf{M}\mathsf{M}^\dagger) = \mathrm{tr}(\sigma_i\mathsf{M}^2) = 2\,\Lambda(\mathsf{M})^0{}_i \ ,$$
$$2\,\Lambda(\mathsf{M})^i{}_j = \mathrm{tr}(\sigma_i\mathsf{M}\sigma_j\mathsf{M}^\dagger)$$
$$= \mathrm{tr}(\cosh^2(\alpha/2)\sigma_i\sigma_j + \sinh\alpha[\sigma_i(n^b\sigma_b)\sigma_j + \sigma_i\sigma_j(n^d\sigma_d)]/2 + n^b n^d\sinh^2(\alpha/2)\sigma_i\sigma_b\sigma_j\sigma_d)$$
$$= 2\delta_{ij}\cosh^2(\alpha/2) + \mathrm{i}\,\sinh\alpha\overbrace{(n^b\epsilon_{ibj} + n^d\epsilon_{ijd})}^{\text{zero}} + 2\,\sinh^2(\alpha/2)n^b n^d(\delta_{ib}\delta_{jd} - \delta_{ij}\delta_{bd} + \delta_{id}\delta_{jb})$$
$$= 2\,\delta_{ij} + 4\,\sinh^2(\alpha/2)n^i n^j = 2\,\delta_{ij} + 2n^i n^j(\cosh\alpha - 1) \ .$$

(ii)

$$2\,\Lambda(\mathsf{M})^i{}_j = \mathrm{tr}(\sigma_i\mathsf{M}\sigma_j\mathsf{M}^\dagger)$$
$$= \mathrm{tr}(\cos^2(\phi/2)\sigma_i\sigma_j - (\mathrm{i}/2)\sin\phi[\sigma_i(n^b\sigma_b)\sigma_j - \sigma_i\sigma_j(n^c\sigma_c)] + \sin^2(\phi/2)n^b n^c\sigma_i\sigma_b\sigma_j\sigma_c)$$
$$= 2\delta_{ij}\cos^2(\phi/2) + 2\,\sin\phi\,\epsilon_{ibj}n^b + 2\,\sin^2(\phi/2)n^b n^c(\delta_{ib}\delta_{jc} - \delta_{ij}\delta_{bc} + \delta_{ic}\delta_{jb})$$
$$= 2\,\delta_{ij}\,\cos\phi + 2(1 - \cos\phi)n^i n^j - 2\,\sin\phi\,\epsilon_{ijk}n^k \quad .$$

S10.3 (i) We have

$$[\cos(\phi_1/2) - \mathrm{i}\,\hat{\mathbf{n}}_1\cdot\sigma\sin(\phi_1/2)][\cos(\phi_2/2) - \mathrm{i}\,\hat{\mathbf{n}}_2\cdot\sigma\sin(\phi_2/2)]$$
$$= \cos(\phi/2) - \mathrm{i}\,\hat{\mathbf{n}}\cdot\sigma\sin(\phi/2)$$

where

$$\cos(\phi/2) = \cos(\phi_1/2)\cos(\phi_2/2) - \mathbf{n}_1\cdot\mathbf{n}_2\sin(\phi_1/2)\sin(\phi_2/2) \quad ,$$
$$\hat{\mathbf{n}}\sin(\phi/2) = \hat{\mathbf{n}}_1\sin(\phi_1/2)\cos(\phi_2/2) + \hat{\mathbf{n}}_2\sin(\phi_2/2)\cos(\phi_1/2)$$
$$+(\hat{\mathbf{n}}_1\times\hat{\mathbf{n}}_2)\sin(\phi_1/2)\sin(\phi_2/2) \quad .$$

(ii) For $\hat{\mathbf{n}}_1 = (1, 0, 0)$, $\hat{\mathbf{n}}_2 = (0, 1, 0)$, $\phi_1 = \phi_2 = \pi/2$, (i) gives $\cos(\phi/2) = 1/2$, $\phi = 2\pi/3$, $\hat{\mathbf{n}} = (\hat{\mathbf{e}}_1 + \hat{\mathbf{e}}_2 + \hat{\mathbf{e}}_3)/\sqrt{3}$. Hence $R_{ij} = (-\delta_{ij} + 1 - \epsilon_{ijk})/2$ ($k \neq i$ and

$k \neq j$), $R_{11} = R_{12} = R_{22} = R_{23} = R_{31} = R_{33} = 0$, $R_{13} = R_{21} = R_{32} = 1$. This agrees with

$$\mathsf{R}_1(\pi/2)\ \mathsf{R}_2(\pi/2) = \begin{pmatrix} 1 & 0 & 0 \\ 0 & 0 & -1 \\ 0 & 1 & 0 \end{pmatrix} \begin{pmatrix} 0 & 0 & 1 \\ 0 & 1 & 0 \\ -1 & 0 & 0 \end{pmatrix} = \begin{pmatrix} 0 & 0 & 1 \\ 1 & 0 & 0 \\ 0 & 1 & 0 \end{pmatrix}$$

$(\mathsf{x} = \mathsf{R}\mathsf{x}' : \ x' = y, y' = z, z' = x)$.

S10.4

$$\text{Left side} \rightarrow \left(\sqrt{\frac{\gamma_0+1}{2}} + \sigma_1\sqrt{\frac{\gamma_0-1}{2}}\right)\left(\sqrt{\frac{\gamma_0+1}{2}} + \sigma_2\sqrt{\frac{\gamma_0-1}{2}}\right)$$

$$= [(\gamma_0+1) + (\sigma_1+\sigma_2)\sqrt{\gamma_0^2-1} + i\sigma_3(\gamma_0-1)]/2$$

$$\text{Right side} \rightarrow \left(\sqrt{\frac{\gamma_0^2+1}{2}} + \left[\frac{\gamma_0}{\sqrt{1+\gamma_0^2}}\sigma_1 + \frac{1}{\sqrt{1+\gamma_0^2}}\sigma_2\right]\sqrt{\frac{\gamma_0^2-1}{2}}\right)$$

$$\frac{1}{\sqrt{2(\gamma_0^2+1)}}[(\gamma_0+1) + i\sigma_3(\gamma_0-1)]$$

$$= [(\gamma_0+1) + (\sigma_1+\sigma_2)\sqrt{\gamma_0^2-1} + i\sigma_3(\gamma_0-1)]/2$$

S10.5 The generator for $x^\mu \rightarrow x^\mu + \delta x^\mu$ is $\delta G = p_\nu \delta x^\nu$. In fact,

$$[x^\mu, \delta G] = [x^\mu, p_\nu]\delta x^\nu = g^\mu_{\ \nu}\, \delta x^\nu = \delta x^\mu .$$

If $L_{\mu\nu} = x_\mu p_\nu - x_\nu p_\mu$, the generators of space rotations are

$$J_i = (1/2)\epsilon_{ijk} L^{jk},\ J_1 = L_{23} \text{ etc.}$$

In fact,

$$\delta x^1 = [x^1, J_3]\delta\phi = -x_2\delta\phi = x^2\delta\phi,\ \delta x^2 = [x^2, J_3]\delta\phi = x_1\delta\phi = -x^1\delta\phi,$$

$$\delta p_1 = [p_1, J_3]\delta\phi = p_2\delta\phi,\ \delta p_2 = [p_2, J_3]\delta\phi = -p_1\delta\phi.$$

The generators of Lorentz transformations are $K_i = L_{0i} = x_0 p_i - x_i p_0$. For instance, for an infinitesimal Lorentz transformation along the 1-axis,

$$x'^0 = x^0 + \delta x^0 = x^0 + [x^0, K_1]\epsilon = x^0 + \epsilon x^1,$$

$$x'^1 = x^1 + \delta x^1 = x^1 + [x^1, K_1]\epsilon = x^1 + \epsilon x^0.$$

(ii) The Poisson brackets are

$$[p_\mu, p_\nu] = 0,\ [p_0, J_i] = 0,\ [p_i, J_j] = -\epsilon_{ijk} p_k,$$

$$[p_0, K_i] = -p_i,\ [p_i, K_j] = -p_0\delta_{ij},\ [J_1, J_2] = -J_3 \text{ etc.},$$

$$[J_i, K_j] = -\epsilon_{ijk} K_k,\ [K_i, K_j] = \epsilon_{ijk} J_k.$$

Verify that this algebra is realized by putting $p_\mu = -\partial/\partial x^\mu$ in all the above generators and replacing Poisson brackets by commutators.

S10.6 (i) $\mathrm{d}C_\mu/\mathrm{d}s = \dot{p}_\mu - (q/2c)F_{\mu\nu}\dot{x}^\nu = \mathrm{d}[m\dot{x}_\mu + (q/2c)x^\nu F_{\nu\mu}]/\mathrm{d}s - (q/2c)F_{\mu\nu}\dot{x}^\nu = m\ddot{x}_\mu - (q/c)F_{\mu\nu}\dot{x}^\nu = 0$
(ii) $C_0 = p_0 - (q/2c)F_{0\nu}x^\nu = mc\dot{t} - (q/c)Ez = (\gamma mc^2 - qEz)/c$ (total energy divided by c), $C_3 = p_3 - (q/2c)F_{3\nu}x^\nu = -m\dot{z} + qEt = (-m\dot{z})_{t=0}$, $C_1 = p_1$, $C_2 = p_2$
(iii) Since $H = mc^2/2$, we have

$$m^2c^2 = (p_\mu - qA_\mu/c)(p^\mu - qA^\mu/c) = (C_\mu - 2qA_\mu/c)(C^\mu - 2qA^\mu/c) \quad \text{etc.}$$

S10.7 (i) The equations

$$m\frac{\mathrm{d}}{\mathrm{d}t'}\left(\gamma'\frac{\mathrm{d}x'}{\mathrm{d}t'}\right) = \frac{qB'}{c}\frac{\mathrm{d}y'}{\mathrm{d}t'} \quad , \quad m\frac{\mathrm{d}}{\mathrm{d}t'}\left(\gamma'\frac{\mathrm{d}y'}{\mathrm{d}t'}\right) = -\frac{qB'}{c}\frac{\mathrm{d}x'}{\mathrm{d}t'}$$

imply γ' =constant. A first integration gives

$$m\gamma'\frac{\mathrm{d}x'}{\mathrm{d}t'} = \frac{qB'}{c}\,y' - m\gamma' v' \quad , \quad m\gamma'\frac{\mathrm{d}y'}{\mathrm{d}t'} = -\frac{qB'}{c}\,x' \quad ,$$

satisfying the initial condition for the velocity components. These can be written in the form $\mathrm{d}x'/\mathrm{d}t' = \omega' y' - v'$, $\mathrm{d}y'/\mathrm{d}t' = -\omega' x'$, with $\omega' = qB'/mc\gamma'$, from which $x' = r\,\cos(\omega' t')$, $y' = r[1 - \sin(\omega' t')]$ with $r = v'/\omega' = mc\gamma' v'/qB'$.
(ii) $x'^2 + (y' - r)^2 = r^2$ becomes $(x - v't)^2 + (y - r)^2/\gamma'^2 = r^2/\gamma'^2$. Since

$$v' = cE/B,\ 1/\gamma'^2 = 1 - v'^2/c^2 = 1 - E^2/B^2,$$

we have

$$r = mc\gamma' v'/qB' = mc^2E/qB^2(1 - E^2/B^2).$$

S10.8 In the rest system $u^\nu = (c, 0, 0, 0)$, and so $S'_0 = 0$ because $\epsilon_{00\rho\sigma} = 0$. Only the spatial component S'_3 is different from zero: $S'_3 = S'\epsilon_{3012} = -S'$, $S'^3 = S'$.

In the lab system, $u^0 = \gamma c$, $u^1 = \gamma u_x$, etc., $r_1^0 = \gamma u_x/c$, $r_2^0 = \gamma u_y/c$, $r_1^1 = 1 + (\gamma - 1)u_x^2/u^2$, $r_1^2 = (\gamma - 1)u_y u_x/u^2$, $r_1^3 = (\gamma - 1)u_z u_x/u^2$, $r_2^1 = (\gamma - 1)u_x u_y/u^2$, $r_2^2 = 1 + (\gamma - 1)u_y^2/u^2$, $r_2^3 = (\gamma - 1)u_z u_y/u^2$.
Using these, one finds $S_0 = \gamma S' u_z/c$, $S^1 = \gamma^2(S' u_z)u_x/c^2(\gamma + 1)$, $S^2 = \gamma^2(S' u_z)u_y/c^2(\gamma + 1)$, $S^3 = S' + \gamma^2(S' u_z)u_z/c^2(\gamma + 1)$.

Some authors, for instance J.L. Anderson, *Principles of Relativity Physics* (Academic Press,1967) p. 250, define the "polarization vector"

$$w_\mu = (1/2)\epsilon_{\mu\nu\rho\sigma}s^{\nu\rho}p^\sigma \ ,$$

where p^σ is a component of the momentum, and $s^{\nu\rho}$ is a skewsymmetric spin tensor with the property $s_{\mu\nu}u^\nu = 0$. Therefore in the rest system only the space components s_{ij} are different from zero. But for trivial factors, w_μ is equal to our S_μ if

$$s^{\mu\nu} = (r_1^\mu r_2^\nu - r_1^\nu r_2^\mu)/2 \ .$$

S10.9 Expressing $S_\nu du^\nu/ds$ in terms of lab quantities, we have

$$S_\nu \frac{du^\nu}{ds} = S_0 c \frac{d\gamma}{ds} - \mathbf{S} \cdot \frac{d\gamma\mathbf{u}}{ds}$$

$$= \overbrace{(cS_0 - \mathbf{S}\cdot\mathbf{u})}^{\text{zero}} \frac{d\gamma}{ds} - \gamma\mathbf{S}\cdot\frac{d\mathbf{u}}{ds} \quad .$$

The rest is trivial.

S10.10 $\mathbf{S}$ and $\mathbf{u}$ in plane of motion,
$\mathbf{S} = (|\mathbf{S}|\cos\varphi, |\mathbf{S}|\sin\varphi)$, $\mathbf{u} = (0, |\mathbf{u}|)$, $\delta\mathbf{u} = (-|\mathbf{u}|\sin\alpha, -|\mathbf{u}|(1-\cos\alpha))$.
From section 10.5, regarding α as small, we have
$\delta\mathbf{S} = (\gamma/c)^2(\mathbf{S}\cdot\delta\mathbf{u})\mathbf{u}$, $\delta S_x = 0$, $\delta S_y \simeq -(\gamma/c)^2|\mathbf{S}||\mathbf{u}|^2\sin\alpha\ \cos\varphi$.
From $\delta\tan\varphi = \delta(S_y/S_x)$ we find $\delta\varphi \simeq -(\gamma/c)^2|\mathbf{u}|^2\sin\alpha\cos^2\varphi$. This gives $\delta\varphi = -\gamma^2\beta^2\sin\alpha$ for $\varphi = 0$, and $\delta\varphi = 0$ for $\varphi = \pm\pi/2$.

Repeating the work for φ' one finds $\delta\varphi' = -(\gamma-1)\sin\alpha$.

S10.11 Multiply (10.104) by $\mathbf{u}$, finding $\mathbf{S}\cdot\mathbf{u} = \gamma(\mathbf{S}'\cdot\mathbf{u})$. Express $\mathbf{S}'$ in terms of $\mathbf{S}$ and $\mathbf{u}$, obtain $d\mathbf{S}'/dt$ in terms of $d\mathbf{S}/dt$, $\mathbf{u}$, and $d\mathbf{u}/dt$, use (10.95) and (10.104).

S10.12 (i)

$$d\mathbf{S}'/dt = 0$$

(ii)

$$d(m\gamma\mathbf{u})/dt = 0,\ d\mathbf{S}'/dt = (gq/2mc\gamma)\mathbf{S}'\times\mathbf{B}$$

(iii) $d(m\gamma\mathbf{u})/dt = (q/c)\mathbf{u}\times\mathbf{B}$ gives $d\mathbf{u}/dt = \mathbf{u}\times\omega_L$, the particle moves on a circle in a plane normal to $\mathbf{B}$ with the Larmor angular velocity $\omega_L = q\mathbf{B}/mc\gamma$. Since $\mathbf{S}'\times(\mathbf{u}\times(\mathbf{u}\times\mathbf{B})) = -[c^2(\gamma^2-1)/\gamma^2]\mathbf{S}'\times\mathbf{B}$, we have

$$\frac{d\mathbf{S}'}{dt} = \frac{q}{mc\gamma}\left(1 + \frac{g-2}{2}\gamma\right)\mathbf{S}'\times\mathbf{B} \quad .$$

This equation is important for the measument of the $g-2$ anomaly. For $g=2$ $\mathbf{u}$ and $\mathbf{S}'$ are in phase, $d\mathbf{u}/dt = \mathbf{u}\times\omega_L$ and $d\mathbf{S}'/dt = \mathbf{S}'\times\omega_L$.

S10.13 The correct result is obtained from $d\mathbf{S}'/dt = [\mathbf{S}', H]$ and the Poisson brackets $[S'_x, S'_y] = S'_z$ etc. Can you justify these latter?

Chapter 11

CONTINUOUS SYSTEMS

To illustrate the application of analytical mechanics to continuous systems, we present two case studies, the uniform string and the ideal non-viscous fluid.

11.1 Uniform string

We assume the reader is familiar with the equation

$$\frac{\partial^2 y}{\partial x^2} - \frac{1}{v^2}\frac{\partial^2 y}{\partial t^2} = 0 \tag{11.1}$$

for small transversal displacements $y(x,t)$ from the equilibrium configuration (the x-axis). The phase velocity is $v = \sqrt{\tau/\mu}$, where τ and μ are the tension and the linear density, both constant.

The general solution is

$$y = f(\xi) + g(\eta) \quad , \tag{11.2}$$

where $\xi = x - vt$ and $\eta = x + vt$, while f and g are twice differentiable, but otherwise arbitrary.

If f and g are simple-harmonic, the wave number k and the angular frequency ω are related by $\omega = vk$, and there is no dispersion.

The vibrating string provides a one-dimensional model for the propagation of light in vacuum. Later on we will write equation (11.1) in the form

$$(\partial_0^2 - \partial_1^2)y = 0 \quad , \tag{11.3}$$

where $\partial_0 = \partial/\partial x^0$, $\partial_1 = \partial/\partial x^1$, $x^0 = vt$, and $x^1 = x$. More concisely

$$\partial_\mu \partial^\mu y = 0 \quad , \tag{11.4}$$

where $\partial^\mu = g^{\mu\nu}\partial_\nu$ ($g^{00} = 1$, $g^{11} = -1$, $g^{10} = g^{01} = 0$).

Note at once the invariance under x^0 and x^1 translations and the Lorentz-like transformation

$$x'^0 = \cosh\alpha\, x^0 - \sinh\alpha\, x^1 \ , \ x'^1 = \cosh\alpha\, x^1 - \sinh\alpha\, x^0 \quad ,$$

$$\partial_0 = \cosh\alpha\, \partial'_0 - \sinh\alpha\, \partial'_1 \ , \ \partial_1 = \cosh\alpha\, \partial'_1 - \sinh\alpha\, \partial'_0 \quad .$$

In all these cases, y transforms as a scalar, $y'(x'^0, x'^1) = y(x^0(x'), x^1(x'))$.

With $\partial_t = \partial/\partial t$ and $\partial_x = \partial/\partial x$, define the quantities

$$\begin{aligned} T_{tt} &= [\mu(\partial_t y)^2 + \tau(\partial_x y)^2]/2 \ , \ T_{xx} = T_{tt} \quad , \\ T_{xt} &= -\tau\, \partial_x y\, \partial_t y \ , \ T_{tx} = -\mu\, \partial_t y\, \partial_x y = v^{-2} T_{xt} \quad . \end{aligned} \tag{11.5}$$

Using the wave equation, we find

$$\partial_t T_{tt} = \mu\, \partial_t y\, \partial_{tt} y + \tau\, \partial_x y\, \partial_{tx} y$$

$$= \tau(\partial_t y\, \partial_{xx} y + \partial_x y\, \partial_{tx} y) = \partial_x(\tau \partial_x y\, \partial_t y) = -\partial_x T_{xt} \quad ,$$

$$\partial_t T_{tt} + \partial_x T_{xt} = 0 \quad . \tag{11.6}$$

This is a continuity equation describing energy conservation. T_{tt} is the energy density of transversal motion.

Integrating equation (11.6) with respect to x, we have

$$\frac{\mathrm{d}}{\mathrm{d}t} \int_a^b T_{tt}\, \mathrm{d}x = T_{xt}(a) - T_{xt}(b) \quad . \tag{11.7}$$

Thus $T_{xt}(a)$ and $T_{xt}(b)$ can be interpreted as the rates of energy flow into and out of the interval (a, b). Of course, they may be negative.

For a wave propagating in the positive x direction, $y = f(\xi)$, we have

$$T_{xt} = -\tau \left(\frac{\mathrm{d}f}{\mathrm{d}\xi}\right) \left(-v \frac{\mathrm{d}f}{\mathrm{d}\xi}\right) = \tau v \left(\frac{\mathrm{d}f}{\mathrm{d}\xi}\right)^2 > 0 \ , \tag{11.8}$$

while for one propagating in the negative x direction, $y = g(\eta)$, we have

$$T_{xt} = -\tau \left(\frac{\mathrm{d}g}{\mathrm{d}\eta}\right) \left(v \frac{\mathrm{d}g}{\mathrm{d}\eta}\right) = -\tau v \left(\frac{\mathrm{d}g}{\mathrm{d}\eta}\right)^2 < 0 \quad . \tag{11.9}$$

T_{xt} is the work per unit time done by the portion of the string left of x on the portion right of x. In fact (see figure 1.1) $T_{xt} = (-\tau\, \partial_x y)\partial_t y = F_y v_y$, where $F_y = -\tau\, \sin\theta \simeq -\tau\, \tan\theta = -\tau\, \partial_x y$, $v_y = \partial_t y$.

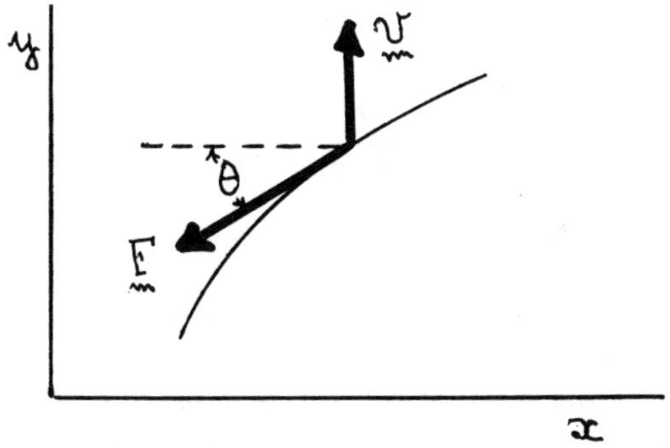

Figure 11.1: Interpreting T_{xt}

Consider now T_{tx}. We have

$$\partial_t T_{tx} = -\mu\ \partial_{tt}y\ \partial_x y - \mu\ \partial_t y\ \partial_{tx}y = -\mu v^2 \partial_{xx}y\ \partial_x y - \mu\ \partial_t y\ \partial_{tx}y$$

$$= -\partial_x[\tau(\partial_x y)^2 + \mu(\partial_t y)^2]/2 = -\partial_x T_{xx} \quad ,$$

$$\partial_t T_{tx} + \partial_x T_{xx} = 0 \quad . \tag{11.10}$$

The quantity $T_{tx} = -\mu\ \partial_t y\ \partial_x y$ is interpreted as a longitudinal momentum density. It is positive/negative for propagation in the positive/negative x direction.

Integration of equation (11.10) over an interval (a, b) gives

$$\frac{\mathrm{d}}{\mathrm{d}t}\int_a^b T_{tx}\mathrm{d}x = T_{xx}(a) - T_{xx}(b) \quad . \tag{11.11}$$

Lagrangian

Consider the action

$$A = \int_{t_1}^{t_2}\mathrm{d}t \int_a^b \mathrm{d}x\ \mathcal{L}(y, \partial_x y, \partial_t y) = v^{-1}\int_D \mathrm{d}x^0 \mathrm{d}x^1 \mathcal{L} \quad , \tag{11.12}$$

where

$$\mathcal{L} = [\mu(\partial_t y)^2 - \tau(\partial_x y)^2]/2 = \tau(\partial_\mu y)(\partial^\mu y)/2 \tag{11.13}$$

is a Lagrangian density.

Assume that $\delta A = 0$ under $y \to y + \delta y$ with $\delta y = 0$ for $t = t_1$, $t = t_2$ (any x) and $x = a$, $x = b$ (any t), or, more briefly, on the boundary of D. From this assumption follows the Lagrange equation

$$\partial_t \frac{\partial \mathcal{L}}{\partial(\partial_t y)} + \partial_x \frac{\partial \mathcal{L}}{\partial(\partial_x y)} = \frac{\partial \mathcal{L}}{\partial y} \tag{11.14}$$

or, more concisely,

$$\partial_\mu \frac{\partial \mathcal{L}}{\partial(\partial_\mu y)} = \frac{\partial \mathcal{L}}{\partial y} \quad . \tag{11.15}$$

It is easy to see that this equation reduces to (11.4)

$$\mu\, \partial_t(\partial_t y) - \tau\, \partial_x(\partial_x y) = 0 \ , \ \partial_{tt} y - (\tau/\mu)\partial_{xx} y = 0 \ , \ \partial^\mu \partial_\mu y = 0 \quad .$$

Invariance properties of $\mathcal{L}$ and conservation laws

Consider a Lagrangian density of the form $\mathcal{L}(y, \partial_\mu y)$, where $y = y(x^0, x^1)$. Assume that $\mathcal{L}$ is invariant under $x'^\mu = x^\mu + \delta x^\mu$ (δx^μ infinitesimal) up to terms of the second order in δx^μ. This means

$$\delta \mathcal{L} = \mathcal{L}(y', \partial'_\mu y') - \mathcal{L}(y, \partial_\mu y) = 0 \quad , \tag{11.16}$$

where $y'(x'^0, x'^1) = y(x^0(x'), x^1(x'))$.

For the string, $\mathcal{L} = (\tau/2)\partial_\mu y\ \partial^\mu y$ is invariant under finite x^0 and x^1 translations, and under finite Lorentz-like transformations. We are interested only in the respective infinitesimal cases, ($x'^0 = x^0 - \epsilon, x'^1 = x^1$), ($x'^0 = x^0, x'^1 = x^1 - \epsilon$), and ($x'^0 = x^0 - \epsilon x^1, x'^1 = x^1 - \epsilon x^0$).

Now $\delta y = y'(x') - y(x) \simeq \bar{\delta} y + (\partial_\mu y)\delta x^\mu$, where

$$\bar{\delta} y = y'(x') - y(x') \simeq y'(x) - y(x)$$

is the change in form of y. Of course, $\delta y = 0$ in the present case in which the "field" is a scalar field. Equation (11.16) ($\delta \mathcal{L} = 0$) gives

$$\frac{\partial \mathcal{L}}{\partial y}\delta y + \frac{\partial \mathcal{L}}{\partial(\partial_\mu y)}\ \delta \partial_\mu y = 0 \quad ,$$

$$\frac{\partial \mathcal{L}}{\partial y}(\bar{\delta} y + (\partial_\nu y)\delta x^\nu) + \frac{\partial \mathcal{L}}{\partial(\partial_\mu y)}(\bar{\delta}\partial_\mu y + (\partial_\nu \partial_\mu y)\delta x^\nu) = 0 \quad .$$

Since in the first order $\bar{\delta}\partial_\mu y = \partial_\mu \bar{\delta} y$, we have

$$\bar{\delta}\mathcal{L} + \frac{\partial \mathcal{L}}{\partial x^\nu}\ \delta x^\nu = 0 \quad ,$$

where

$$\bar{\delta}\mathcal{L} = \frac{\partial \mathcal{L}}{\partial y}\ \bar{\delta} y + \frac{\partial \mathcal{L}}{\partial(\partial_\mu y)}\ \partial_\mu \bar{\delta} y = 0 \quad .$$

By Lagrange's equation (11.15) this can be expressed in the form

$$\partial_\mu \left[\frac{\partial \mathcal{L}}{\partial(\partial_\mu y)}\bar{\delta} y\right] + (\partial_\mu \mathcal{L})\delta x^\mu = 0 \quad .$$

Assuming $\partial_\mu \delta x^\mu = 0$, which is trivially true for x^0 and x^1 translations and for Lorentz-like transformations, we have

$$\partial_\mu \left[\frac{\partial \mathcal{L}}{\partial(\partial_\mu y)} \bar{\delta} y + \mathcal{L}\, \delta x^\mu \right] = 0 \quad . \tag{11.17}$$

For x^0 translations $x'^0 = x^0 - \epsilon$ $(\delta x^0 = -\epsilon)$, $y'(x'^0, x'^1) = y(x^0, x^1)$,
$y'(x^0 - \epsilon, x^1) = y(x^0, x^1)$, $y'(x^0, x^1) = y(x^0 + \epsilon, x^1)$, $\bar{\delta} y = \epsilon\, \partial_0 y$.
Similarly for x^1 translations $\bar{\delta} y = \epsilon\, \partial_1 y$.

Equation (11.17) gives

$$\partial_\mu T^{\mu\nu} = 0 \quad , \tag{11.18}$$

where

$$T^{\mu\nu} = \frac{\partial \mathcal{L}}{\partial(\partial_\mu y)} \partial^\nu y - \mathcal{L} g^{\mu\nu} \quad . \tag{11.19}$$

For the string Lagrangian (11.13) this formula yields

$$T^{00} = T_{tt} \ , \ T^{11} = T_{xx} = T_{tt} \ ,$$

$$T^{01} = v T_{tx} \ , \ T^{10} = v^{-1} T_{xt} \ ,$$

where the quantities T_{tt}, T_{xx}, T_{tx}, and T_{xt} are given by equations (11.5). Note that $T^{01} = T^{10}$, while $T_{tx} \neq T_{xt}$.

The equation $\partial_0 T^{00} + \partial_1 T^{10} = 0$ is equivalent to equation (11.6) $(\partial_t T_{tt} + \partial_x T_{xt} = 0)$, and $\partial_0 T^{01} + \partial_1 T^{11} = 0$ is equivalent to equation (11.10) $(\partial_t T_{tx} + \partial_x T_{xx} = 0)$.

For the Lorentz-like transformation $\delta x^0 = -\epsilon x^1$, $\delta x^1 = -\epsilon x^0$,
$\bar{\delta} y = y(x^0 + \epsilon x^1, x^1 + \epsilon x^0) - y(x^0, x^1) = \epsilon(x^1 \partial_0 y + x^0 \partial_1 y)$,
we find

$$\partial_0 \left[\frac{\partial \mathcal{L}}{\partial(\partial_0 y)} \epsilon(x^1 \partial_0 y + x^0 \partial_1 y) + (-\epsilon x^1)\mathcal{L} \right]$$

$$+\partial_1 \left[\frac{\partial \mathcal{L}}{\partial(\partial_1 y)} \epsilon(x^1 \partial_0 y + x^0 \partial_1 y) + (-\epsilon x^0)\mathcal{L} \right] = 0 \quad ,$$

$$\partial_0[-x^0 T^{01} + x^1 T^{00}] + \partial_1[x^1 T^{10} - x^0 T^{11}] = 0 \quad . \tag{11.20}$$

Since $\partial_\alpha T^{\alpha 0} = \partial_\alpha T^{\alpha 1} = 0$, this gives $T^{10} - T^{01} = 0$. Thus invariance of $\mathcal{L}$ under x^0 and x^1 translations, combined with invariance under Lorentz-like transformations, implies $T^{\mu\nu} = T^{\nu\mu}$.

Integrating (11.20) and assuming that $T^{\mu\nu}$ vanishes for $x = \pm\infty$, we have

$$\frac{\mathrm{d}}{\mathrm{d}t} \int (-v^2 t T_{tx} + x T_{tt}) \mathrm{d}x = 0 \ ,$$

$$\int T_{tt}x \ \mathrm{d}x = v^2 t \int T_{tx} \ \mathrm{d}x + \text{constant} = t \int T_{xt} \ \mathrm{d}x + \text{constant} \ ,$$

where the integrals are over the whole x axis.

Since $\int T_{tt} \mathrm{d}x = $ constant, we find for the "energy center"

$$\langle x \rangle = \frac{\int T_{tt} x \ \mathrm{d}x}{\int T_{tt} \ \mathrm{d}x} = \left(\frac{\int T_{xt} \ \mathrm{d}x}{\int T_{tt} \ \mathrm{d}x} \right) t + \text{const} = \frac{P}{E} \ t + \text{const} \quad ,$$

where P is the total longitudinal momentum.

Hamiltonian

Define

$$\pi(x,t) = \frac{\partial \mathcal{L}}{\partial(\partial_t y(x,t))} \tag{11.21}$$

and the Hamiltonian density

$$\mathcal{H}(\pi, y, \partial_x y) = \pi(x,t)\partial_t y(x,t) - \mathcal{L}(y, \partial_t y, \partial_x y) \quad . \tag{11.22}$$

Hamilton's equations can be derived from the invariance of the action

$$A = \int \mathcal{L} \ \mathrm{d}t\mathrm{d}x = \int (\pi \partial_t y - \mathcal{H}) \ \mathrm{d}t\mathrm{d}x \tag{11.23}$$

under variations of y and π vanishing on the boundary of the integration region. In fact, $\delta A = 0$ gives

$$0 = \int \left(\partial_t y \ \delta\pi + \pi \ \delta\partial_t y - \frac{\partial \mathcal{H}}{\partial \pi} \delta\pi - \frac{\partial \mathcal{H}}{\partial y} \ \delta y - \frac{\partial \mathcal{H}}{\partial(\partial_x y)} \ \delta\partial_x y \right) \mathrm{d}t\mathrm{d}x \quad .$$

Using $\delta\partial_t y = \partial_t \delta y$, $\delta\partial_x y = \partial_x \delta y$, and integrating by parts, we have

$$\int \left[\left(\partial_t y - \frac{\partial \mathcal{H}}{\partial \pi} \right) \delta\pi + \left(-\partial_t \pi - \frac{\partial \mathcal{H}}{\partial y} + \partial_x \frac{\partial \mathcal{H}}{\partial(\partial_x y)} \right) \delta y \right] \mathrm{d}t\mathrm{d}x = 0 \quad .$$

Assuming the independence of $\delta\pi$ and δy, we find Hamilton's equations

$$\partial_t y = \frac{\partial \mathcal{H}}{\partial \pi} \quad , \quad \partial_t \pi = -\frac{\partial \mathcal{H}}{\partial y} + \partial_x \frac{\partial \mathcal{H}}{\partial(\partial_x y)} \quad . \tag{11.24}$$

For the string, we have

$$\pi = \mu \ \partial_t y \quad ,$$

$$\mathcal{H} = [\mu(\partial_t y)^2 + \tau(\partial_x y)^2]/2 = \pi^2/2\mu + \tau(\partial_x y)^2/2 \quad ,$$

and Hamilton's equations

$$\partial_t y = \pi/\mu \quad , \quad \partial_t \pi = \tau \ \partial_{xx} y \quad , \quad \mu \ \partial_{tt} y = \tau \ \partial_{xx} y \quad .$$

Poisson brackets etc.

Let A and B be two functionals of $\pi(x,t)$ and $y(x,t)$. Define the Poisson bracket of A and B as

$$[A,B] = \int \mathrm{d}x \left[\frac{\delta A}{\delta y(x,t)} \frac{\delta B}{\delta \pi(x,t)} - \frac{\delta A}{\delta \pi(x,t)} \frac{\delta B}{\delta y(x,t)} \right] \quad , \tag{11.25}$$

where the integral is over the whole x axis and $\delta A/\delta y(x,t)$ etc. denote a "functional derivative" as defined below.

Let $F[f|t]$ be a functional of $f(x,t)$, namely a correspondence between the function $f(x,t)$ of x and t and the function $F[f|t]$ of t alone. The Lagrangian $L = \int \mathcal{L}(y, \partial_t y, \partial_x y)\, \mathrm{d}x$ is a functional of $y(x,t)$.

Assume that for $y(x,t) \to y(x,t) + \delta y(x)$ (not $\delta y(x,t)$)

$$\delta F[f|t] = \int A(x,t)\, \delta y(x)\, \mathrm{d}x \quad .$$

We then say that the functional derivative of F is

$$\frac{\delta F[f|t]}{\delta y(x,t)} = A(x,t) \quad .$$

For the above Lagrangian

$$\frac{\delta L}{\delta y(x,t)} = \frac{\partial L}{\partial y(x,t)} - \partial_x \frac{\partial L}{\partial(\partial_x y(x,t))}$$

and the Lagrange equation (11.15) can be written in the form

$$\partial_t \frac{\partial \mathcal{L}}{\partial(\partial_t y(x,t))} = \frac{\delta L}{\delta y(x,t)} \quad .$$

Similarly

$$\partial_t y(x,t) = \frac{\partial \mathcal{H}}{\partial \pi(x,t)} = \frac{\delta H}{\delta \pi(x,t)} \quad ,$$

$$\partial_t \pi(x,t) = -\frac{\partial \mathcal{H}}{\partial y(x,t)} + \partial_x \frac{\partial \mathcal{H}}{\partial(\partial_x y(x,t))} = -\frac{\delta H}{\delta y(x,t)} \quad ,$$

where

$$H = \int \mathcal{H}\, \mathrm{d}x \quad .$$

Writing $y(x',t) = \int y(x,t)\delta(x-x')\mathrm{d}x$, we find at once that

$$\frac{\delta y(x',t)}{\delta y(x,t)} = \delta(x-x') \quad .$$

Similarly

$$\frac{\delta \pi(x',t)}{\delta \pi(x,t)} = \delta(x-x') \quad .$$

Thus we obtain the Poisson brackets

$$[y(x,t),\pi(x',t)] = \int \mathrm{d}x''[\delta(x-x'')\delta(x'-x'') - \text{zero}] = \delta(x-x') \quad , \tag{11.26}$$

$$[y(x,t),y(x',t)] = 0 \quad , \quad [\pi(x,t),\pi(x',t)] = 0 \quad . \tag{11.27}$$

Since

$$[y(x,t),H] = \int \mathrm{d}x' \left(\frac{\delta y(x,t)}{\delta y(x',t)} \frac{\delta H}{\delta \pi(x',t)} - \frac{\delta y(x,t)}{\delta \pi(x',t)} \frac{\delta H}{\delta y(x',t)} \right) \quad ,$$

$$[y(x,t),H] = \frac{\delta H}{\delta \pi(x,t)} \quad ,$$

we have

$$\partial_t y(x,t) = [y(x,t),H] \quad . \tag{11.28}$$

Similarly

$$\partial_t \pi(x,t) = [\pi(x,t),H] \quad . \tag{11.29}$$

In general, if F is a functional of $y(x,t)$ and $\pi(x,t)$, possibly explicitly dependent on t, we have

$$\frac{\mathrm{d}F}{\mathrm{d}t} = \frac{\partial F}{\partial t} + [F,H] \quad . \tag{11.30}$$

For the Hamiltonian

$$\frac{\mathrm{d}H}{\mathrm{d}t} = \frac{\partial H}{\partial t} \quad .$$

11.2 Ideal fluids

One-dimensional incompressible ideal fluid

In the absence of external forces, the velocity $v(x,t)$ of the fluid at position x and time t obeys the equation

$$\frac{\partial v}{\partial t} + v\,\frac{\partial v}{\partial x} + \frac{1}{\mu}\,\frac{\partial p}{\partial x} = 0 \quad , \tag{11.31}$$

where $\mu =$ constant is the linear density and p is the "pressure" (with the dimensions of a force).

We refer the reader to intermediate mechanics textbooks for the three-dimensional Euler equation

$$\frac{\partial \mathbf{v}}{\partial t} + (\mathbf{v}\cdot\nabla)\mathbf{v} + \frac{1}{\rho}\,\nabla p = \frac{1}{\rho}\,\mathbf{f} \quad , \tag{11.32}$$

where $\mathbf{f}$ is the force per unit volume and $\mathbf{f}/\rho$ the force per unit mass. Using this together with the continuity equation

$$\frac{\partial \rho}{\partial t} + \nabla \cdot (\rho \mathbf{v}) = 0 \tag{11.33}$$

one finds

$$f_i = \frac{\partial(\rho v_i)}{\partial t} + \partial_j(\rho v_i v_j + p\delta_{ij}) \quad . \tag{11.34}$$

Within the limited scope of this book, we wish to discuss the Lagrangian formulation of this problem. We follow in part D.E. Soper, *Classical Field Theory* (Wiley, 1976) ch.4, where we learned the existence of the original work by G. Herglotz, *Ann. Phys.*,**36**(1911)493.

We must switch from Euler's description (observation of fluid at a given position) to Lagrange's description (follow fluid particles in their motion).

If Euler observes a particle at x at time t, let $X(x,t)$ be the position the particle had at $t = 0$. Therefore $X(x,0) = x$. Let $M(X)\mathrm{d}X$ be the total mass of particles with coordinates between X and $X+\mathrm{d}X$ at $t = 0$. At time t, these particles will be in the interval $(x, x+\mathrm{d}x)$. Since $\mathrm{d}X = (\partial X/\partial x)\mathrm{d}x$, the linear density at x at time t is

$$\mu(x,t) = M(X)\,\frac{\partial X}{\partial x} \quad . \tag{11.35}$$

Note now that a particle at x at time t will be at $x + v(x,t)$ dt at time $t+$ dt. When it was at x its X was $X(x,t)$, when it is at $x + v$ dt its X is $X(x + v\,\mathrm{d}t)$. The corresponding dX is

$$\mathrm{d}X = \frac{\partial X}{\partial x}\,v\,\mathrm{d}t + \frac{\partial X}{\partial t}\,\mathrm{d}t \quad .$$

Since X is just a label for the particle, must be d$X = 0$, and so [1]

$$v = -\frac{\partial X/\partial t}{\partial X/\partial x} \quad , \tag{11.36}$$

and the current is

$$J = \mu v = \left[M(X)\frac{\partial X}{\partial x}\right]\left[-\frac{\partial X/\partial t}{\partial X/\partial x}\right] = -M(X)\frac{\partial X}{\partial t} \, . \tag{11.37}$$

The continuity equation is automatically satisfied:

$$\frac{\partial \mu}{\partial t} + \frac{\partial J}{\partial x} = \frac{\mathrm{d}M}{\mathrm{d}X}\left(\frac{\partial X}{\partial t}\frac{\partial X}{\partial x} - \frac{\partial X}{\partial x}\frac{\partial X}{\partial t}\right) + M\left(\frac{\partial^2 X}{\partial t \partial x} - \frac{\partial^2 X}{\partial x \partial t}\right) = 0 \, .$$

[1] For uniform motion we would have $x = X+vt$, $X = x-vt$, $\partial X/\partial x = 1$, $\partial X/\partial t = -v$, $v = -(\partial X/\partial t)/(\partial X/\partial x)$.

Equation (11.31) can be derived from the Lagrangian density

$$\mathcal{L} = \mathcal{L}(X, \partial_t X, \partial_x X) = \mu v^2/2 - \mu\, U(\lambda) \quad , \tag{11.38}$$

where $\lambda = 1/\mu$ ("volume" of the unit mass) and we assume that the "equation of state" is $p = -\mathrm{d}U(\lambda)/\mathrm{d}\lambda$.
Remembering that

$$\mathcal{L} = M(X)\left[\frac{1}{2}\frac{(\partial_t X)^2}{\partial_x X} - (\partial_x X)\, U\left(\frac{1}{M(X)\partial_x X}\right)\right] \quad , \tag{11.39}$$

a fairly long calculation using the continuity equation yields

$$\begin{aligned} 0 &= \partial_t \left(\frac{\partial \mathcal{L}}{\partial(\partial_t X)}\right) + \partial_x \left(\frac{\partial \mathcal{L}}{\partial(\partial_x X)}\right) - \frac{\partial \mathcal{L}}{\partial X} \\ &= -M(X)\left(\partial_t v + v\,\partial_x v + \frac{1}{\mu}\,\partial_x p\right) \quad . \end{aligned}$$

The following is another proof of the one-dimensional version of Euler's equation (11.31),

$$\partial_t v + v\,\partial_x v + \frac{1}{\mu}\,\partial_x p = 0 \quad .$$

Using (11.19) with y replaced by X, and the Lagrangian (11.38,39), we find $(x_0 = x^0 = t, x^1 = x = -x_1)$

$$T^{00} = \mu v^2/2 + \mu U(\lambda)\ ,\ T^{01} = \mu v\ ,\ T^{10} = v(T^{00} + p)\ ,\ T^{11} = \mu v^2 + p\ .$$

Then (11.18) for $\nu = 0$ $(\partial_0 T^{00} + \partial_1 T^{10} = 0)$ using the continuity equation $\partial_t \mu + \partial_x(\mu v) = 0$, gives

$$\mu v(\partial_t v + v\partial_x v + \mu^{-1}\partial_x p) = 0$$

and for $\nu = 1$ $(\partial_0 T^{01} + \partial_1 T^{11} = 0)$

$$\mu(\partial_t v + v\partial_x v + \mu^{-1}\partial_x p) = 0 \quad .$$

Two-dimensional incompressible fluid
With $\mathbf{x} = (x^1, x^2)$ and $\mathbf{X} = (X^1, X^2)$ and a natural extension from the one-dimensional case, we express the density in the form

$$\rho(\mathbf{x}, t) = \frac{\partial(\mathbf{X})}{\partial(\mathbf{x})} M(\mathbf{X}) \quad , \tag{11.40}$$

where

$$\frac{\partial(\mathbf{X})}{\partial(\mathbf{x})} = \begin{vmatrix} \partial X^1/\partial x^1 & \partial X^1/\partial x^2 \\ \partial X^2/\partial x^1 & \partial X^2/\partial x^2 \end{vmatrix} . \tag{11.41}$$

Then

$$\rho(\mathbf{x}, t)\ \mathrm{d}^2x = M(\mathbf{X})\ \mathrm{d}^2X \quad . \tag{11.42}$$

The analogue of the one-dimensional equation (11.36) ($v\partial_x X + \partial_t X = 0$) is

$$\begin{cases} (\partial_1 X^1)v^1 + (\partial_2 X^1)v^2 = -\partial_t X^1 \ , \\ (\partial_1 X^2)v^1 + (\partial_2 X^2)v^2 = -\partial_t X^2 \ , \end{cases} \tag{11.43}$$

where $\partial_i = \partial/\partial x^i$ ($i = 1, 2$) and $\partial_t = \partial/\partial t$.
Solving for v^1 and v^2 we find $v^i = J^i/\rho$ ($i = 1, 2$), where

$$\begin{cases} J^1 = -M(\mathbf{X})[(\partial_t X^1)(\partial_2 X^2) - (\partial_t X^2)(\partial_2 X^1)] \ , \\ J^2 = -M(\mathbf{X})[(\partial_t X^2)(\partial_1 X^1) - (\partial_t X^1)(\partial_1 X^2)] \ . \end{cases} \tag{11.44}$$

Continuity equation

Let $\epsilon^{\alpha\beta\gamma}$ be skewsymmetric in all indices which take the values $(0, 1, 2)$, and $\epsilon^{012} = 1$. The density and the current components can be written

$$\begin{cases} \rho = M\epsilon^{0\alpha\beta}(\partial_\alpha X^1)(\partial_\beta X^2) \ , \\ J^i = M\epsilon^{i\alpha\beta}(\partial_\alpha X^1)(\partial_\beta X^2) \ . \end{cases} \tag{11.45}$$

These satisfy the continuity equation. In fact

$$\partial_t\rho + \partial_1 J^1 + \partial_2 J^2 = \partial_\mu(M\epsilon^{\mu\alpha\beta}(\partial_\alpha X^1)(\partial_\beta X^2))$$

$$= \sum_{i=1}^{2}(\partial M/\partial X^i)\epsilon^{\mu\alpha\beta}(\partial_\mu X^i)(\partial_\alpha X^1)(\partial_\beta X^2) + M\partial_\mu(\epsilon^{\mu\alpha\beta}(\partial_\alpha X^1)(\partial_\beta X^2)) = 0+0 = 0 \ .$$

Lagrangian

From the Lagrangian density

$$\mathcal{L} = (\rho/2)[(v^1)^2 + (v^2)^2] - \rho U(\lambda) \tag{11.46}$$

($\lambda = 1/\rho$) one might try to derive the two-dimensional version of (11.32) with $\mathbf{f} = 0$. However one can proceed more simply as follows.

From

$$T^{\mu\nu} = \sum_{m=1}^{2} \frac{\partial\mathcal{L}}{\partial(\partial_\mu X^m)}\partial^\nu X^m \ - \mathcal{L}g^{\mu\nu} \tag{11.47}$$

we have

$$T^{01} = \rho v^1 \ , \ T^{11} = \rho(v^1)^2 + p \ , \ T^{21}\rho v^1 v^2 \ ,$$

$$0 = \partial_0 T^{01} + \partial_1 T^{11} + \partial_2 T^{21} = \rho(\partial_t v^1 + v^1 \partial_1 v^1 + v^2 \partial_2 v^1 + \rho^{-1} \partial_1 p) \ .$$

Note that from

$$\partial_0 T^{00} + \partial_1 T^{10} + \partial_2 T^{20} = 0$$

with

$$T^{00} = (\rho/2)[(v^1)^2 + (v^2)^2] + \rho U(\lambda) \ , \ T^{i0} = v^i(T^{00} + p) \ ,$$

we would find

$$\rho(\mathbf{v} \cdot \partial_t \mathbf{v} + v^i \mathbf{v} \cdot \partial_i \mathbf{v} + \rho^{-1} v^i \partial_i p) = 0 \ .$$

Circulation theorem

We derive the theorem in two dimensions from the assumption that the Lagrangian density $\mathcal{L}(X^i, \partial_\mu X^i)$ is invariant under an infinitesimal transformation $X^i \to X^i + \epsilon \xi^i(\mathbf{X})$ defined below. Then

$$0 = \epsilon \left(\frac{\partial \mathcal{L}}{\partial(\partial_\mu X^i)} \partial_\mu \xi^i + \frac{\partial \mathcal{L}}{\partial X^i} \right) \quad ,$$

and, using Lagrange's equations,

$$0 = \partial_\mu \left(\frac{\partial \mathcal{L}}{\partial(\partial_\mu X^i)} \xi^i \right) \quad . \tag{11.48}$$

Therefore

$$\int \mathrm{d}^2 x \frac{\partial \mathcal{L}}{\partial(\partial_t X^i)} \xi^i = \text{constant} \quad . \tag{11.49}$$

The Lagrangian density (11.46) is invariant under a transformation $\mathbf{X} \to \mathbf{X}'$ such that

$$M(\mathbf{X}') = M(\mathbf{X}) \, \frac{\partial(X^1, X^2)}{\partial(X'^1, X'^2)} \quad , \tag{11.50}$$

where $M(\mathbf{X}')$ is obtained from $M(\mathbf{X})$ by replacing X^i by X'^i $(i = 1, 2)$.

In fact ρ, J^1, and J^2 are invariant:

$$\rho(\mathbf{x}, t) = \frac{\partial(\mathbf{X})}{\partial(\mathbf{x})} M(\mathbf{X}) = \frac{\partial(\mathbf{X}')}{\partial(\mathbf{x})} \frac{\partial(\mathbf{X})}{\partial(\mathbf{X}')} M(\mathbf{X}) = \frac{\partial(\mathbf{X}')}{\partial(\mathbf{x})} M(\mathbf{X}') \ ,$$

$$\begin{aligned} J^1(\mathbf{x}, t) = -M(\mathbf{X}) \frac{\partial(X^1, X^2)}{\partial(t, x^2)} &= -M(\mathbf{X}') \frac{\partial(X'^1, X'^2)}{\partial(X^1, X^2)} \frac{\partial(X^1, X^2)}{\partial(t, x^2)} \\ &= -M(\mathbf{X}') \frac{\partial(X'^1, X'^2)}{\partial(t, x^2)} \ , \end{aligned}$$

and similarly for J^2.

For infinitesimal transformations $X'^i = X^i + \epsilon\ \xi^i(\mathbf{X})$ the condition (11.50) gives $M(\mathbf{X})(1 - \epsilon\ \partial\xi^i/\partial X^i) = M(\mathbf{X}) + \epsilon(\partial M(\mathbf{X})/\partial X^i)\xi^i$,

$$\frac{\partial(M(\mathbf{X})\xi^i)}{\partial X^i} = 0 \quad .$$

This is satisfied by

$$M(\mathbf{X})\ \xi^i(\mathbf{X}) = -\int_0^1 \mathrm{d}\tau\ \frac{\mathrm{d}\Omega^i(\tau)}{\mathrm{d}\tau}\ \delta^2(\mathbf{X} - \boldsymbol{\Omega}(\tau)) \quad , \quad \boldsymbol{\Omega}(1) = \boldsymbol{\Omega}(0)$$

(see D.E. Soper, *loc. cit.*, p.51, but this is a little more general, $M(\mathbf{X})$ is not regarded as constant).

The conserved quantity is

$$\begin{aligned} Q(t) &= -\int_0^1 \mathrm{d}\tau \int \mathrm{d}^2x\ \frac{\partial\mathcal{L}}{\partial(\partial_t X^i)}\ \frac{\mathrm{d}\Omega^i}{\mathrm{d}\tau}\ \frac{1}{M(\mathbf{X})}\ \delta^2(\mathbf{X} - \boldsymbol{\Omega}(\tau)) \\ &= -\int_0^1 \mathrm{d}\tau\ \frac{\mathrm{d}\Omega^i}{\mathrm{d}\tau} \left[\frac{\partial(\mathbf{x})}{\partial(\mathbf{X})}\ \frac{1}{M(\mathbf{X})}\ \frac{\partial\mathcal{L}}{\partial(\partial_t X^i)}\right]_{\mathbf{x}=\mathbf{x}(\mathbf{X}=\boldsymbol{\Omega}(\tau))} \\ &= -\int_0^1 \mathrm{d}\tau\ \frac{\partial x^j}{\partial X^k}\ \frac{\mathrm{d}\Omega^k}{\mathrm{d}\tau} \left[M(\mathbf{X})\frac{\partial(\mathbf{X})}{\partial(\mathbf{x})}\right]^{-1}\ \frac{\partial\mathcal{L}}{\partial(\partial_t X^i)}\ \partial_j X^i \quad , \end{aligned}$$

where $\mathbf{X} = \boldsymbol{\Omega}(\tau)$ and $\mathbf{x} = \mathbf{x}(\boldsymbol{\Omega})$. By this rearrangement we have $(\partial x^j/\partial\Omega^k)(\mathrm{d}\Omega^k/\mathrm{d}\tau) = \mathrm{d}x^j/\mathrm{d}\tau$, while $(\partial\mathcal{L}/\partial(\partial_t X^i))\partial_j X^i = -\rho v^j$, as can be shown by a little calculation. Therefore

$$Q(t) = -\oint \mathrm{d}x^j\ \frac{1}{\rho}\ (-\rho v^j) = \oint \mathbf{v}\cdot\mathrm{d}\mathbf{x} \quad .$$

Three-dimensional ideal fluids

Rather than extending the two-dimensional Lagrangian calculations, we summarize the standard treatment of the three-dimensional fluid.

Define the "vorticity" $\boldsymbol{\Omega} = \nabla\times\mathbf{v}$. The Euler equation for a non-viscous fluid acted upon by conservative forces ($\mathbf{f} = -\rho\nabla\Phi$) can be expressed in the form

$$\frac{\partial\mathbf{v}}{\partial t} + \frac{1}{2}\ \nabla v^2 - \mathbf{v}\times\boldsymbol{\Omega} + \frac{1}{\rho}\ \nabla p + \nabla\Phi = 0 \quad .$$

Note that for steady irrotational flow of an incompressible non-viscous fluid ($\boldsymbol{\Omega} = 0$, $\rho =$ constant) this equation reduces to

$$\frac{\partial}{\partial x^i}\left(\frac{\rho}{2}v^2 + p + \rho\Phi\right) = 0 \ \text{(Bernoulli equation)} \quad .$$

For a non-viscous fluid the substantial derivative of the vorticity,

$$\frac{\mathrm{D}\boldsymbol{\Omega}}{\mathrm{D}t} = \frac{\partial \boldsymbol{\Omega}}{\partial t} + (\mathbf{v}\cdot\nabla)\boldsymbol{\Omega} \quad ,$$

obeys the equation

$$\frac{\mathrm{D}\boldsymbol{\Omega}}{\mathrm{D}t} = (\boldsymbol{\Omega}\cdot\nabla)\mathbf{v} - \boldsymbol{\Omega}(\nabla\cdot\mathbf{v}) + \frac{1}{\rho^2}\nabla\rho\times\nabla p \quad .$$

In fact, taking the rotation of the Euler equation we have

$$\frac{\partial \boldsymbol{\Omega}}{\partial t} - \nabla\times(\mathbf{v}\times\boldsymbol{\Omega}) - \frac{1}{\rho^2}\,\nabla\rho\times\nabla p = 0 \quad .$$

Since $\nabla\cdot\boldsymbol{\Omega} = 0$, we have

$$\nabla\times(\mathbf{v}\times\boldsymbol{\Omega}) = -\boldsymbol{\Omega}(\nabla\cdot\mathbf{v}) + (\boldsymbol{\Omega}\cdot\nabla)\mathbf{v} - (\mathbf{v}\cdot\nabla)\boldsymbol{\Omega} \quad .$$

If the fluid is incompressible (ρ = constant, $\nabla\cdot\mathbf{v} = 0$) this equation reduces to

$$\frac{\mathrm{D}\boldsymbol{\Omega}}{\mathrm{D}t} = (\boldsymbol{\Omega}\cdot\nabla)\mathbf{v} \quad . \tag{11.51}$$

Let $\mathrm{d}A$ be the area of an infinitesimal surface element, $\hat{\mathbf{n}}$ a unit vector normal to it, and $\mathrm{d}\mathbf{S} = \hat{\mathbf{n}}\,\mathrm{d}A$. In problem 11.6 we shall prove that

$$\frac{\mathrm{D}\mathrm{d}S_i}{\mathrm{D}t} = (\nabla\cdot\mathbf{v})\mathrm{d}S_i - \frac{\partial v^j}{\partial x^i}\,\mathrm{d}S_j \quad . \tag{11.52}$$

Using equations (11.51,52) one finds that

$$\frac{\mathrm{D}(\boldsymbol{\Omega}\cdot\mathrm{d}\mathbf{S})}{\mathrm{D}t} = 0 \quad . \tag{11.53}$$

From equation (11.53) one infers the Helmoltz theorem for a non-viscous fluid

$$\frac{\mathrm{D}}{\mathrm{D}t}\int_S \boldsymbol{\Omega}\cdot\mathrm{d}\mathbf{S} = 0 \quad . \tag{11.54}$$

If S is a surface having the curve C as a boundary, since $\boldsymbol{\Omega} = \nabla\times\mathbf{v}$ one obtains the Thomson (Kelvin) theorem

$$\frac{\mathrm{D}}{\mathrm{D}t}\int_C \mathbf{v}\cdot\mathrm{d}\mathbf{r} = 0 \quad . \tag{11.55}$$

11.3 Chapter 11 problems

11.1 (i) Show that the continuity equation (11.33) can be expressed in the form

$$\frac{\partial}{\partial t}(\rho\tilde{\omega}^3) + \tilde{d}[\rho\tilde{\omega}^3(\bar{V})] = 0 \quad ,$$

where $\tilde{\omega}^3(\bullet,\bullet,\bullet) = \tilde{d}x \wedge \tilde{d}y \wedge \tilde{d}z$, $\bar{V} = v_x\partial/\partial x + v_y\partial/\partial y + v_z\partial/\partial z$, and
$\tilde{\omega}^3(\bar{V}) = \tilde{\omega}^3(\bar{V},\bullet,\bullet) = \rho(v_x\tilde{d}y \wedge \tilde{d}z - v_y\tilde{d}x \wedge \tilde{d}z + v_z\tilde{d}x \wedge \tilde{d}y$.
(ii) Show also that the continuity equation can be expressed in terms of Lie derivatives as

$$(\partial/\partial t \ + L_{\bar{V}})(\rho\tilde{\omega}^3) = 0 \quad .$$

11.2 Using problem 11.1 (i) derive the continuity equation in curvilinear coordinates,

$$\frac{\partial\rho}{\partial t} + \frac{1}{\sqrt{g}}\frac{\partial}{\partial x^i}\left(\rho\sqrt{g}\,V^i\right) = 0 \quad .$$

11.3 (i) Show that the Euler equation

$$\frac{\partial \mathbf{v}}{\partial t} + (\mathbf{v}\cdot\nabla)\mathbf{v} + \frac{1}{\rho}\,\nabla p = -\nabla\Phi$$

can be expressed in the form

$$\frac{\partial\tilde{V}}{\partial t} + \frac{1}{2}\,\tilde{d}(\mathbf{v}^2) + (\tilde{d}\tilde{V})(\bar{V}) + \frac{1}{\rho}\,\tilde{d}p + \tilde{d}\Phi = 0 \quad ,$$

where $\bar{V} = v_x\partial/\partial x + v_y\partial/\partial y + v_z\partial/\partial z$ and $\tilde{V} = v_x\tilde{d}x + v_y\tilde{d}y + v_z\tilde{d}z$.
We are using Cartesian coordinates ($v^1 = v_1 = v_x, \ldots$).
(ii) Show that the Euler equation can also be expressed as

$$\left(\frac{\partial}{\partial t} + L_{\bar{V}}\right) + \frac{1}{\rho}\,\tilde{d}p + \tilde{d}\left(\Phi - \frac{1}{2}\mathbf{v}^2\right) = 0 \quad .$$

11.4 From the equation $\partial\tilde{V}/\partial t + (1/2)\tilde{d}(\mathbf{v}^2) + \tilde{d}\tilde{V}(\bar{V}) + (1/\rho)\tilde{d}p + \tilde{d}\Phi = 0$ derive Euler's equation in curvilinear coordinates

$$\rho\left[\frac{\partial V_i}{\partial t} - \left(\frac{\partial V_j}{\partial x^i} - \frac{\partial V_i}{\partial x^j}\right)V^j\right] + \frac{\partial p}{\partial x^i} + \rho\frac{\partial}{\partial x^i}\left(\frac{1}{2}\,V_jV^j + \Phi\right) = 0 \quad .$$

11.5 Show that under a general coordinate transformation $V'_i W'^i = V_i W^i$ and

$$\left(\frac{\partial V'_i}{\partial x'^j} - \frac{\partial V'_j}{\partial x'^i}\right) V'^j = \frac{\partial x^k}{\partial x'^i}\left[\left(\frac{\partial V_k}{\partial x^m} - \frac{\partial V_m}{\partial x^k}\right) V^m\right] \quad .$$

11.6 Show that the continuity equation

$$\frac{\partial \rho}{\partial t} + \frac{1}{\sqrt{g}}\frac{\partial}{\partial x^i}(\rho\sqrt{g}\, V^i) = 0$$

is invariant under general coordinate transformations in 3-space.

11.7 (i) Consider an infinitesimal $\mathrm{d}\mathbf{r} = (\mathrm{d}x^1, \mathrm{d}x^2, \mathrm{d}x^3)$ and the particles of fluid that are on it at the time t. Show that at the time $t + \mathrm{d}t$ the same particles will be found on $\mathrm{d}\mathbf{r}' = (\mathrm{d}x'^1, \mathrm{d}x'^2, \mathrm{d}x'^3)$,

$$\mathrm{d}x'^i = \mathrm{d}x^i + \frac{\partial v^i}{\partial x^j}\,\mathrm{d}x^j \quad .$$

(ii) Show that the substantial derivative of the oriented-area vector $\mathrm{d}\mathbf{S} = \mathrm{d}^{(1)}\mathbf{r} \times \mathrm{d}^{(2)}\mathbf{r}$, $\mathrm{d}S_i = \epsilon_{ijk}\mathrm{d}^{(1)}x^j\mathrm{d}^{(2)}x^k$, is

$$\frac{\mathrm{D}\mathrm{d}S_i}{\mathrm{D}t} = \frac{\partial v^j}{\partial x^j}\,\mathrm{d}S_i - \frac{\partial v^j}{\partial x^i}\,\mathrm{d}S_j \quad .$$

11.8 Derive the equation $\partial\mathbf{\Omega}/\partial t - \nabla \times (\mathbf{v} \times \mathbf{\Omega}) - (1/\rho^2)\nabla\rho \times \nabla p = 0$ from the form version of Euler's equation,

$$\left(\frac{\partial}{\partial t} + L_{\tilde{V}}\right)\tilde{V} + \frac{1}{\rho}\,\tilde{d}p + \tilde{d}\left(\Phi - \frac{1}{2}\,\mathbf{v}^2\right) = 0 \quad .$$

11.9 In three dimensions define

$$J^\mu(\mathbf{x}, t) = M(\mathbf{X})\epsilon^{\mu\alpha\beta\gamma}(\partial_\alpha X^1)(\partial_\beta X^2)(\partial_\gamma X^3) \quad ,$$

where $\mathbf{X} = (X^1, X^2, X^3)$ and the Greek indices run from zero to three. Verify that

$$\rho = J^0 = M(\mathbf{X})\,\frac{\partial(\mathbf{X})}{\partial(\mathbf{x})} \quad , \quad v^k = \frac{J^k}{\rho} \quad ,$$

and that the continuity equation is satisfied,

$$\partial_\mu J^\mu = 0 \quad .$$

11.10 (i) A massless scalar field (scalar photons) is described by the Lagrangian

$$L = (1/2) \int \partial_\mu \varphi \partial^\mu \varphi \, \mathrm{d}^3 x,$$

yielding the wave equation $\partial_\mu \partial^\mu \varphi = 0$.

Show that the energy-momentum tensor

$$T^{\mu\nu} = \partial^\mu \varphi \partial^\nu \varphi - g^{\mu\nu} (\partial_\lambda \varphi \partial^\lambda \varphi)/2$$

satisfies $\partial_\mu T^{\mu\nu} = 0$, and derive energy and momentum conservation.

(ii) Let

$$A = \int \mathrm{d}^3 x \, \mathrm{d}t [(1/2) \partial_\mu \varphi \partial^\mu \varphi + (M/2) \dot{\mathbf{X}}^2 \, \delta(\mathbf{x} - \mathbf{X}(t)) + f \varphi u(\mathbf{x} - \mathbf{X}(t))]$$

be the action for the above field interacting with a particle of mass M, position $\mathbf{X}(t)$. Write the Lagrange equations.

(iii) Show that the energy

$$E = (1/2) \int [(\partial_t \varphi)^2 / c^2 + (\nabla \varphi)^2] \mathrm{d}^3 x + (M/2) \dot{\mathbf{X}}^2 - f \int \varphi \, u(\mathbf{x} - \mathbf{X}) \, \mathrm{d}^3 x$$

and the momentum

$$\mathbf{P} = -c^{-2} \int \partial_t \varphi \, \nabla \varphi \, \mathrm{d}^3 x + M \dot{\mathbf{X}}$$

are conserved.

Solutions to ch. 11 problems

S11.1 (i)

$$\tilde{d}[\rho\tilde{\omega}^3(\bar{V})] = \partial(\rho v_x)/\partial x\ \tilde{d}x \wedge \tilde{d}y \wedge \tilde{d}z - \partial(\rho v_y)/\partial y\ \tilde{d}y \wedge \tilde{d}x \wedge \tilde{d}z$$
$$+\partial(\rho v_z)/\partial z\ \tilde{d}z \wedge \tilde{d}x \wedge \tilde{d}y$$
$$= \nabla \cdot (\rho \mathbf{v})\ \tilde{\omega}^3$$

(ii) Using the general formula $L_{\bar{A}}\tilde{\omega} = \tilde{d}[\tilde{\omega}(\bar{A})] + (\tilde{d}\tilde{\omega})(\bar{A})$ we have $L_{\bar{V}}(\rho\tilde{\omega}^3) = \tilde{d}[\rho\tilde{\omega}^3(\bar{V})] + (\tilde{d}(\rho\tilde{\omega}^3))(\bar{V})$. The second term is zero since there cannot be forms of order greater than 3 in 3-dimensional space.
Using (i) we have $\nabla \cdot (\rho\mathbf{v}) = \tilde{d}[\rho\tilde{\omega}^3(\bar{V})] = L_{\bar{V}}(\rho\tilde{\omega}^3)$.

S11.2

$$\tilde{\omega}^3 = \tilde{d}x \wedge \tilde{d}y \wedge \tilde{d}z = \frac{\partial x}{\partial x^i}\frac{\partial y}{\partial x^j}\frac{\partial z}{\partial x^k}\ \tilde{d}x^i \wedge \tilde{d}x^j \wedge \tilde{d}x^k$$
$$= J\ \tilde{d}x^1 \wedge \tilde{d}x^2 \wedge \tilde{d}x^3 \quad ,$$

where

$$J = \epsilon_{ijk}\frac{\partial X^i}{\partial x^1}\frac{\partial X^j}{\partial x^2}\frac{\partial X^k}{\partial x^3} \quad (X^1 = x, X^2 = y, X^3 = z) \quad .$$

Let $g = \det(\mathsf{g})$, where g is the matrix with elements

$$g_{ij} = \frac{\partial x}{\partial x^i}\frac{\partial x}{\partial x^j} + \frac{\partial y}{\partial x^i}\frac{\partial y}{\partial x^j} + \frac{\partial z}{\partial x^i}\frac{\partial z}{\partial x^j} \quad .$$

If J is the matrix with elements $\partial X^i/\partial x^j$, we have

$$\mathsf{g} = \mathsf{J}\mathsf{J}^{\mathrm{T}},\ g = \det(\mathsf{J})\ \det \mathsf{J}^{\mathrm{T}} = J^2.$$

Hence $\tilde{\omega}^3 = \sqrt{g}\ \tilde{d}x^1 \wedge \tilde{d}x^2 \wedge \tilde{d}x^3$. Now with $\bar{V} = V^i\partial/\partial x^i$, we have

$$\tilde{d}[\rho\sqrt{g}\ \tilde{d}x^1 \wedge \tilde{d}x^2 \wedge \tilde{d}x^3(\bar{V})] = (\partial(\rho\sqrt{g}\ V^i)/\partial x^i)\ \tilde{d}x^1 \wedge \tilde{d}x^2 \wedge \tilde{d}x^3 \quad .$$

The final result follows at once.

S11.3 (i)

$$\tilde{d}(\mathbf{v}^2) = (\partial\mathbf{v}^2/\partial x^i)\tilde{d}x^i,$$
$$(\tilde{d}\tilde{V})(\bar{V}) = [(\nabla \times \mathbf{v}) \times \mathbf{v}]_i\tilde{d}x^i = -(1/2)(\partial\mathbf{v}^2/\partial x^i)\tilde{d}x^i + v_j(\partial v_i/\partial x^j)\tilde{d}x^i$$

(ii) The general formula $L_{\bar{A}}\tilde{\omega} = \tilde{d}[\tilde{\omega}(\bar{A})] + (\tilde{d}\tilde{\omega})(\bar{A})$ gives

$$L_{\bar{V}}\tilde{V} = \tilde{d}[\tilde{V}(\bar{V})] + (\tilde{d}\tilde{V})(\bar{V}) = \tilde{d}(\mathbf{v}^2) + (\tilde{d}\tilde{V})(\bar{V}).$$

Thus (i) reduces to (ii).

S11.4 With $\tilde{V} = V_j\tilde{d}x^j$, we have

$$\tilde{d}\tilde{V}(\bar{V},\bar{W}) = (\tilde{d}[V_j\tilde{d}x^j])(\bar{V},\bar{W}) = \left(\frac{\partial V_j}{\partial x^i}\ \tilde{d}x^i \wedge \tilde{d}x^j\right)(\bar{V},\bar{W})$$
$$= \frac{\partial V_j}{\partial x^i}(V^iW^j - V^jW^i) = \left(\frac{\partial V_j}{\partial x^i} - \frac{\partial V_i}{\partial x^j}\right)V^iW^j \quad \text{etc.}$$

Note that one can also start from $\partial\mathbf{v}/\partial t + (\mathbf{v}\cdot\nabla)\mathbf{v} + (1/\rho)\nabla p = -\nabla\Phi$ using the formula $(\mathbf{v}\cdot\nabla)\mathbf{v} = (1/2)\nabla\mathbf{v}^2 - \mathbf{v}\times(\nabla\times\mathbf{v})$. Then in Cartesian coordinates for which $v_i = v^i$, one finds the proposed equation with $V_i \to v_i$ and $V^i \to v^i$. The transition to curvilinear coordinates is then made by using the results of the following problem.

S11.5 $\mathrm{d}x'^i = (\partial x'^i/\partial x^j)\mathrm{d}x^j$, $V'^i = (\partial x'^i/\partial x^j)V^j$, $\partial/\partial x'^i = (\partial x^j/\partial x'^i)\partial/\partial x^j$, $V'_i = (\partial x^j/\partial x'^i)V_j$,

$$V'_i W'^i = V_j \frac{\partial x^j}{\partial x'^i}\frac{\partial x'^i}{\partial x^k} W^k = V_j \delta^j{}_k W^k = V_j W^j \quad .$$

Since the second term in

$$\frac{\partial V'_i}{\partial x'^j} = \frac{\partial}{\partial x'^j}\left(\frac{\partial x^k}{\partial x'^i} V_k\right) = \frac{\partial x^m}{\partial x'^j}\frac{\partial x^k}{\partial x'^i}\frac{\partial V_k}{\partial x^m} + \frac{\partial^2 x^k}{\partial x'^j \partial x'^i} V_k$$

is symmetric in i and j, it cancels out when we interchange i and j and subtract. The required result follows at once.

S11.6 We must show that

$$\frac{1}{\sqrt{g'}}\frac{\partial}{\partial x'^i}(\rho\sqrt{g'}\,V'^i) = \frac{1}{\sqrt{g}}\frac{\partial}{\partial x^i}(\rho\sqrt{g}\,V^i) \quad .$$

From eq. (4.4.2) of S. Weinberg, *Gravitation and Cosmology: Principles and Applications of the General Theory of Relativity* (Wiley, 1972), we have

$$g'_{ij} = \frac{\partial x^m}{\partial x'^i} g_{mn} \frac{\partial x^n}{\partial x'^j} \quad ,$$

from which one finds at once that $g' = J^2 g$, where J is the Jacobian determinant. A brief calculation gives

$$\frac{1}{\sqrt{g'}}\frac{\partial}{\partial x'^i}(\rho\sqrt{g'}V'^i) = \frac{1}{\sqrt{g}}\frac{\partial}{\partial x^i}(\rho\sqrt{g}V^i) + \frac{\rho V^j}{J}\frac{\partial x^k}{\partial x'^i}\frac{\partial}{\partial x^k}\left(\frac{\partial x'^i}{\partial x^j}J\right) \quad .$$

We must show that the second term vanishes. We have

$$\frac{1}{J}\frac{\partial x^k}{\partial x'^i}\frac{\partial}{\partial x^k}\left(\frac{\partial x'^i}{\partial x^j}J\right) = \frac{\partial \ln J}{\partial x^j} + \frac{\partial x^k}{\partial x'^i}\frac{\partial}{\partial x^k}\left(\frac{\partial x'^i}{\partial x^j}\right)$$

$$= \frac{\partial \ln J}{\partial x^j} + \frac{\partial x^k}{\partial x'^i}\frac{\partial}{\partial x^j}\left(\frac{\partial x'^i}{\partial x^k}\right)$$

$$= \frac{\partial \ln \det \mathsf{J}}{\partial x^j} + \mathrm{trace}\left(\mathsf{J}\frac{\partial \mathsf{J}^{-1}}{\partial x^j}\right) = \frac{\partial \ln \det \mathsf{J}}{\partial x^j} - \mathrm{trace}\left(\mathsf{J}^{-1}\frac{\partial \mathsf{J}}{\partial x^j}\right) \quad ,$$

where J is the matrix with elements $\partial x^k/\partial x'^i$. This is zero by virtue of Weinberg's equation (4.7.5),

$$\mathrm{trace}\left(\mathsf{M}^{-1}(x)\frac{\partial \mathsf{M}(x)}{\partial x^i}\right) = \frac{\partial \ln \det \mathsf{M}(x)}{\partial x^i} \quad .$$

S11.7 (i) During the time $\mathrm{d}t$, the end points $\mathbf{a}$ and $\mathbf{b}$ of a vector $\mathbf{b}-\mathbf{a}$ will be transported by the fluid motion to $\mathbf{a}'=\mathbf{a}+\mathbf{v}(\mathbf{a})\mathrm{d}t$ and $\mathbf{b}'=\mathbf{b}+\mathbf{v}(\mathbf{b})\mathrm{d}t$, and the vector $\mathbf{b}-\mathbf{a}$ will be transported to $\mathbf{b}'-\mathbf{a}'=\mathbf{b}-\mathbf{a}+(\mathbf{v}(\mathbf{b})-\mathbf{v}(\mathbf{a}))\mathrm{d}t$. For $\mathbf{a}=\mathbf{r}$, $\mathbf{b}=\mathbf{r}+\mathrm{d}\mathbf{r}$, $\mathbf{a}'=\mathbf{r}'$, $\mathbf{b}'=\mathbf{r}'+\mathrm{d}\mathbf{r}'$, we have $\mathrm{d}\mathbf{r}'=\mathrm{d}\mathbf{r}+(\mathbf{v}(\mathbf{r}+\mathrm{d}\mathbf{r})-\mathbf{v}(\mathbf{r}))\mathrm{d}t$, yielding at once the required formula. Thus the substantial derivative of $\mathrm{d}x^i$ is

$$\frac{\mathrm{Dd}x^i}{\mathrm{D}t}=\frac{\partial v^i}{\partial x^j}\,\mathrm{d}x^j \quad .$$

(ii)

$$\frac{\mathrm{Dd}S_i}{\mathrm{D}t}=\epsilon_{ijk}\left(\frac{\mathrm{Dd}^{(1)}x^j}{\mathrm{D}t}\,\mathrm{d}^{(2)}x^k+\mathrm{d}^{(1)}x^j\,\frac{\mathrm{Dd}^{(2)}x^k}{\mathrm{D}t}\right)$$

$$=\epsilon_{ijk}\frac{\partial v^j}{\partial x^m}\,(\mathrm{d}^{(1)}x^m\mathrm{d}^{(2)}x^k-\mathrm{d}^{(1)}x^k\mathrm{d}^{(2)}x^m)=\epsilon_{ijk}\,\frac{\partial v^j}{\partial x^m}\,\epsilon^{rmk}\,\mathrm{d}S_r$$

$$=(\delta_i{}^r\delta_j{}^m-\delta_i{}^m\delta_j{}^r)\frac{\partial v^j}{\partial x^m}\,\mathrm{d}S_r \quad \text{etc.}$$

S11.8 Acting with the operator $\tilde{d}$, we find

$$\frac{\partial\tilde{d}\tilde{V}}{\partial t}+\tilde{d}L_{\bar{V}}\tilde{V}-\frac{1}{\rho^2}\,\tilde{d}\rho\wedge\tilde{d}p=0 \quad .$$

Since $\tilde{d}\tilde{V}=(1/2)\epsilon_{ijk}\Omega^i\tilde{d}x^j\wedge\tilde{d}x^k$, $\tilde{d}L_{\bar{V}}\tilde{V}=\tilde{d}[\tilde{d}\tilde{V}(\bar{V})]=\tilde{d}[(\mathbf{\Omega}\times\mathbf{v})_j\tilde{d}x^j]$ $=(1/2)\epsilon_{ijk}[\nabla\times(\mathbf{\Omega}\times\mathbf{v})]^i\tilde{d}x^j\wedge\tilde{d}x^k$, $\tilde{d}\rho\wedge\tilde{d}p=(1/2)\epsilon_{ijk}(\nabla\rho\times\nabla p)^i\tilde{d}x^j\wedge\tilde{d}x^k$, we only have to take the coefficients of $\epsilon_{ijk}\tilde{d}x^j\wedge\tilde{d}x^k$.

S11.10 (i) Proving that $\partial_\mu T^{\mu\nu}=0$ is straightforward using the wave equation. Space integration gives

$$\frac{1}{c}\frac{\mathrm{d}}{\mathrm{d}t}\int T^{0\nu}\mathrm{d}^3x+\int\partial_iT^{i\nu}\mathrm{d}^3x=0 \quad .$$

Assuming that φ vanishes at infinity, we find

$$\frac{\mathrm{d}}{\mathrm{d}t}\int T^{0\nu}\mathrm{d}^3x=0 \quad ,$$

$$E=\int T^{00}\mathrm{d}^3x=(1/2)\int[(\partial_t\varphi)^2/c^2+(\nabla\varphi)^2]\mathrm{d}^3x=\text{constant} \quad ,$$

$$P^i=(1/c)\int T^{0i}\mathrm{d}^3x=-c^{-2}\int\partial_t\varphi\partial_i\varphi\;\mathrm{d}^3x=\text{constant}^i \quad .$$

(ii)

$$\partial_0\frac{\partial\mathcal{L}}{\partial(\partial_0\varphi)}+\partial_i\frac{\partial\mathcal{L}}{\partial(\partial_i\varphi)}=\frac{\partial\mathcal{L}}{\partial\varphi}$$

gives

$$(\partial_0^2 - \nabla^2)\varphi = f\ u(\mathrm{x} - \mathrm{X}) \ ,$$

whereas

$$\frac{\partial}{\partial t}\frac{\partial \mathcal{L}}{\partial \dot{X}^i} = \frac{\partial \mathcal{L}}{\partial X^i}$$

gives

$$M\ddot{\mathbf{X}} = \nabla_{\mathbf{X}} \int f\varphi\, u(\mathbf{x} - \mathbf{X})\mathrm{d}^3x$$

$$= -f \int \varphi \nabla_{\mathbf{x}} u(\mathbf{x} - \mathbf{X})\mathrm{d}^3x = f \int u(\mathbf{x} - \mathbf{X})\nabla_{\mathbf{x}}\varphi\, \mathrm{d}^3x \ .$$

(iii) The field energy-momentum tensor $T^{\mu\nu}$ (see (i)) now obeys the equation

$$\partial_\mu T^{\mu\nu} = (\partial_\mu \partial^\mu \varphi)\partial^\nu \varphi = f\ u(\mathbf{x} - \mathbf{X})\partial^\nu \varphi \ .$$

Therefore

$$\frac{\mathrm{d}}{\mathrm{d}t}\int T^{00}\mathrm{d}^3x = f \int u(\mathbf{x} - \mathbf{X})\partial_t \varphi\, \mathrm{d}^3x$$

$$= f\frac{\mathrm{d}}{\mathrm{d}t}\int u(\mathbf{x}-\mathbf{X})\varphi\, \mathrm{d}^3x + f\int (\dot{\mathbf{X}}\cdot\nabla_{\mathbf{x}}u)\varphi\, \mathrm{d}^3x = \ldots - \dot{\mathbf{X}}\cdot(f\int u\nabla_{\mathbf{x}}\varphi\, \mathrm{d}^3x) = \ldots - M\dot{\mathbf{X}}\cdot\ddot{\mathbf{X}} \ ;$$

$$\frac{\mathrm{d}}{\mathrm{d}t}\int T^{0i}\mathrm{d}^3x = -\int f\ u(\mathbf{x} - \mathbf{X})\frac{\partial \varphi}{\partial x^i}\, \mathrm{d}^3x = -M\ddot{\mathbf{X}}^i \ .$$

Index